Nicholas Procter Burgh

A Practical Treatise on Boilers and Boiler-Making

Nicholas Procter Burgh

A Practical Treatise on Boilers and Boiler-Making

ISBN/EAN: 9783744743600

Printed in Europe, USA, Canada, Australia, Japan

Cover: Foto ©berggeist007 / pixelio.de

More available books at **www.hansebooks.com**

A

PRACTICAL TREATISE

ON

BOILERS

AND

BOILER-MAKING.

BY

N. P. BURGH,

MEM. INST. MECH. ENG., ASSC. CIV. ENG., ETC.

ILLUSTRATED WITH 1163 ENGRAVINGS AND 50 PLATES.

LONDON:
E. & F. N. SPON, 48, CHARING CROSS.
NEW YORK: 446, BROOME STREET.
1873.

EXPLANATORY REMARKS.

THIS work—my twelfth—was commenced in the early part of the year 1872, and, in setting about its construction, I classified the subjects in the following order :—

MARINE BOILERS,
LOCOMOTIVE BOILERS,
PORTABLE LAND BOILERS,
STATIONARY LAND BOILERS,
CYLINDRICAL BOILERS,
VERTICAL BOILERS,
TUBE BOILERS,
CIRCULAR BOILERS,
WATER TUBE BOILERS,
FIRE ENGINE BOILERS,
FURNACE BOILERS,
GAS FUEL BOILERS,
OIL FUEL BOILERS,
MECHANICAL COAL-FEEDING FURNACES,
MECHANICAL OIL-FEEDING FURNACES,
SELF-ACTING NON-FEED ALARM APPARATUS,
SAFETY VALVES, WITH LOADS AND SELF-ADJUSTING SPRINGS,
FEED PUMPS,
FEED INJECTORS,
SELF-ACTING FEED-WATER APPARATUS,
SMOKE-CONSUMING APPARATUS,
CONSTRUCTION,
PROPORTIONS,
SETTING,
RULES,
REPAIRS,
EXPLOSIONS,
NATURE OF METALS USED,
GALVANIC ACTIONS AND INCRUSTATION.

I then arranged the chapters as will be seen in the Table of Contents, and as those two accomplishments lie in a literary nutshell before my readers, they may, perhaps, be interested in knowing how I compressed the vast amount of matter into such a small space.

To begin with, after arranging preliminaries, I gathered together in our Patent Office a list of the total amount of boilers that had been patented since the year 1663 to 1873, or a period of two hundred and ten years, and discovered that over three thousand examples had been offered for legal protection.

My next task was the "picking out" the possible from the impossible arrangements that could be constructed ; and, as careful as I have tried to be, I am afraid some of the doubtful characters have appeared, while at the same time the selection has been as fair as I could then judge.

In Chapter I., vertical land boilers are fully discussed, to the extent of one hundred and ninety examples, by two hundred and seventy-four illustrations, contained in one hundred and two pages of letterpress. Those engineers and others who are conversant with that class of boiler will recognise the popular as well as the generally unknown productions, and that the same arrangement has been favoured by two or three claimants for the invention, to which circumstance in all cases I have directed attention.

In Chapter II., horizontal land boilers are illustrated by one hundred and sixty examples; and it will be noticed, also, that where the setting forms a medium for working the boiler, it is shown; in fact, I have made this chapter a guide for boiler-setting as well as a record of inventions. The relation of the several arrangements as to similarity in design I have pointed out also, and clearly proved the looseness of our present patent law, as indeed I have in nearly all the chapters.

The tube boiler I have thoroughly investigated in Chapter III., and I have proved that the first high-pressure tube boiler proper was invented in the year 1824 by an Englishman, "Moore," by name, shown on page 164, thus upsetting the general belief that Dr. Alban first introduced it: he certainly proved it, but that was nearly twenty years after. The other boiler worthy of notice here is Perkins's, shown on page 168; in this case vertical and horizontal tubes connected at right angles form the arrangement, which is a contrast to Moore's cage boiler. I cannot help remarking how very little it is understood that the best and most economical method of raising steam is by enveloping the boiler as Moore did, and yet he knew it nearly fifty years ago, while a large number of engineers do not even know it now.

This Chapter III. contains fifty-six examples, and very contrasted they are too; for there are coil tubes, angular tubes, cross tubes, vertical tubes, ring tubes, annular tubes, syphon tubes, hanging tubes, and unit tubes. I may add, in passing, one inventor did, in the year 1872, partially put the boiler in the fire, as shown by Fig. 489, on page 188.

In the fourth chapter I have settled the impracticability of injection boilers, from the fact that if the steam is raised by a gush of feed water, the power must be uneven, and therefore the motion of the engine is unsteady; and so well is this understood, that only twenty-eight examples have been selected as inventors' claims.

Marine boilers next engaged my attention, which I put into Chapter V., making one hundred and five examples; and I was rather surprised to find that there has been but little improvement in this class of boilers during the last five years; but in mitigation of that fact is the practical reason that shipowners prefer to use a type of boiler that they know has answered, and are tardy in admitting new arrangements, because a ship on her voyage means capital afloat. The time, undoubtedly, is close at hand when the "drum" cylindrical boiler will give place to the small tube arrangement, and surface condensers, in connection with distilling apparatus, will prepare the feed water.

Liquid fuel having received some attention lately, I of course fully investigated the boilers in use and proposal for that purpose, and discovered that the interest was of the feeblest kind in the inventive world, as only twenty-five examples were to be found, and those I illustrated in Chapter VI. Of the practicability of burning liquid fuel, I have not the least doubt; but there remains at present to know, first, how to prepare the oil for combustion; second, how to burn it; and, third, to provide an apparatus not liable to get out of order. The present difficulty with liquid fuel is that the deposit of carbon is so rapid that the caloric properties of the flame after a time become almost ineffective for evaporation.

Chapter VII. I devoted to locomotive boilers, of which I selected ninety-five examples out of about three hundred locomotives; in fact, the engines have received a vast amount of attention, while the boilers have not been so specially favoured. But in my case I have considered the arrangement of the fire boxes and tubes as demanding most attention; and a careful comparison of the examples I have illustrated will repay any one interested in the subject.

Having waded through thus far, and accomplished what I did, I turned with full vigour to the details commenced in Chapter VIII., and turned over five hundred blue books on that matter, selecting ninety-three safety-valves, seventy-one alarm valves, and fifty-four feed pump engines, known commonly as "donkey pumps."

Chapter IX. I devoted to the connection of tubes, and illustrated seventy-two methods.

Fire bars next came in my list, and, in Chapter X., I illustrated forty-five different solid, hollow, fixed, and movable bars, selecting, of course, the most practical kind.

Following that, in proper order, came "mechanical feed fuel apparatus," of which, in Chapter XI., I illustrated twenty-four different arrangements; and of fire box doors, seventeen examples.

Thus far the mechanical matter was completed, but the main questions were still before me for digestion, i.e., the ignition of coal, combustion, action of flame, raising of steam, etc.

Chapter XII., on the "ignition of coal," next appeared, and as that sentence is a volume in itself, and equally as words can but hardly explain it, I illustrated the question in seven different stages, which conveys to the mind at once what language might fail to express.

Next I explained "combustion" in Chapter XIII., beginning with the constituents, and thoroughly described the different gases that arise during the process of their release, giving in detail their properties, value, and relative volume. Following that, the amount of air per lb. of coal I discussed, and, as a comparison to the ignition, some new light has been shown.

On the smoke question and its cause, I quoted two well-known authorities on that subject, while, at the same time, I expressed my own views. I then collated a series of experiments on combustion, evaporation, and other valuable matter relating to the powers of stationary locomotive and marine boilers.

The "action of flame and raising of steam" I considered too worthy a subject to be treated by words alone, consequently, I illustrated it fully; and in no work that I know of has such a lucid pictorial explanation been given.

Although boilers are made pretty strong now, unfortunately they explode sometimes; and, therefore, in Chapter XV., the causes of boiler explosions are digested and illustrated. And as a contrast to that subject I followed it with boiler-making, in Chapter XVI., which I illustrated, not only with woodcuts, but with three plates, A, B, and C. I set more value on them than any of the others, from the fact of their being more practically instructive in detail; and general thanks are due to their donors.

My last chapter, XVII., is, perhaps, as valuable as any, because it gives information and data for future practice, in the form of "Tables, Rules, and Memoranda for Boiler-making," and as such, becomes important as a key to the whole preceding it. The practitioner will find information refreshing to his memory, while the student will obtain knowledge of use to him in due course; in fact, I have endeavoured to give such matter as is really in immediate connection with boiler-making, and scientifically requiring attention; for, as I stated at the beginning of Chapter XVI., "Boiler-making requires more primary consideration, talent, scientific education, and practical knowledge in construction than any other branch of engineering yet known."

The plates contained at the end of this work are, without doubt, the best of their kind; and no small amount of gratitude is due to their contributors—I, of course, exclude my own—because the attention paid to their completion and correctness must have been extreme.

My epilogue is now ended, and I, like other actors on this stage of life, retire for a time, to appear, I trust, refreshed to act again, and meet with a repetition of those substantial plaudits I have already received.

N. P. BURGH.

78 WATERLOO BRIDGE ROAD, LONDON, S.E.
October 1st, 1873.

CONTENTS.

CHAPTER I.
LAND STATIONARY VERTICAL BOILERS . . . 1 to 102

CHAPTER II.
LAND STATIONARY HORIZONTAL BOILERS . 103 to 163

CHAPTER III.
LAND STATIONARY TUBE BOILERS . . . 164 to 189

CHAPTER IV.
INJECTION BOILERS . . . . 190 to 204

CHAPTER V.
MARINE BOILERS . . . . . . 205 to 236

CHAPTER VI.
LIQUID FUEL BOILERS . . . . . . 237 to 250

CHAPTER VII.
LOCOMOTIVE BOILERS . . . . . . 251 to 278

CHAPTER VIII.
BOILER STEAM SAFETY VALVES AND GEAR 279 to 288
BOILER ALARM SAFETY VALVES AND GEAR 288 to 296
BOILER FEED PUMPS AND ENGINES . . 296 to 305

CHAPTER IX.
SECURING AND CONNECTING TUBES . . . 306 to 307
SECURING WATER CIRCULATING TUBES. . 307 to 310
SECURING BRANCH PIECES TO CIRCULATING
TUBES . . . . 310 to 312
CONNECTION OF TUBES AT RIGHT ANGLES,
ETC. . . . . 312 to 313

CHAPTER X.
PERFORATED FIRE BARS 314
SOLID AND HOLLOW FIRE BARS 315
WATER FIRE BARS . . . . . . 315 to 316
MOVABLE FIRE BARS . . . . . . 316 to 317

CHAPTER XI.
MECHANICAL FEED FUEL APPARATUS . . 318 to 323
FIRE-BOX DOORS . . . . 323 to 326

CHAPTER XII.
IGNITION OF COAL 327

CHAPTER XIII.
THE CAUSE AND EFFECT OF COMBUSTION . 328 to 350

CHAPTER XIV.
ACTION OF FLAME AND RAISING OF STEAM 351 to 355

CHAPTER XV.
CAUSES OF BOILER EXPLOSIONS . . . . 356 to 360

CHAPTER XVI.
BOILER-MAKING . . . . 361 to 363

CHAPTER XVII.
TABLES, RULES, AND MEMORANDA FOR
BOILER-MAKING . . . . . . 364 to 372

INDEX OF PLATES . . . . 373 to 374
INDEX OF ILLUSTRATIONS . . . . . 375 to 389
INDEX OF SUBJECT-MATTER . . . . . 390 to 391
ADVERTISEMENTS . . . . . at the end of the Plates

NOTICE TO THE BINDER.—Insert Plates A, B, and C between pages 362 and 363; and the remainder of the plates in consecutive order at the end of the Index of Subject-Matter.

STEAM BOILERS AND BOILER-MAKING.

CHAPTER I.

LAND STATIONARY VERTICAL BOILERS.

THE upright or vertical boiler being the first kind in use, according to chronological records of scientific occurrences, we therefore begin with that subject, which originated as far back as the year A.D. 1663, when the Marquis of Worcester is said to have invented the boiler and apparatus in connection, for lifting water by steam, as illustrated by Fig. 1; from which it is

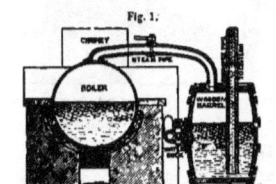

Fig. 1.

The first Globular Boiler and Water Cask used as a "Steam Water-lift." Invented in the year 1663, by the Marquis of Worcester.

apparent that the boiler was globular, set in brickwork, and the steam passed through a pipe into a cask or wooden barrel, which, when fed with water, condensed the steam thereby, and thus a vacuum occurred sufficient to prompt the water below to rise into the cask while the cock was opened; but directly it was closed condensation ceased, and the steam then from the boiler forced the water in the cask up through the discharge pipe into the tank above. This undoubtedly was a very clever idea for a pumping engine; but in

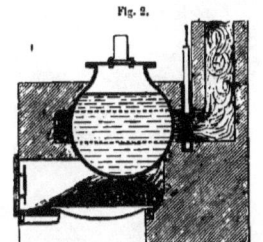

Fig. 2.

Savery's Steam Boiler, used in connection with a Water-lift in the year 1698.

practice the use of steam by condensation was so extreme over what was requisite to force the water, that the working expenses became a matter of grave consideration. But that, however, did not prevent Savery from patenting and erecting a boiler as Fig. 2 represents, which was in connection with a very similar condensing apparatus, of metal, as the Marquis of Worcester's. The Marquis's boiler, however, was not set as scientifically as that by Savery; and we therefore illustrate how Newcomen set his globular boiler, as shown by Fig 3, he having had the experience of his

Fig. 3.

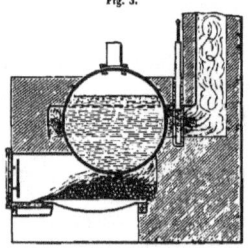

Newcomen's Globular Boiler, set in brickwork, for Steam Engines, as used in the year 1710.

predecessors to guide him, as also shown in plan by Fig. 4, from which it is apparent

Fig. 4.

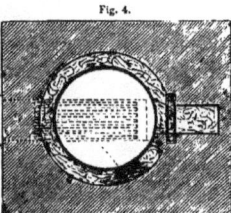

Plan of the Globular Boiler shown by Fig. 3.

that the flame surrounded the boiler just below the water level, and finally passed up the chimney, that was situated opposite the fire-place.

At this period the steam engine began to cause some stir amongst machinists, and the result was, that the few engineers then existing soon found out that the globular boiler, although the strongest of shapes, was not the most economical either in manufacture or working; while as the pressure of steam raised did not extend beyond 5 lbs. on the square inch, there was little difficulty with that, and therefore the cylindrical vertical boiler was next introduced, as shown by Fig. 5, in the

Fig. 5.

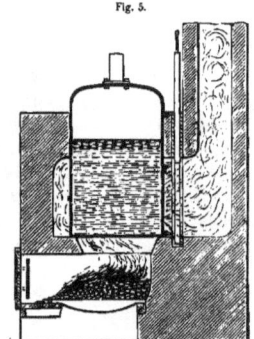

Savery's Cylindrical Vertical Boiler, with flat ends, as used in the year 1711.

year 1711. But this form being only resistable to pressure at the side, the ends soon gave way, and science combined with practice prompted the form of vertical boiler, as shown by Fig. 6. in which the ends and sides partake of the globular shape, or are portions of arcs of a circle. And even that did not satisfy the wishes of the engineers, and thus a third form of cylindrical boiler next made its appearance, as illustrated by Fig. 7, which

Fig. 6.

Newcomen's Cylindrical Vertical Boiler, with curved ends, as used in the year 1714.

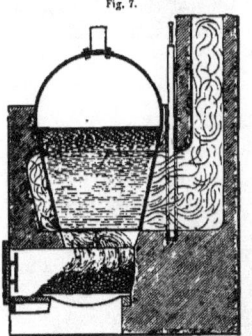

Fig. 7.

Newcomen's Haycock Cylindrical Boiler, as used in the year 1711.

was aptly termed the "haycock boiler;" and the main cause for the sides being angular was that the flame might impinge on it, rather than slide only, as with the vertical side boilers shown by Figs. 5 and 6. After this the haycock boiler was recessed at the sides, to still further absorb the heat, as illustrated

Fig. 8.

Newcomen's Haycock Cylindrical Boiler, with recessed sides, as used in the year 1711.

by Fig. 8, and the bottom end curved upwards to resist the pressure.

From this period, 1711, Newcomen was superseded by Watt, who introduced the Wagon and other long horizontal boilers — which are all fully illustrated in the next chapter — while also Smeaton had not been idle with his improvements; as for example he introduced the boiler illustrated by Fig. 9, on the next page, which was a vast improvement over the prior boilers used.

We next direct attention to a cylindrical boiler with a spiral flue, which was introduced late in the seventeenth century by an inventor whose name unfortunately has not been recorded in the annals of engineering events. This boiler is illustrated by Fig. 10 in sectional elevation and plan, and shows that the flame first enters into a "pot" which is

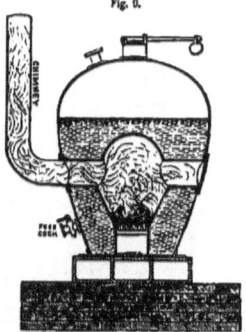

Fig. 9.

Sweaton's Haycock Cylindrical Boiler, with internal fire-box, flues, and iron chimney, being the first Vertical Flued Boiler made and introduced in the year 1769.

in connection with the spiral flue that leads to the chimney, by which means the flame has a very great traverse, and thus time is permitted for the water to absorb a great deal of the heat, also showing that the inventor knew what was requisite—as far as the practice then—to generate steam.

We advance now to the date 1822, because from 1700 to then nothing in vertical boilers

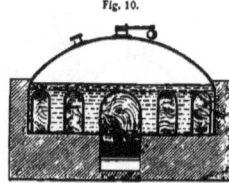

Fig. 10.

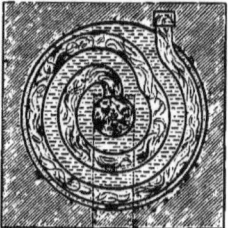

Cylindrical Boiler with Spiral Flue, supposed to have been used in the year 1780. Inventor unknown.

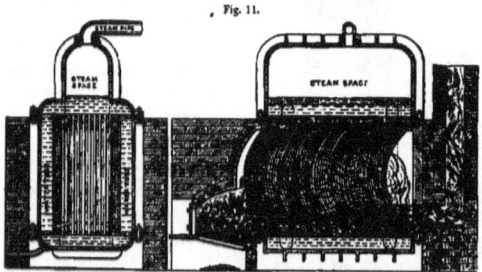

Fig. 11.

Clark's Vertical Tubular Boiler, with top and bottom casings—the first Vertical Tube Boiler proper. Patented in the year 1822.

was carried out sufficient to warrant notice in these pages, and we there next direct attention to Fig. 11, which is a sectional elevation and plan of the first tubular boiler; and to the credit of Mr. Clark we make it known that he invented the *first surface condenser* also, at the same time.

This boiler is composed of a number of curved tubes placed in an erect position, by which a greatly extended surface is exposed to the action of the fire. These tubes are supplied with water from vessels into which their extremities are inserted, the construction of which will be best seen by reference to the illustrations, which are a transverse section of the boiler and furnace, taken through one of the water tubes, and also a longitudinal sectional view, which shows the fireplace, the ash pit, and the furnace containing the tubes. This space is closed above, below, and on the sides with fire brick; and at the top and bottom are cast iron plates through which the tubes pass; and above and below them are the upper and lower parts of the boiler, containing also water, which flows through the water tubes that support the upper part of the boiler, and by means of bolts and flanges the whole is connected together.

The curved tubes for generating the steam, upon the external surface of which the fire acts, are made of sheet copper, about one-tenth of an inch thick, which, when brased or soldered with spelter, form the pipes or tubes, about one inch diameter, and which are to be curved in the form shown, and this curve has for its object the capability of the tubes expanding or contracting without fracture. The holes in the iron tube plates are bored to receive the ends of the tubes; those in the upper plate are something larger than in the lower one, in order to permit the tubes to pass down. The ends of the tubes, being accurately fitted to the holes, are made to attach themselves firmly thereto by means of a mandrel driven in, and, on withdrawing which, a ferrule, of the same metal of which the tubes are composed, is introduced and driven tight. The holes in the ferrules allow the free passage of the water, as will be described hereafter. There is a vacant space between the arch of fire brick and the upper tube plate, to prevent the heat of the furnace injuring the plate.

Two years elapsed, and then a Mr. Moore patented the arrangement of pipes to form a boiler, as shown by Fig 12, where it will be

Fig. 12.

Moore's Vertical Tube Boiler, with horizontal pipe rings, surrounded with brickwork. Patented in the year 1824.

seen that it consisted of three ring pipes that are connected by vertical and angular tubes. The bottom and intermediate ring pipes, and the portions of the tubes from the same level, contain the water, while the pipe and portions of the tubes above receive the steam, which is generated by fuel put on and ignited on a grate that is situated just above the bottom ring pipe. This entire mechanical arrangement is surrounded by brickwork, which forms a kiln or furnace, so that there is ample space for combustion. We have no hesitation in stating that Mr. Moore, although not an engineer in practice, evidently was one in theory, because he put his boiler bodily in the fire, and that is what must be done for high pressure steam, for which only tubular boilers are serviceable.

and the parts are united together so that the internal capacities of those tubes form a receptacle for the water which is to be converted into steam. Each of these combinations of tubes constitute a sort of cylindrical cage, within which the fire is made. The lower part of the small curved tubes perform the office of the bars of a fire grate; viz., to support the burning fuel, and through the spaces or interstices between the tubes fresh air is introduced to maintain the combustion. The direct red heat of the different fires is thus applied with great effect to the water con-

Fig. 13.

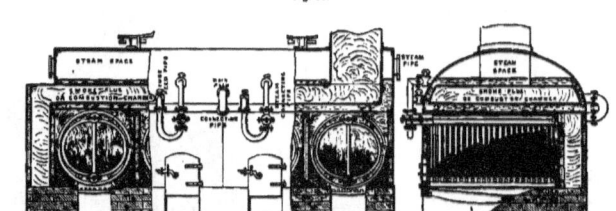

Teissier's Circular and Vertical Tube Boilers, in connection with a top flue and steam chamber. Patented in the year 1825.

At the latter part of the year 1825, another non-professional man named Teissier patented the arrangement of tubes as illustrated by Fig. 13, which is that there are four fireplaces, those spaces being surrounded—except the bottom—with water, which is contained in the narrow spaces between surfaces of metal plates, so that the water shall be spread out thinly, and in a suitable manner receive the heat. Each furnace or fireplace consists of a series of small curved tubes, suitably combined with strait perpendicular tubes, and united by large horizontal tubes situated top and bottom;

tained in the small tubes and in the upper part of the lower horizontal tube. The flame which proceeds from the fires passes through the spaces between the upper parts of the curved tubes, and surrounds the large top horizontal tube; the flame also strikes against the interior surfaces which form the boundaries of the fireplaces. The sides and ends of the boiler are formed of a number of such flat water spaces of suitable dimensions as required, placed in vertical positions, and put together as follows, viz.:—Two spaces are placed in a vertical position to form the ends of the boiler, and three other spaces form the vertical partitions between the fireplaces, for which also two other vertical spaces are placed parallel to each other, one forming the end of the four fireplaces, and the other forming the upright side of the boiler; the interval between the two spaces is the flue to carry off the smoke and heated air. The space included amongst all these vertical flat vessels is covered over by the principal portion, which forms the upper part of the boiler which receives the steam; but there is also a main flue or combustion chamber, which passes horizontally through the lower part of the to the extremities of the horizontal tubes, and from these tubes the water flows freely into all the other tubes. These branches are provided with stopcocks, by which the communication with any of the fireplaces can be intercepted at pleasure. The steam which is generated in each of the combinations of the small tubes, and, by the heat of the fire and flame, rises up and accumulates in the upper horizontal tubes, from whence it passes by the curved branches into the upper part of the boiler, and in each branch is a stopcock to shut the passages when required. Suitable communicating branches are likewise pro-

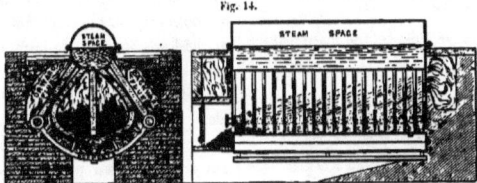

Fig. 14.

Teissier's Vertical Angular and Curved Tubes Boiler, with horizontal water and steam pipes. Patented in the year 1825.

steam space, and occupies so much of its capacity as only to leave thin flat spaces round the flue to contain the water which surrounds every part of the flue.

The water which is to be converted into steam is introduced into the dome or top part of the boiler by the feed pump, and from the lower portion of this part the water is distributed and conveyed into the interior of each of the flat spaces by short connecting pipes.

The combinations of tubes which form the fireplaces are supplied with water through pipes which join from the lower part of the dome vided, in order to convey the steam from each of the vertical flat spaces into the upper part of the large receptacle.

Another arrangement is shown above by Fig. 14, which is that three horizontal pipes are connected to a larger one by angular, vertical, and curved pipes, and the top pipe is sufficiently large to contain a steam space also. The fuel is laid on the lower parts of curved pipes and a portion of the bottom horizontal pipe, so that the ordinary fire bars are not used. The flame passes from the central space to one side forward and backward, on the other to the main flue.

Mr. Teissier having exhausted his ideas of tubular boilers, he then turned his attention to flat water space boilers, an example of which is shown by Fig. 15 in sectional eleva-

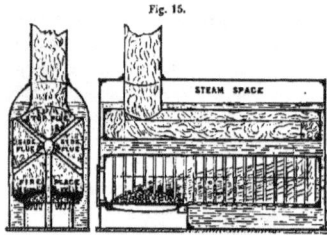

Fig. 15.

Teissier's Vertical and Angular Water-spaced Flue Boiler.
Patented in the year 1825.

tions, and in plan by Fig. 16, the main feature claimed being that the boiler is so formed that

Fig. 16.

Plan of the Boiler shown by Fig. 15.

all the surfaces which are exposed to the fire and flame are inclined in an angle of at least 45° of the vertical, in order that any salt or other sediment which may be formed upon such inclined surfaces may not remain thereupon, but may slide down, and collect into proper receptacles provided at the lower part of the boiler, where no heat is applied, and from which the sediment may be withdrawn when a sufficiency is collected. The fireplace is included within and surrounded on every side with water, so as to communicate heat thereto. The upper surfaces of the fireplace are inclined planes, in the manner of the roof of a house, and sloping so much that the sediment cannot lodge or rest thereupon, but the same will slide down and fall through the water into the receptacles on each side of the fire, but situated lower than the fire, so as not to be heated thereby. In the centre of the fireplace a series of small tubes are fixed in a perpendicular direction. They are joined at their upper ends to the top of the fireplace, so as to be at all times freely supplied with water; and being contained in the midst of the fire, and surrounded thereby on all sides, they receive the direct heat therefrom, and convert the contained water into steam. The tubes also communicate at their lower ends to another sediment receptacle, which is also placed lower than the fire grate, and is adapted to receive the solids which are formed within the tubes and which fall down into the receptacle. The row of perpendicular tubes divides the fireplace into two, as shown in the plan; but the flame can pass between the tubes so as to act on all sides of them.

The flame from the fireplace passes out at the far end, and rises up into a flue or passage, which is of a triangular form, as is shown, the surfaces of two of its sides being placed at an inclination of 45° with the vertical. The other, or third side of the triangular flue is a vertical surface placed opposite to the outside of the boiler, and near thereto, so as to leave but a small space for the water. The side flue turns round at the front end of the boiler, over the fire door, and joins to another triangular flue, which is exactly similar to the other, and conveys the smoke back to far end of the boiler, where the end flue joins to a third triangular flue, which passes along the middle of the boiler, and leads to the perpendicular chimney.

It will be noticed that in that boiler the stays are omitted, and that the construction

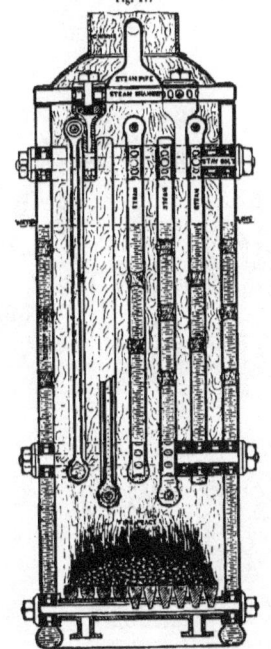

Fig. 17.

Hancock's Vertical Flat Cellular Boiler. Patented in the year 1827.

would be very difficult, while repair means breaking one half to repair the other. A Mr. Hancock, however, in the year 1827, invented a better cellular boiler, as shown in sectional elevation by Fig. 17, and in plan by Fig. 18, the arrangement of which is that a suitable number of the narrow flat cells are put together in parallel vertical planes, or edgeways upwards over the fire, which is made upon the grate, leaving narrow vertical spaces between the several cells, to admit the flame, smoke, and heated air to pass between their adjacent vertical surfaces. The two outside cells are of greater vertical depth than the others, and they descend sufficiently below the others to form a space between them for the fireplace. The narrow upright ends of all

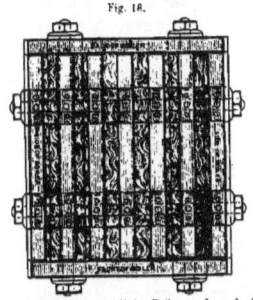

Fig. 18.

Sectional plan of Hancock's Cellular Boiler, as shown by Fig. 17.

the cells apply against the flat surfaces of two other large cells, which form the front and back ends of the boiler, and enclose the space which constitutes the fireplace. The front cell has an opening at the lower part of the fire-door. The tops of all the cells are covered over by a flat cell, which forms the steam chamber to receive and collect the steam which is produced in all the different cells. The several cells, which stand side by side, are united together laterally by strong bolts which extend across them through the apertures, and circular rings are interposed between the several cells to keep them at the proper distances apart, in order to leave the required spaces between them for the fire to act in. The bolts pass through all the external rings, as well as through the corresponding internal rings, which are placed withinside each cell before it is riveted up. Rings of very thin sheet lead or of tinfoil doubled are also applied at each side of each of the external rings, to serve for packings around the margins of the several apertures, and thus form the contact with the plate of which the flat sides of the cells are composed. The bolts have strong screws and nuts on the ends, which are screwed up very tight in order to bind all the several cells very strongly together, side by side; and by compressing the various rings very forcibly towards each other, the metal plate of which the flat vertical sides of the cells are formed is pinched so close between those rings, that, together with the interposed packings of lead or tin, the fittings around the margins of the several apertures are made water-tight and steam-tight. Those apertures through the cells are of corresponding size to the interior of the several rings, but the bolts which pass through the rings and the apertures being much smaller, a sufficient open space is left around the bolts for the water and steam to pass freely along the bolts, in order to flow from one ring into the next; and all the internal rings which are withinside the several cells being perforated with holes from the circumference towards the centre, the water and steam will have free passage from one vessel to another throughout the whole series.

In the year 1834, or twelve years after, Mr. Clark introduced his tube boiler, as shown by Fig. 11 on page 4. A "canny" Scotch engineer, named McDowall, patented the arrangement shown by Fig. 19, which, like

Fig. 19.

McDowall's Vertical Tube Boiler. Patented in the year 1834.

Clark's, is composed of top and bottom chambers that are connected by vertical tubes; but straight in this case and curved in the other; while Mr. McDowall stipulates wrought iron chambers connected, as shown, and Mr. Clark not being so far advanced, proposed cast iron, but there the difference ends. But we think Mr. Clark deserves the greater credit, however, not only by being first in the field, but also for curving his tubes to allow for expansion and contraction — matters that Mr. McDowall omits to provide for.

Next came a Mr. Holmes, who ignored tubes, but preferred a flued boiler, as shown by Fig. 20.

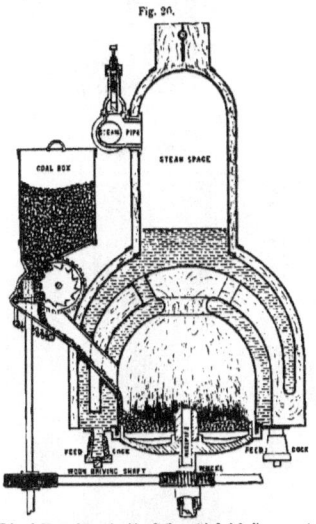

Holmes's Vertical Annular Flue Boiler, with fuel feeding apparatus and revolving grate. Patented in the year 1836.

Fig. 20, added to which was a mechanical fuel feeding apparatus and revolving grate, the arrangement of the whole being as follows.

The boiler is composed of the several parts or sections that are connected together in pairs to form the water chambers, and are firmly secured to the main bottom plates of the boiler; the steam chamber being formed by the dome part, and the whole surrounded by the flame jacket or outer casing.

The fire is supplied with fuel from the hopper, the coal or coke being carried or taken therefrom by the grooved or toothed roller, and caused to descend the pipe on to the fire bars, the fuel being retained or held up by the shelf, which turns with its pin in a hinge joint, and is kept up by the spring applied to the lever on the end of the joint pin, whereby the shelf will be allowed to give way to any large lumps of fuel, and again return to retain the smaller pieces from falling through.

The feed roller is actuated by a worm or endless screw on the upper end of the vertical shaft, working into a toothed wheel on the end of the axis of the roller, this shaft being actuated by an endless screw on the horizontal or worm-driving shaft.

The rotary motion is communicated to the fire bars or grate through this shaft, which carries another endless screw gearing into a toothed wheel mounted on the boss or collar, which is connected to the hollow shaft or air pipe, which supports the frame of the fire bars by a key and slot, so as to cause the shaft to revolve with it; at the same time the shaft with the grate can be raised or lowered without throwing the worm and wheel out of gear, for the purpose of removing the fire when necessary, on any extraordinary pressure of steam in the boiler, or for the purpose of lighting the fire or other purposes.

From the year 1836 to 1851 nothing was done to improve the vertical tubular boiler, and we attribute this fact to the cause of people being afraid to use them, which fear evidently arose from a confidence in horizontal boilers, and a non-understanding of the vertical kind. A Mr. Stenson, however, rushed to the scene of doubt, and patented his "pot" and tubular vertical boiler, as illustrated by Fig. 21, which is a cylindrical shell containing a fire box, with a deep recessed water space in it—commonly known *now* as a "pot." The flame from the grate, directly under it, acts on and around, and then passes out at the top of the annular space, through horizontal tubes, into down-take tubes to a ring chamber that is below the grate, and from there ascends through up-take tubes into the smoke box above, and finally up the chimney.

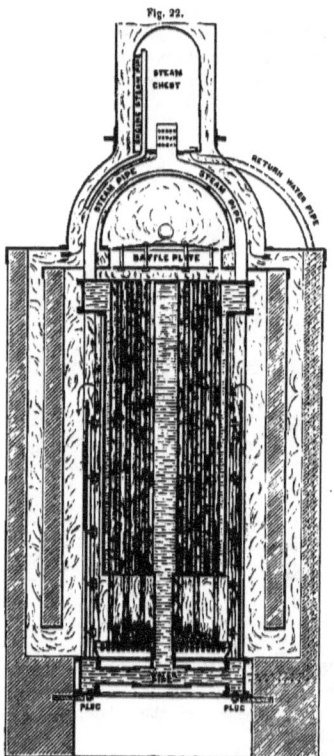

Craddock's Vertical Tubular Boiler, with outside return flues. Patented in the year 1852.

One year from the last date a Mr. Craddock

came forward with his ideas of what a vertical boiler should be, as shown in sectional elevation by Fig. 22, and in plan by Fig. 23.

Fig. 23.

Plan of Craddock's Vertical Tubular Boiler, as shown by Fig. 22.

The boiler rests on the bottom chamber, which is made of wrought iron, as is the top chamber; and into those chambers at top and bottom, the outside or fire box range of tubes are inserted in the same way as the tubes are in locomotive boilers; also plugs are put in from the inside, and screwed up from the outside, into the holes made in the bottom plate of the bottom chamber, which holes are made for the purpose of driving the ferrules into the tubes that are inserted in the bottom plates of the chamber. By taking out any one of those plugs, opposite any tube, the means required for inserting a new tube are obtained; also it will be seen that those plugs are kept clear of the plate on which the boiler rests.

The central tube chamber is joined to the top chamber, and is thus suspended over the grate. From this chamber will be seen passing through the centre of the grate a water pipe, communicating with the bottom chamber, which is for the purpose of opening a communication with the top chamber.

The complete communication of all the steam and water spaces is effected by the two pipes going from the top chamber, by which the steam is conveyed to the steam chest, whilst any water that is carried with the steam into the steam chest passes back to the bottom chamber through the return water pipe, leaving the dry steam to be conveyed to the engine by the steam pipe.

The action of the flame in this boiler is rather peculiar, it being that a portion of the flame passes through the central small tubes, and strikes the baffle plate, and thus acts on the top of the water chamber, and then descends and is met by the other portion of the flame that passes amongst the outer large tubes, above the side flue plates, and from there both portions of the flame pass down the inner flues and up the outer flues to the chimney.

The next example of boiler that deserves notice is illustrated in sectional elevation by Fig. 24, on the next page, being the patented idea of a Mr. Balmorth in the year 1852, the main improvement being the midway division of the length of the tubes by a combustion chamber, so that the flame from the first set of tubes shall expand, and thus intermingle before passing into the second or upper set, which idea is sound in theory and effectual in practice.

In the same year a Mr. Bellford laid claim for a patent of his conception of the best boiler with tubes internally arranged in rows, as illustrated in sectional elevation by Fig. 25, and in sectional plan by Fig. 26; and the only feature in this example worthy of record is that the patentee introduced baffle or priming plates in the steam space; for his arrangement of a large water space and chamber below the fire grate is the worst possible, while the position of the tubes do not permit the flame to act on them with the best advantage.

Following Mr. Bellford—but certainly not copying him—Mr. Cameron came forward with his boiler as illustrated by Fig. 27, which is three cylindrical casings or shells put within the other. The smallest shell is a conical water space, and the next is a furnace and

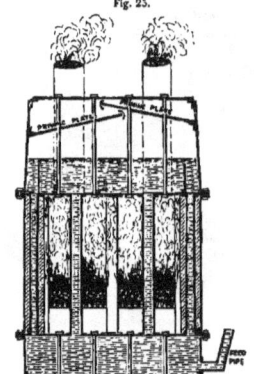

Bellford's Vertical Cylindrical Tubular Boiler, with internal combustion chamber and water spaces below the fire grate. Patented in the year 1852.

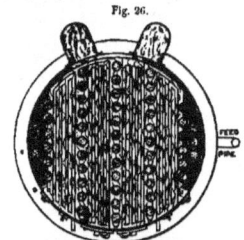

Sectional plan of Bellford's Vertical Tubular Boiler, as shown by Fig. 25.

combustion chamber, while the third, the longest and largest, is the final casing that contains the other two, with the fuel, water, and steam. The boiler is set in brickwork, with surrounding flues that lead to the chimney, but the brickwork can be dispensed with, and iron casing stand in its place, if preferred.

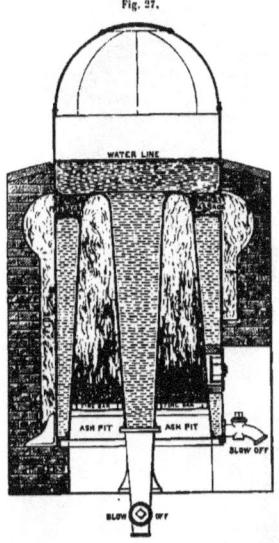

Cameron's Vertical Cone-Central and Cone-Annular Water-spaced Cylindrical Boiler. Patented in the year 1852.

In the same year, 1852, a Mr. Huddart patented the arrangement of boiler as illustrated by Fig. 28, which is a central vertical chamber for containing the fuel to be consumed, and this chamber is provided with furnace bars at its lower end, and it is closed in at top by a hopper, through which the fuel is fed to the furnace, from time to time, by the withdrawal of a slide, which forms the bottom of the hopper. The hopper is itself closed with a lid, to prevent the passage of air downwards with the fuel to the fire. Air for supporting combustion is supplied to the fire chamber or furnace, in part through the spaces between the fire bars, but chiefly through a central air tube, which is in connexion with a fan or blower not shown. This tube is pierced with holes, to allow of the escape of the air

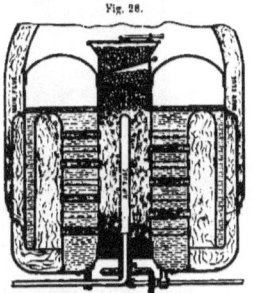

Huddart's Vertical Cylindrical Boiler, with horizontal fire tubes, annular flues, revolving air tube, and top feed fuel apparatus. Patented in the year 1852.

therefrom in streams amongst the burning fuel. It is also furnished with radial pins, which act as stirrers to the fire; and for this purpose it is necessary to give the tube a slow rotary motion by bevel gearing.

Projecting radially from the fire chamber are the horizontal tubes, which pass through an annular water space, and communicate with an annular flue which descends, and thereby forms a second annular water space. Beyond these, at opposite sides thereof, rise the main flues, which conduct away the gaseous products of combustion from the furnace.

Fig. 29.

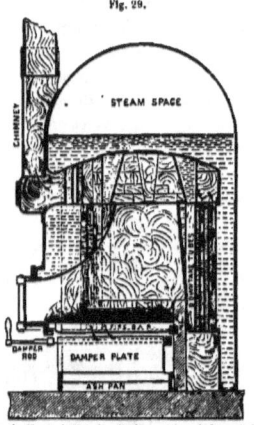

Galloway's Vertical Tubular Boiler, with tubular combustion chamber, return flame tubes, and bottom flue. Patented in the year 1853.

or dome, and the fire box contains two furnaces that are separated by the fire brick partition. The products of combustion from the two fires unite above the partition, and pass through the upper part or throat of the fire box into

Fig. 30.

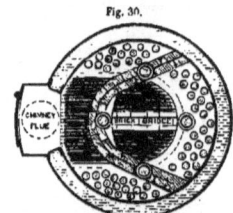

Sectional plan of Galloway's Boiler, shown by Fig. 29.

the combustion chamber, from whence they descend through a number of vertical tubes into the bottom flue, and again ascend through other vertical tubes into the chamber, and pass through the opening to the chimney.

The chamber is formed by one circular box,

Fig. 31.

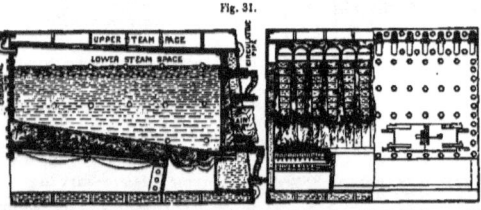

Cowper's Vertical Cellular Boiler, fitted steam and water circulating pipes. Patented in the year 1853.

In the next year, 1853, tubular vertical boilers engaged a great deal of attention from engineers, and Mr. Galloway was the first to make known the earliest novelty of that period, as illustrated by Figs. 29 and 30. This boiler is a cylindrical shell, with a hemispherical top divided into two parts by the fire brick partition, and strengthened by several conical water tubes and a fire brick curb, seen in plan, which protects the angle formed by the junction of the fire box from the direct action of the flame. There is a damper for regulating the admission of air to the fires, and it will be seen that the boiler has an internal furnace, contained within the shell, and surrounded by water; also that the greater part of the shell is curved, and that the weaker parts are strengthened by tubes or stays, so that the whole boiler may be capable of resisting the pressure of the steam.

Next came Mr. Cowper with his novelty, as illustrated in elevations by Fig. 31. The shell of the boiler is formed by flat cells or water spaces—which are connected and properly stayed, as shown—and contain a series of flat cells, made of plate iron or copper, and strengthened by a number of stays, as shown in the transverse sections.

Those cells are made tapering or wedge shaped, being narrower at the bottom than at top, and each of them is connected by a pipe with the steam chamber, and by another pipe with the water chamber.

The feed water is supplied to the water chamber, and flows through pipes connecting the cells, and the steam formed in those cells passes through the front pipes into the steam chamber, and any water which may be carried up with the steam into the chamber, flows back into the water chamber through the back pipes. In order as much as possible to prevent the water from depositing its impurities in and incrusting the interior of the cells, each cell is provided with a second set of pipes, which communicates with the water chamber over the main flue, therefore the effect of the ebullition in the cells is to cause a circulation of the water through those pipes; while the water in the chamber being less exposed to the heat of the fire than the water in the cells, it is in a less agitated state, and the impurities or sediment in that water are thus enabled to subside to the bottom of the chamber, whence they are blown off from time to time through a cock or valve.

The water pipes are each provided with a valve, opening towards the chambers, as shown in section, and when the boiler is at work, those valves are open; but in the event of the bursting of either of the cells, those valves would be closed by the rush of water and steam towards the burst cell, so that the escape of water and steam would be confined to the quantity contained in the burst cell, instead of the whole contents of the boiler escaping at once.

Mr. Kendrick followed Mr. Cowper, but

Fig. 32.

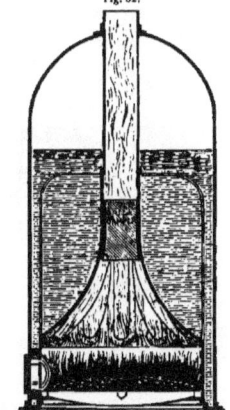

Kendrick's Vertical Radial Flat Water-spaced Boiler, with a central circular flue. Patented in the year 1853.

carried out his views in rather a different manner, by four examples; and Figs. 32 and

LAND STATIONARY VERTICAL BOILERS.

33 are illustrations of the first in sectional elevation and plan, which arrangement is that the shell is fitted with radial flat water spaces, that form also a central vertical flue from the top portion, and to prevent most of the flame

Fig. 33.

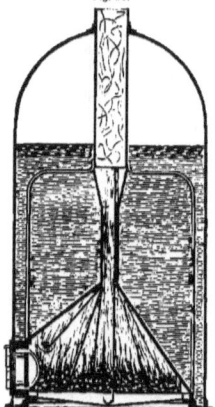

Sectional plan of the boiler shown by Fig. 32.

Fig. 34.

Kendrick's Vertical Radial Flat Water-spaced Boiler. Patented in the year 1853.

passing up through the flue, a brick damper is hung in it.

The second arrangement is illustrated by

Figs. 34 and 35, in sectional elevation and plan, in which the spaces are radially longer by the omission of the vertical flue below their tops, but joining on just above that part. His third arrangement is shown in sectional

Fig. 35.

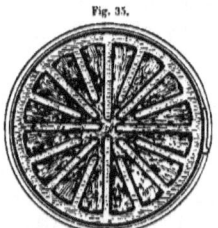

Sectional plan of the boiler shown by Fig. 34.

Fig. 36.

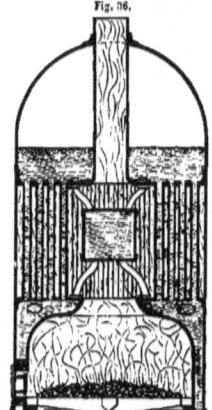

Kendrick's Vertical Cylindrical Boiler, fitted with vertical water tubes and central water box. Patented in the year 1853.

elevation by Fig. 36, which is a cylindrical shell fitted with a water box, surrounded with water tubes, while the flame from the grate below acts on and around their surfaces, and from thence passes up the vertical flue.

The fourth example is shown by Fig. 37,

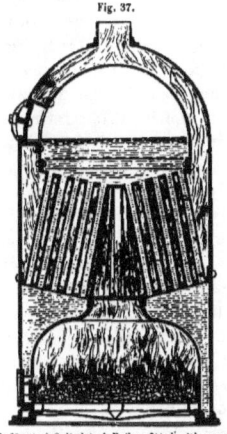

Fig. 37.

Kendrick's Vertical Cylindrical Boiler, fitted with angular water tubes. Patented in the year 1853.

and is the most sensible arrangement, because it is the simplest and easiest to construct. The tubes are angularly arranged in a water chamber over the fire box, and the flame passes from amongst them through an annular space to the chimney.

About the middle of this year Mr. Bellford came forward again with his improved boiler, as shown by Figs. 38 and 39, in sectional elevation and plan; and it will be apparent, on looking at Figs. 25 and 26, on page 14, how far behind he was then, in comparison to what he introduced now, and also how much more complication he introduced to demonstrate his improvement, the arrangement of which is that the outer shell of the boiler consists of a vertical shell, within which is a smaller shell, the two shells being united by

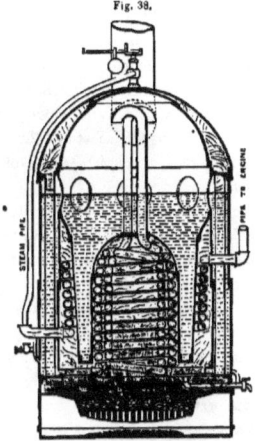

Fig. 38.

Bellford's Vertical Cylindrical Shell Water-spaced Boiler, fitted with water and steam coils, and vertical flue pipes. Patented in the year 1853.

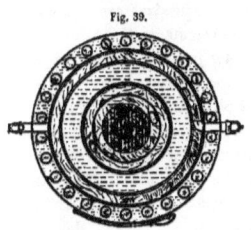

Fig. 39.

Sectional plan of Bellford's boiler shown by Fig. 29.

annular plates, at top and bottom, and thereby form an outer water jacket. Within the outer water jacket there is an inner water jacket, formed by a third shell as a hollow frustrum of a cone, and a fourth shell, united at their bottoms, which are a little above the bottoms of the first and second shells.

The smallest shell is open at its bottom, and terminates at the top in a dome, which is the fire box. Also the second shell terminates in a dome, which forms a steam chamber.

There is a steam and water communication between the outer water jacket and the interior of the second shell, through a series of holes, as shown.

The boiler is supported by a circular base, upon which the outer shell rests; this contains the circular fire grate and the ash pit. The fire grate is so arranged as to leave a shelf all around it, between which shelf and the bottom of the outer water jacket there is a space, which may be considered as a circular flue. A series of tubes passing vertically through the outer water jacket, and terminating in the top annular plates, form a communication between the lower flue and the upper annular flue, which is formed around the dome.

Within the fire box are two water coils of tubing in communication with the lower part of the outer jacket, and whose upper ends pass through the dome, and then connect with two upright pipes, which reach nearly to the top of the dome, into which they turn downwards.

A steam coil of a similar nature is placed in the annular space between the outer and inner water jackets. The products of combustion, after acting in the fire box, then descend, and pass off into the circular flue, from whence they escape through the vertical tubes to the dome flue, and from thence to the chimney. The steam generated by the contact of the water with all these heating surfaces rises into the dome or steam chamber, from which it is taken off for use by the steam coil.

The arrangement and connections of the water jackets and water coils have the effect of preserving a water level in the jackets, but not in the coils; because the steam is generated in and rises through the coils of pipe with such rapidity, that constant streams of water are driven forcibly through them into the steam chamber so long as any water is left in the boiler. The water thus driven up descends again when it leaves the pipes, to be again carried up, and thus a constant circulation of water is kept up in every part of the boiler. In case the level of the water becomes low in the water jackets, the water that is forced through the coils of pipe into the steam chamber has the effect of keeping the heating surfaces moist, and thereby obviates the danger of explosion, and prevents the plates from burning, while the forcible circulation of water through the coils prevents any accumulation of impurities.

From this period, till the year 1855, nothing was done in the improvement of the vertical tubular boiler, and the earliest next in the field was Mr. Chaplin, with three examples of vertical cylindrical boilers, the first of which is shown by Fig. 40. The outer shell is a vertical cylinder, with a hemispherical top, and with an inside cylindrical fire box in its lower part. From the crown of this fire box a series of flue tubes pass vertically upwards to a top tube plate, through which the flue currents pass into a curved funnel-shaped smoke box or draught chamber communicating with the chimney. The boiler water surrounds the outside of the fire box and the tubes, and the steam passes off from the annular space in the top of the boiler, and encircling the smoke box. The flue tubes are conical, or tapered with their small ends upwards, and they are inserted in their tube plates by being passed up through the fire box crown from below. The upper ends of the tubes are screwed externally to receive nuts above the top plate, by screwing which nuts the tubes are made to bear well up into their supporting holes in the plates. The tubes may also be tightened up by means of the nuts if they become slack after use, and they serve as longitudinal stays for the boiler. When the tubes are to be taken out of the boiler, the top nuts are screwed off them, and the tubes, being tapped on their tops by a hammer, are easily removed through the fire box.

The next boiler, represented by Fig. 41, differs from that represented in Fig. 40, just described, in as far as regards the shape of the

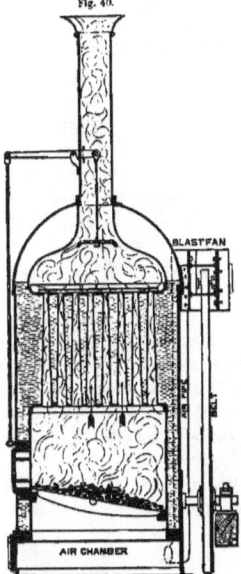

Fig. 40.

Chaplin's Vertical Cylindrical Boiler, with conical fire tubes, internal smoke box, and blast fan. Patented in the year 1855.

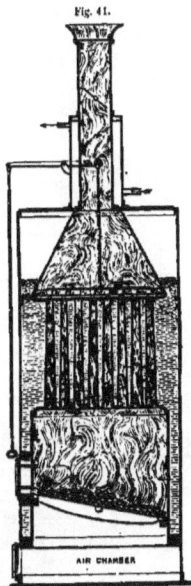

Fig. 41.

Chaplin's Vertical Cylindrical Boiler, with internal smoke box and tube stop plate. Patented in the year 1855.

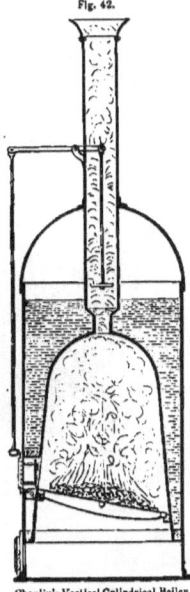

Fig. 42.

Chaplin's Vertical Cylindrical Boiler, with parabolic fire box, central chimney, and adjustable damper. Patented in the year 1855.

top of the boiler and of the smoke chamber above the tubes. The cylindrical portion of this boiler is carried up to the top, and is surmounted by a flat plate, whilst the smoke box is made conical, and larger than in Fig. 40, but still leaving a sufficient steam space between it and the top and sides of the boiler.

The third boiler, represented by Fig. 42, is the same externally as that represented by

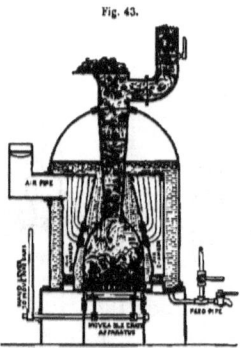

Fig. 43.

Atkinson's Vertical Cylindrical Water Tube Boiler, fitted with a central fire box, fed from the top with fuel through the central flue, and an oscillating fire grate moved by hand. Patented in the year 1855.

Fig. 40, but internally the tubes are omitted, and the fire box is more of a parabolic shape.

The boiler shown by Fig. 40 also illustrates Chaplin's system of forced combustion in the fire, and for this purpose the boiler shell is continued down below the grate bar level, so as to form a closed chamber or ashpit beneath the grate. Access to this chamber when necessary, is by a door, such as represented. Atmospheric air is forced into this closed chamber beneath the grate by means of a small fan fixed to the side of the boiler, the discharge pipe of the fan passing down the side of the boiler and entering the chamber. The fan is represented as driven by a belt from the fly wheel upon the engine shaft, the fly wheel being made with a suitable rim to receive the belt. The current of gases from the fire passing off to the chimney is governed by a damper, adjustable by means of a lever and rod, so that by closing this valve more or less, a greater or less air pressure may be kept up beneath the grate bars, thus forcing the combustion of the fuel thereon.

A Mr. Atkinson followed Mr. Chaplin with the arrangement illustrated by Fig. 43, the fire box of which is nearly parabolic also, and surrounded by a conical-shaped hollow vessel, thus forming an annular water space, the fire box having an opening at the top, from whence proceeds a taper flue, the upper part of which protrudes through the crown of the dome that encloses the boiler and the steam chest. The upper part of the flue has a cover hinged thereto, by opening which the fire below may be fed or charged with fuel.

The chimney is connected to the fuel feeding box, and has a damper, to regulate the draught in the flue.

The outer casing of the boiler is cylindrical and concentric to the inner conical casing, thus leaving an annular space or chamber between those two casings for the passage of atmospheric air, which enters through openings in the base plate. An outlet air pipe is connected to the air chamber, and furnished with a damper, for regulating the quantity of air admitted into the chamber.

The water tubes have their lower ends connected to the conical casing, and the upper ends to the top of the air chamber, thus exposing their surface to the action of the air heat in the chamber, and keep up a circulation of water in the boiler, together with the series of short pipes that connect the water spaces at the bottom.

The fire grate is so constructed as to oscillate upon pivots, by means of a rod and levers and handle affixed thereto for that purpose, as shown.

Novelty being always interesting, we next direct attention to a very novel boiler, shown by Fig. 44, which shows a conical combustion chamber, surrounded by a screw-shaped casing, and the spaces between them contain the water, while a pipe on the top carries off the steam. The flame acts on the hollow part of the screw, and in the chamber also, and thus the water is exposed to its effect on both sides. We need scarcely add that this arrangement is the only one of its kind, and is likely to remain so.

Mr. Golding having wound up the year

Fig. 44.

Golding's Screw Flued Vertical Boiler. Patented in the year 1853.

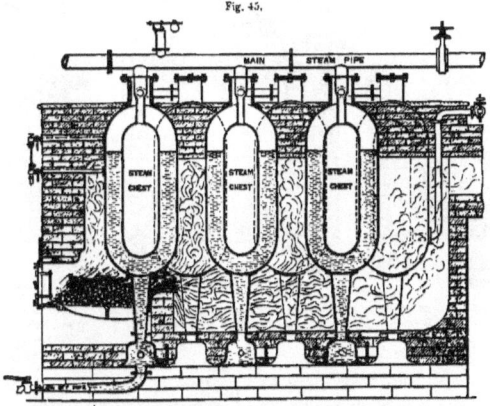

Fig. 45.

Ferinhough's Vertical Water-leg Boiler, with internal steam chests. Patented in the year 1856.

1855 with his scheme, we next show how a Mr. Ferinhough began the ensuing year with his ideas as illustrated by Fig. 45, which shows

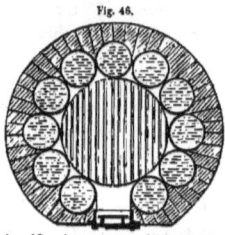

Fig. 45.

Sectional plan of Dunn's arrangement of Cylindrical Semi-globular End Vertical Boiler, fitted with a central steam and water dome, connected by pipes, as shown by Figs. 47 and 48. Patented in the year 1856.

internal pipe, which is connected to the main steam pipe on the top.

Mr. Dunn came suddenly on Mr. Golding with his vertical semi-globular end boilers, the arrangement being shown by Fig. 46 in sectional plan, and in elevation by Figs. 47 and 48, the difference in those two figures being only in the form of the recesses in the base of the dome for the flame to act on.

Fig. 49 is a similar arrangement of conical boilers, semi-globular at their larger ends, and secured at their smaller ends to the base of the steam and water domes, therefore the connecting pipes shown in the two former figures are omitted.

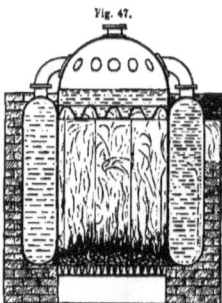  

Fig. 47.   Fig. 48.   Fig. 49.

Dunn's Vertical Semi-globular End Boiler, with corrugated recesses in the bottom of the steam and water dome for the flame to act on. Patented in the year 1856.

Dunn's Vertical Semi-globular End Boiler, with square sectioned recesses in the bottom of the steam and water dome for the flame to act in. Patented in the year 1856.

Dunn's Angular Conical Semi-globular One End Boiler, connected direct to a steam and water dome, having angular recesses in its base for the flame to act on. Patented in the year 1856.

a series of semi-globular end boilers, supported on water legs, which are connected by pipes at their base, as are also the tops of the boiler: and to heat the steam and keep it so, each boiler has an internal steam chest in it of the same form as the boiler, suspended by an

It is apparent that the flame in those arrangements acted only on about half of the surface of the vertical boilers, and consequently the remaining half was not exposed to the heat, and thus the evaporation reduced it. Therefore, to obviate that, Mr. Dunn introduced a series of arrangements of larger cylindrical shell vertical boilers, having semiglobular ends, and fitted within with fire grates, tubes, and flues; three examples of which are shown in sectional elevation and plan by Figs. 50 to 55 inclusive. In each case the flame descends

Fig. 50.

Dunn's Vertical Boiler, fitted with an internal fire box, combustion chamber, two rows of water tubes, and a main flue. Patented in the year 1856.

Fig. 52.

Dunn's Vertical Boiler, fitted with internal twin fire boxes and combustion chambers, and a cylindrical central main flue, with one row of water tubes. Patented in the year 1856.

Fig. 54.

Dunn's Vertical Boiler, fitted with internal twin fire boxes and combustion chambers, and a flat central main flue, with four rows of water tubes. Patented in the year 1856.

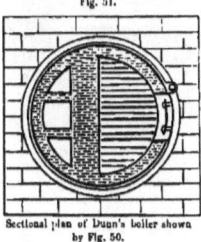

Fig. 51.

Sectional plan of Dunn's boiler shown by Fig. 50.

Fig. 53.

Sectional plan of Dunn's boiler shown by Fig. 52.

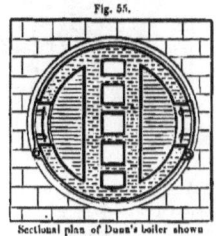

Fig. 55.

Sectional plan of Dunn's boiler shown by Fig. 54.

and passes through a flue below the fire grate, instead of ascending as in the former examples.

and 57 are a sectional elevation and plan of an arrangement in which the fire box and combustion chamber are on the same level,

Fig. 56.

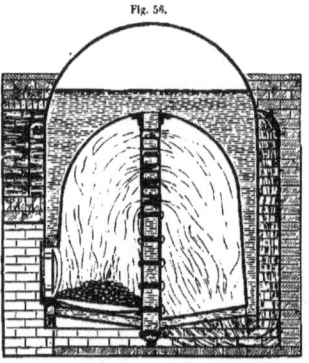

Holt's Vertical Cylindrical Boiler, fitted with the fire box and combustion chamber on the same level, divided by a water space and horizontal tubes. Patented in the year 1856.

Fig. 58.

Holt's Vertical Cylindrical Boiler, fitted with the combustion chamber over the fire box, divided by a side water space and vertical tubes. Patented in the year 1856.

Fig. 57.

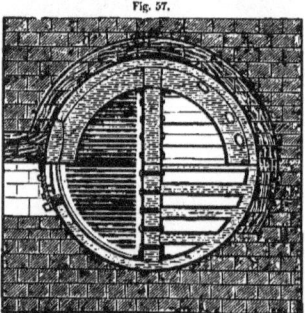

Sectional plan of Holt's boiler shown by Fig. 56.

Fig. 59.

Sectional plan of Holt's boiler shown by Fig. 58.

A Mr. Holt came next, and evidently borrowed a little from Mr. Dunn, as the following illustrations display on comparison. Figs. 56

and alike in shape, but divided by a water space.

The flame from the fire box passes through

LAND STATIONARY VERTICAL BOILERS.

horizontal tubes, and thence down into the flue under the boiler, up the back flue, splitting into the circular flues, and out at the front flue to the chimney. The box and chamber are enclosed in a cylindrical shell with a dome top, and the water spaces at the bottom are connected by pipes situated under the fire bars.

Figs. 58 and 59 are similar views of a boiler boiler is fitted with a duplicate set of details, and the main flue is in the centre.

The next example of Mr. Holt's vertical boilers is shown in sectional elevation by Fig. 62, which it will be seen is nearly arranged as in Fig. 60, the difference being one row of tubes less, and the central flue passing through the dome, so that the flame ascends entirely in this case.

Fig. 60.

Holt's Vertical Cylindrical Boiler, fitted with duplicate fire boxes, horizontal tubes, and central combustion chamber. Patented in the year 1856.

Fig. 61.

Holt's Vertical Cylindrical Boiler, fitted with duplicate vertical tubes, and central main flue. Patented in the year 1856.

Fig. 62.

Holt's duplicate arrangement of Vertical Boiler, with the central flue passing through the dome, and radial water and flame tubes. Patented in the year 1856.

with vertical tubes for the flame to pass through into a combustion chamber, and from thence through a twin set of tubes into the chamber below, after which the circuit is the same as in the other case, as also is the mechanical arrangement.

The next illustrations, Figs. 60 and 61, are precisely as Figs. 56 and 58 in the principle of their arrangements, but in this case the

To enable these views to be fully appreciated, we illustrate a sectional plan of them by Fig. 63, on the next page, which represents that the flame and water tubes are radially arranged rather than direct across, as in Figs. 57 and 59 on page 26, also indicating that Mr. Holt's ideas of mechanical arrangements were so concentrated, it very nearly prevented an obvious difference in them.

The year 1856 being nearly at an end, Mr. Bougleux thought the best way to finish it was to patent his ideas of a vertical boiler, as illustrated by Fig. 64 in sectional elevation, other. The shell is cylindrical with flat ends, the top end being stayed by a single rod, and the bottom by angle iron rings.

Fig. 63.

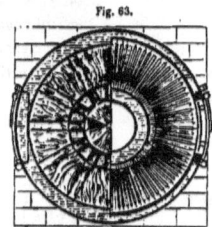

Sectional plan of Holt's boilers shown by Figs. 60 and 62.

Fig. 65.

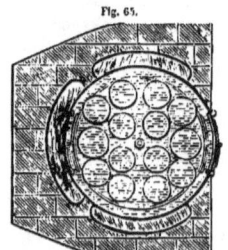

Sectional plan of Bougleux's boiler shown by Fig. 64.

Fig. 64.

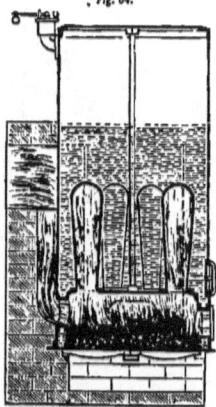

Bougleux's Vertical Cylindrical Boiler, fitted with vertical flame pots on the top of the fire box. Patented in the year 1856.

Fig. 66.

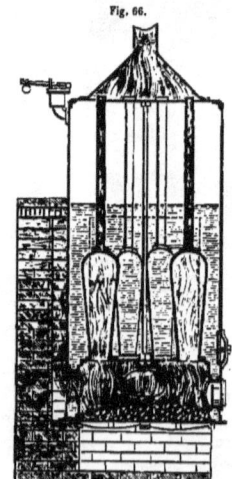

Bougleux's Vertical Cylindrical Boiler, fitted with vertical tubes on the flame pots. Patented in the year 1857.

and in sectional plan by Fig. 65, which is an arrangement of vertical flame pots secured on the top of a cylindrical fire box, having the fire door and the flame opening opposite each

The intended action of the flame in this

boiler is, that after rising from the grate it should fill the "pots," and then descend and pass up the flues; but doubtless the residue of the flame only would ascend into the pots, and thus their duties were reduced.

If that were not the case, why did the same Mr. Bougleux begin the next year, 1857, with an improvement on that boiler, as illustrated in similar views by Figs. 66 and 67; the improvement being the addition of tubes connecting the pots and the top of the boiler, and therefrom the flame could ascend in them for certain.

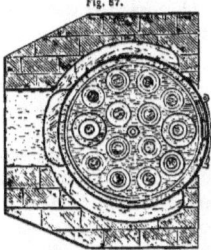

Fig. 67.

Sectional plan of Bougleux's boiler shown by Fig. 66.

A Mr. Fowler then introduced his notion of a vertical boiler, as illustrated by Fig. 68. The boiler is cylindrical, and has a flue through its middle, connected at the top by the funnel or tube to the chimney. It is supported upon hollow or tubular legs connected with its lower end. Through these legs water is admitted to the boiler, and sediment discharged by means of valves or cocks, or any convenient mode. The furnace is placed below the bottom end of the boiler and between the tubular legs. Between the boiler and the surrounding brickwork is an open space nearly as high as the water level. The width of the space must depend on the draft of the chimney and other circumstances. An annular flue is built in the brickwork, which opens directly into the chimney. It is connected with the space by means of the openings through which the heated products of combustion, after having acted on the surface of the boiler, pass to the side chimney, while the flue allows a

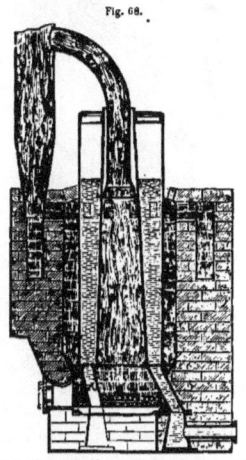

Fig. 68.

Fowler's Vertical Cylindrical Boiler, with an internal flue. Patented in the year 1857.

portion of the heat to pass to the central chimney through the interior of the boiler.

When the year 1858 began, a Mr. Soames invented the arrangement of boiler as shown by Fig. 69, on the next page, which is a conical fire box, surrounded by a conical water space, and that by a flame space which is enclosed by a water space also, and then a second flame space encloses it, above which is the water contained in the shell of the boiler, and the connections of all those spaces are by tubes, as shown.

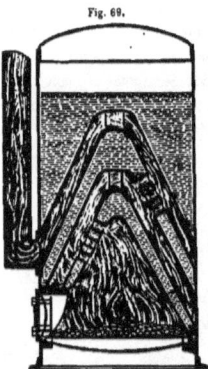

Fig. 69.

Soame's Vertical Cylindrical Boiler, fitted with a conical fire box, and conical flame and water spaces above it. Patented in the year 1858.

The action of the flame in this boiler is that it rises from the fire box through the tubes to the first flame space, and from there to the tubes leading to the second flame space. It ascends next to the top, and then descends to the chimney opening above the fire box door. This boiler would require a fierce draught to work it, and in case of repair must be taken entirely to pieces.

We now direct attention to a very simple vertical boiler which is illustrated by Fig. 70, in sectional elevation, and in sectional plan by Fig. 71, patented by a Mr. Bowman in the year 1858; the arrangement being that over the fire grate, and contained in the fire box, there are two cross tubes that connect the annular spaces surrounding the fire box. But ten years before that a Mr. Millward invented and constructed a vertical boiler as shown by Fig. 72 in sectional elevation, and in sectional plans by Figs. 73 and 74. From which it is evident that Mr. Millward anticipated Mr. Bowman in the use of the cross tube, and also Mr. Cameron in the use of the conical central water box. See Fig. 27, page 15. The action of the flame on Millward's boiler is that after being split by the cross tube, it ascends to the four openings leading to the top flues, and then descends to the lower flues, surrounds the boiler, and passes out at the main opening to the chimney; while in Bowman's boiler the flame ascends direct to the chimney situated on the top of the fire box.

Early in the ensuing year, 1859, Mr. Chaplin, whose former productions are illustrated on page 21, patented an arrangement of vertical tubular boilers, in which the main feature was that the tubes were made conical for about half their length, and parallel for the remainder, or two tubes of those shapes were joined together to form one of unequal form.

The illustration, Fig. 75, on page 32, shows a sectional elevation of Chaplin's vertical cylindrical boiler with a conical shell, and fitted with tubes as described, over the fire box.

The next example is shown by Fig. 76, on page 32, which is a central contracted shell, containing a conical fire box and tubes of the same order as before.

Mr. Chaplin next stretched his scheme to the arrangement illustrated by Fig. 77, on page 33, showing the tubes arranged at the side of the fire box, instead of over it, and he introduced a water fire bridge also in the place of the ordinary brick bridge.

Not being exhausted, Mr. Chaplin extended

his ideas to the curved tube system as arranged for vertical boilers, with the flame passing through them, and illustrated by Fig. 78, p. 33.

The first example is shown in sectional elevation by Fig. 79, and in plan by Fig. 80, on page 34, the arrangement of which is that the

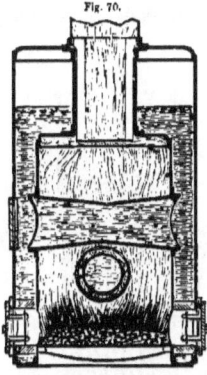

Fig. 70.

Bowman's Vertical Cylindrical Boiler, the fire box being fitted with cross tubes. Patented in the year 1858.

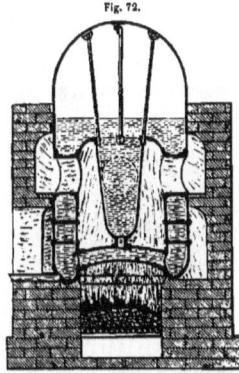

Fig. 72.

Millward's Vertical Cylindrical Boiler, fitted with a "pot" and cross tube over the fire grate. Constructed in the year 1848.

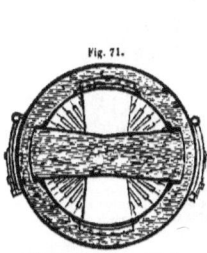

Fig. 71.

Sectional plan of Bowman's Boiler, shown by Fig. 70.

Fig. 73.

Sectional plan through the top flues of Millward's Boiler, shown by Fig. 72.

Fig. 74.

Sectional plan through the cross tube and lower flue of Millward's boiler, shown by Fig. 72.

The year 1860 began inauspiciously for Mr. Rowan, because he patented an arrangement of vertical tubular boilers which were difficult to manufacture, and worse to repair. shell is cylindrical, and is fitted internally with two horizontally placed rings, at the top and bottom, which are connected by vertical tubes equidistantly placed, as shown in the plan.

The bottom ring is connected also by horizontal tubes to a large central vertical tube which extends to the chimney. This tube is conical at the lower part where the fire grate surrounds it just above the ring; and the parallel portion of the tube directly above the cone is surrounded by two horizontal rings, that are connected to two upper rings, on a level with the largest top rings, by tubes, as in the other case.

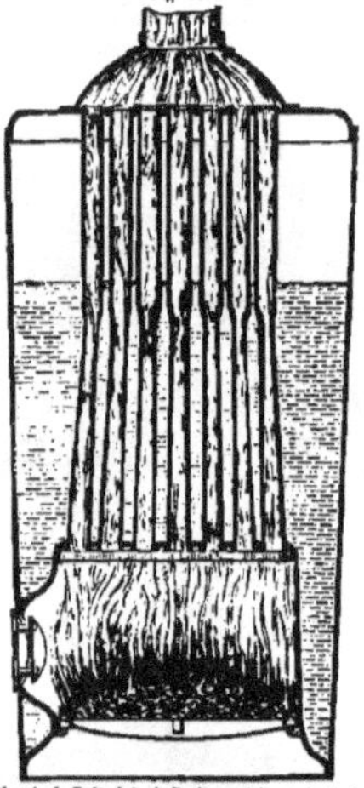

The water, it will be seen, reaches nearly to the top of the tubes, the space above being for the steam, and the rings are connected to the central large tube by curved pipes as shown in the elevation, by which connection the steam space is sufficiently increased.

Fig. 75.

Chaplin's Vertical Cylindrical Boiler, with conical shell, central contracted tubes, and cylindrical fire box. Patented in the year 1859.

The flame surrounds the tubes and rings above the grate, and superheats the steam in passing up the annular space in the chimney.

The next example by Mr. Rowan is still more complicated, the addition being internal

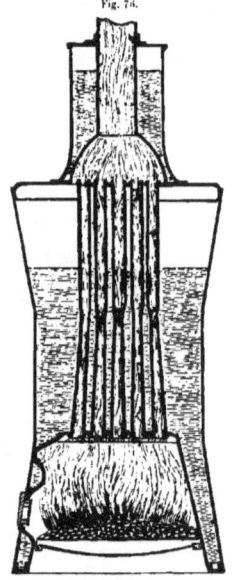

Fig. 74.

Chaplin's Vertical Cylindrical Boiler, with central contracted shell and tubes, conical fire box, and feed water heater surrounding the chimney. Patented in the year 1859.

flame tubes put in the short water tubes, to increase the effect of the flame, as illustrated in sectional elevation and plan by Figs. 81 and 82, on page 34.

The third example is shown similarly by Figs. 83 and 84, on page 34, in which case the rings are square in section, as also are the outer vertical tubes; and the inner short tubes are smaller and more in number comparatively.

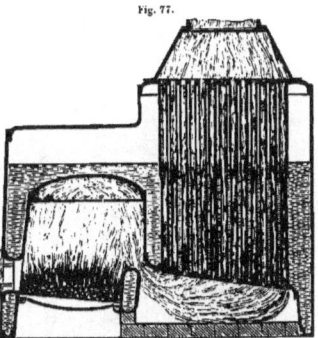

Fig. 77.

Chaplin's Vertical Boiler, fitted with a cylindrical fire box, water bridge, and central contracted tubes arranged at the side of the fire box. Patented in the year 1859.

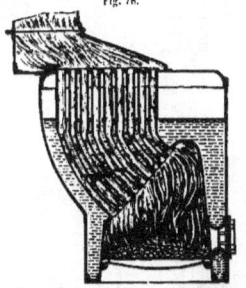

Fig. 78.

Chaplin's Vertical Boiler, fitted with curved central contracted tubes. Patented in the year 1859.

After Mr. Rowan came Mr. Pullan, with a much simpler arrangement of vertical tubular boilers, extending, too, to five examples; the first of which is illustrated by Fig. 85, on page 35. The shell is cylindrical, as also is all the internal fittings, arranged as follows:— The annular water space is suspended in the upper part of the fire box, by being connected at the bottom with a cross tube that extends from side to side of the fire box, and is connected with the main water space of the boiler. The object in this arrangement is to ensure water being kept in the annular water space; and also the sediment is carried down to the lower part of the boiler, thus preventing the plates over the fire from being burnt. Tubes connect the outer flame space with the central fire box, so as to cause a current of flame to pass through from the outer space to the inner one, and from thence to the main flue tubes of the boiler.

The main flue tubes are curved in the annular water space, to cause the flame to take a more circuitous route instead of passing direct to the chimney.

The second example is shown by Fig. 86, and in this case there are two annular water spaces over the fire box, and thus the flame is split, and extends more amongst the water receptacles. The annular water spaces are connected at the bottom by tubes to the shell water spaces. The circuit of the flame is the same in principle in this case as in the other, but more extensively carried out here.

The third example is shown by Fig. 87, and the internal arrangement consists of a horizontal barrel or chamber extending from side to side of the fire box, and communicating with the main water spaces of the boiler. In this chamber are vertical chambers projecting up into the water space, so as to obtain a

LAND STATIONARY VERTICAL BOILERS.

large amount of direct heating surface in a small space. Before arriving at the chimney the heated gases will have to travel round the horizontal barrel, or chamber, and the vertical

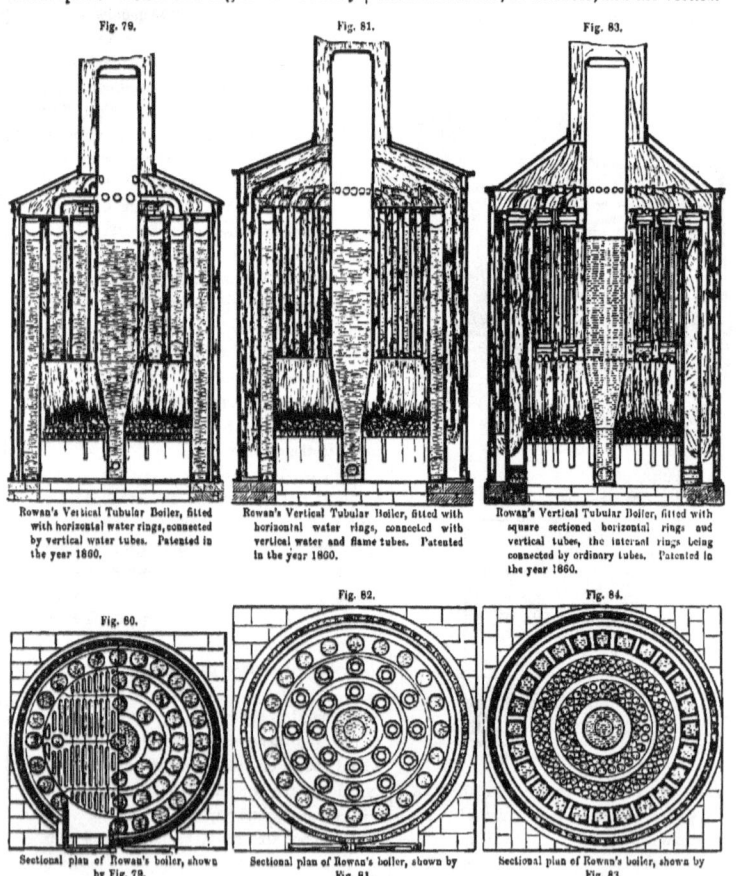

Fig. 79. Fig. 81. Fig. 83.

Rowan's Vertical Tubular Boiler, fitted with horizontal water rings, connected by vertical water tubes. Patented in the year 1860.

Rowan's Vertical Tubular Boiler, fitted with horizontal water rings, connected with vertical water and flame tubes. Patented in the year 1860.

Rowan's Vertical Tubular Boiler, fitted with square sectioned horizontal rings and vertical tubes, the internal rings being connected by ordinary tubes. Patented in the year 1860.

Fig. 80. Fig. 82. Fig. 84.

Sectional plan of Rowan's boiler, shown by Fig. 79.

Sectional plan of Rowan's boiler, shown by Fig. 81.

Sectional plan of Rowan's boiler, shown by Fig. 83.

chambers, and some portion up through the short flame pipes.

The fourth example differs in arrangement from the others, as is shown by Fig. 88, on the next page. This boiler is also cylindrical, and surrounded with a casing lined with fire bricks, thus forming a flame space. The fire box is cylindrical, and is enclosed by an annular chimney. The narrow part of the water space is continued downwards under the fire box, so as to collect the sediment. A steam pipe is connected in the upper part of the steam space of the boiler, and is continued in a coil round the boiler in the flue space, where it is heated for the purpose of super-heating the steam.

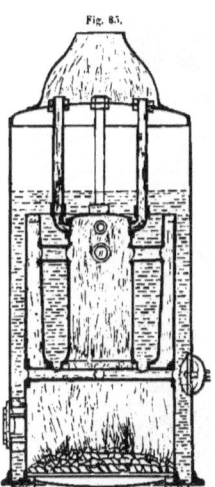

Fig. 85.

Pullan's Vertical Cylindrical Boiler, fitted with central flame box, annular flame space, and flame tubes. Patented in the year 1860.

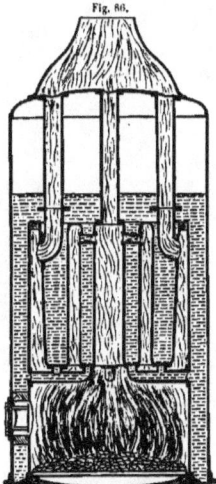

Fig. 86.

Pullan's Vertical Cylindrical Boiler, fitted with central flame box, two annular flame spaces, and flame tubes. Patented in the year 1860.

Fig. 87.

Pullan's Vertical Cylindrical Boiler, fitted with a cross chamber over the fire grate, containing two vertical chambers. Patented in the year 1860.

water space at the sides. Short flue tubes pass through the annular water space of the boiler, and communicate with the flue space between the outside of the boiler and the brickwork. By this arrangement the flame and gases from the fire box are made to pass over the upper exterior of the boiler before arriving at the

Mr. Pullan's fifth arrangement is illustrated by Fig. 89, and is a vertical boiler fitted with a horizontal tubular chamber or water space, extending from side to side of the fire box, and having tubes passing horizontally through the same and fastened into the outer shell of the boiler; thereby opening a communication between the smoke box on each side of the barrel. The outer casings of the smoke boxes are connected with the boiler. In order to connect the fire box with the smoke box, short tubes are passed through the water space of the shell, and the heated gases will, in escaping from the fire box, pass through the tubes into the first smoke box, and back through the long tubes to the second smoke box, and thence to the chimney.

A Mr. Giles then appeared with his notion of what a vertical tubular boiler should be, as illustrated in sectional elevation and plan by

Fig. 89.

Pullan's Vertical Cylindrical Boiler, with one annular water space surrounding the fire box, and an annular flame space surrounding the shell. Patented in the year 1860.

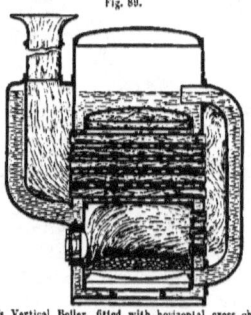

Fig. 89.

Pullan's Vertical Boiler, fitted with horizontal cross chamber, tubes, and outer flues. Patented in the year 1860.

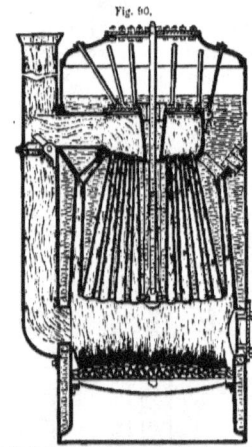

Fig. 90.

Giles's Vertical Cylindrical Boiler, fitted with angular conical tubes, connecting the fire box and internal combustion chamber, and side chimney open to both. Patented in the year 1860.

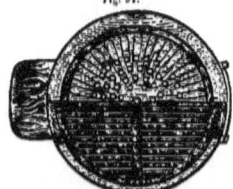

Fig. 91.

Sectional plan of Giles's boiler, shown by Fig. 90.

Figs. 90 and 91, the main feature of which is that the shell, being cylindrical, contains a fire box and combustion chamber of a similar shape, that are connected by conical tubes angularly situated, and a side chimney is secured to the shell connecting the chamber and box, the flame from the latter being regulated in its exit over the grate by a damper in the chimney, which regulation also tends to affect the amount of flame passing above through the tubes and combustion chamber.

Very soon after Mr. Giles had made known his ideas, a Mr. Burch laid a claim to publish his also, and began according to the arrangement illustrated in sectional elevation and plan by Figs. 92 and 93. The fire box is cylindrical, with a convex top. And opposite to the fire door, above the grate, is the flame opening that communicates with the annular flues, that are narrow, and encircle the fire box, with which in height and depth they about correspond, and are of sufficient diameter to form narrow water spaces between them and the furnace. The flame on entering the flues divides, as shown in plan, and meets again at the opposite side over the fire box door at the exit opening, where it passes out through the shell of the boiler to the chimney. The vertical sides of the shell of the boiler enclose a narrow water space between it and the flues, and below the fire grate.

Another method of carrying out this arrangement is shown in sectional elevation and

Fig. 92.

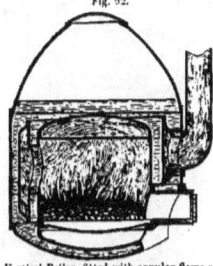

Burch's Vertical Boiler, fitted with annular flame and water spaces. Patented in the year 1860.

Fig. 93.

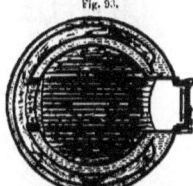

Sectional plan of Burch's boiler, shown by Fig. 92.

Fig. 94.

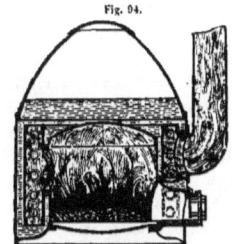

Burch's Vertical Boiler, fitted with circular flame tubes. Patented in the year 1860.

Fig. 95.

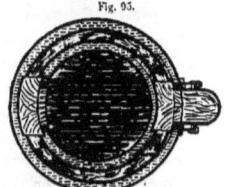

Sectional plan of Burch's boiler, shown by Fig. 94.

plan by Figs. 94 and 95, which illustrate that, in the place of the annular flues, a series of tubes are used for the flame to pass through; but the line of the action of the flame is the same as in the previous case.

But Mr. Burch did not leave off inventing there, but strained himself so much that he schemed to put three of his boilers in one shell, on the top of each other, as illustrated in sectional elevation and plan by Figs. 96 and 97, and shows for what purpose some people exert their talents.

Early in the month of January, 1861, a varnish manufacturer, named Matheson, gave birth to an idea of his, which is illustrated in sectional elevation and plan by Figs. 98 and 99. The coil system, however, was introduced by Mr. Bellford in the year 1853, as shown on page 19. Mr. Matheson's arrangement consists of a square casing that encloses the water coils in communication by their lower ends with a chamber which forms a mud space. The upper ends of the coiled pipes pass through the bottom of a close vessel or chamber, which forms a water reservoir above the casing; and those ends of the pipes being open, steam escapes therefrom into the space in the dome or upper part of the vessel, which is provided with a safety valve. The ends of the pipes are also covered by a cap, to prevent a sudden rush of steam from improperly acting on the safety valve, and also to prevent priming.

The fire is fed with fuel from above through the opening, which is closed by a door as in ordinary furnaces. The furnace or fireplace is situated in the central space that is surrounded by and enclosed within the coils.

An ashpit is placed below the fireplace, and is closed in front by a door as usual. As the fire is situated in the centre of the smallest

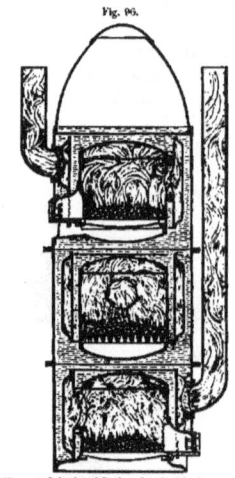

Burch's Vertical Cylindrical Boiler, fitted with three separate fire boxes, annular flues, and chimneys. Patented in the year 1860.

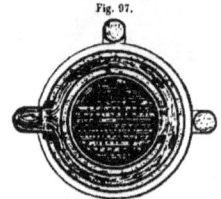

Sectional plan of Burch's boiler, shown by Fig. 96.

coil, those pipes are close together at the top and bottom, to prevent the fuel from passing between them; but at the same time have sufficient spaces between for the flame and heated gases from the fuel to pass through and act on the coils of the outer pipes surrounding the inner.

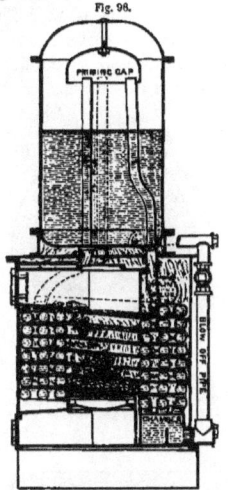

Matheson's Vertical Boiler, fitted with water coils, and an overhead water and steam chamber. Patented in the year 1861.

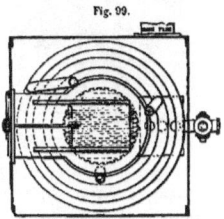

Sectional plan of Matheson's boiler, shown by Fig. 98.

The fireplace is closed in at the top by a hollow cover, to the upper part of which a damper is adapted, and is worked by a rod, for the purpose of creating a draught when the fire is first lighted; and by opening this damper the gaseous products of combustion are allowed to pass direct from the fireplace to the main flue, and will thereby create a strong draught through the fire; but when the damper is closed these gases will pass between the coils, that are left more open at the central part for the purpose.

Upon referring to the plan, it will be seen that, as the pipes are arranged in circular coils within the square casing, there will be space for the gases to circulate in the chamber before passing to the main flue. The water reservoir or top chamber having been supplied with water to the proper level, the stop-cock on the blow-off pipe is opened, and water is allowed to flow from that chamber down the pipe into the lower chamber, from whence it passes up into the coiled pipes, and fills them. As fast as steam is generated in the pipes, it passes out at their upper ends, which for the free exit of the steam are extended above the level of the water in the reservoir, and water from the reservoir will enter at the lower end only of the pipes, and a constant circulation will be thus maintained in the pipes.

An engineer named Vavasseur in the next month introduced his arrangement of tubes, connecting the fire box and smoke box, as illustrated by Figs. 100 and 101 in sectional elevation and plan. This is a conical water box in the fire box, the top of which, and the cone part, are connected by flame tubes to the combustion chamber over the steam space.

Next on the list of inventors is Mr. Bremner, a worthy Scotchman, who considered that the best way to evaporate water and form steam was to make passages for the flame to pass through, and the water to surround their plates and tubes; and he carried out that in one form, as illustrated in sectional elevation by Fig. 102, and in sectional plan by Fig. 103, from which it is seen that the plan of this boiler is square in section, with rounded corners, and is arranged with two furnaces in the width, formed of double lengths of fire bars running from the front plate to the back, but having one common ashpit. These fire grates have each a furnace door passing through the water space in the front of the boiler, which admits of charging or firing the furnaces alternately; for the better burning of the gaseous products of combustion, the smoke evolved from the fresh fuel meets the flame of the fire, and is ignited at the bridge which divides the two fires, as shown in the elevation. All round this rectangular basement, or combined double furnace, there is formed

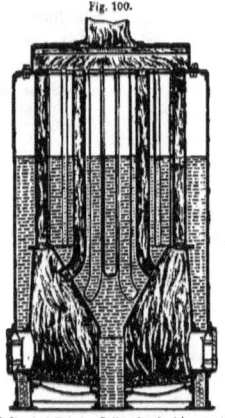

Fig. 100.

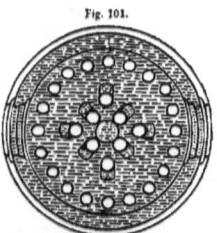

Fig. 101.

Vavasseur's Vertical Tubular Boiler, fitted with a conical water box, connected by flame tubes to the smoke box. Patented in the year 1861.

Sectional plan of Vavasseur's boiler, shown by Fig. 100.

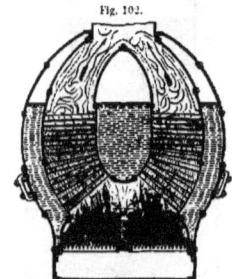

Fig. 102.

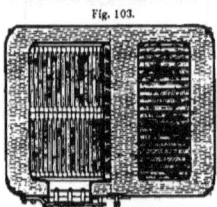

Fig. 103.

Bremner's Vertical Square Boiler, fitted with horizontal angular water tubes, central and side water and steam spaces. Patented in the year 1861.

Sectional plan of Bremner's boiler, shown by Fig. 102.

by the outer sides, and an inner shell or casing, an outer water and steam space. This water space begins close to the level of the furnace bars, and at the front and back extends upwards in a vertical direction to above the height of the tubes also. At the sides this space is considerably enlarged over the furnace, and bulges outwards in a curved or crescent form as it extends upwards; while the inner and outer casings are contracted gradually all round, from about the water level, in the form of a dome, till they are joined at the crown, leaving a circular aperture to receive the chimney.

A water space is also formed in the centre of the boiler over the two furnaces by an inside casing between the back and front inner shells, and joining the water spaces at the two sides. The sides of this casing are somewhat of a pear shape in elevation, but are parallel to each other in plan, and divide the space over the furnaces into two rectangular flues, which conduct the flame and gases of combustion up the two sides of the boiler, and they intermix with each other at the crown, just before entering the chimney; and across those two rectangular flues are arranged a series of tubes in rows, radiating at right angles to the curved surfaces of the two inner shells of the boiler, which form the tube plates for the inner ends of the tubes. The lower rows of tubes are so arranged that the spaces between them are wider than those above, so as to allow the flame easier access to the upper tubes. The flame from each furnace passes up the flues, and impinges on each row in succession, until it reaches the top row, where it acts upon the arched dome.

On second consideration, evidently, Mr. Bremner thought that the horizontal arrangement of tubes might not be the best, and he consequently devised a vertical arrangement for them, as illustrated by Fig. 104. The

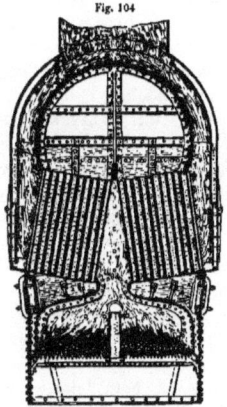

Fig. 104

Bremner's Vertical Square Boiler, fitted with vertical angular water tubes, and water pockets below them. Patented in the year 1861.

two fire grates are in this instance formed by three lengths of bars, and are separated by a feed-water chamber, perforated and arranged with tubes passing through the front and back water spaces, and have mud hole doors at the four corners of the lower part of these water spaces, as in the former example. The back and front water spaces extend up from the furnace in a vertical direction, till they merge into the large water and steam space above the upper tube plates. The inner shells of the side water spaces are each curved inwards partly over the fire grate, and are then turned upwards, and riveted to a flat tube plate. This plate is at a slight angle, inclined upwards, and extends outwards to join the outer shell, that is extended until it meets the outer plate, and is riveted to it, thus enclosing a water space over each furnace, which space extends from the front to the back water spaces. The crown of the fire box is arched over by a stay riveted to the outer shell, and to this stay is firmly riveted the upper tube plates, lying parallel to the lower tube plates.

Those two pairs of upper and lower tube plates enclose two flue spaces, diverging to the right and left into smoke boxes, and from this upwards into two flues, carried round the arched crown of the boiler to the foot of the chimney.

The smoke boxes are enclosed at the back and front ends by a plate projecting from the body of the boiler in a line with the inner lining, and at the sides by double plate doors, for the purpose of getting at the tubes to clean them. The tubes are arranged in a nearly vertical manner, at right angles to their tube plates, and in parallel rows, filling the whole space enclosed by the flue. The outer covering of the flue above the smoke box doors is formed of a double casing of boiler plates enclosing a superheating space, supported or attached to the top of the boiler by brackets.

The arched portion of the boiler, which forms the steam space, is made large enough for the withdrawal or introduction of the tubes, and is fitted with a pipe for conveying the steam from this part to the superheating chamber, from whence the steam is conveyed by a suitable pipe to the engine. A man-hole door is formed at the lower part of the chimney, and another, opening into each of the water spaces over the fire places, for the purpose of making the end joints of the tubes.

It will have been noticed that on page 40,

Fig. 105.

Burch's Vertical Boiler, fitted with water tubes, internal flues, and revolving fire grate. Patented in the year 1861.

Fig. 106.

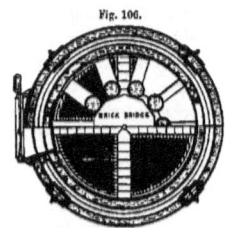

Sectional plan of Burch's boiler, shown by Fig. 105.

we illustrated, by Figs. 100 and 101, an arrangement of vertical boiler with a central water box connected by flame tubes to the smoke box, and that those tubes were curved at their lower ends. We now direct a comparison with the two views illustrated by Figs. 105 and 106, in sectional elevation and plan, where the tubes are also curved, but contain water, with the flame surrounding them, showing that both inventors differed extremely in the methods for raising steam; and that Mr. Burch, who closed the year 1860 so laboriously with his three-furnaced boiler, shown on page 38, recovered himself again in 1861, in the month of May, and introduced the arrangement referred to, which consists of an annular water space around the fire box, and above the grate are tubes connecting the annular space to the space above the crown of the fire box, to cause a circulation of the two portions of water contained therein, which by the way was also done by Mr. Atkinson, in the year 1855, as shown on page 22.

The flame surrounds the tubes, but is prevented from a direct circuit to the flues by a horizontal brick bridge, suspended by brackets secured to the tubes, and a curved bridge secured on the tubes directly opposite the flue opening, which extends around the fire box to a vertical flue, over the fire door, that leads to an angular flue extending through the steam space to the chimney.

Mr. Cater also in the same month put forth his arrangement of a vertical tubular boiler, as shown in sectional plan and elevation by Figs. 107 and 108. The fire box is conical, with a mushroom-shaped top, in which are water tubes, also flame tubes that are connected to an annular flue below the fire grate, leading to the chimney at the side.

In this year, 1861, spiral shape boilers were much in fashion amongst inventors, and after Mr. Matheson introduced his arrangement, as shown by Figs. 98 and 99, on page 39, a Mr. Hughes thought he could improve on it, according as illustrated by Figs. 109, 110, and 111, on page 44, in which is shown a sectional elevation and two plans of a central water tube in connection with four coils arranged in this way. The water tube rests on a stone in the centre of the ashpit, and extends to the top of the furnace, and there connects with

Fig. 107.

Cater's Vertical Cylindrical Boiler, fitted with flame and water tubes, and side chimney. Patented in the year 1861.

Fig. 108.

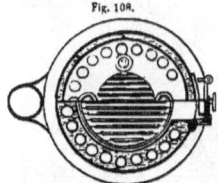

Sectional plan of Cater's boiler, shown by Fig. 107.

a steam chest fitted with flame tubes to superheat the steam; and above this chest is a steam dome, having a steam pipe and safety valve secured on it, the man-hole being about midway of the dome. The coils are situated in the furnace, and are angularly placed, that the flame may act fully on them; and as the water line is above the coils, the water in them evaporates into the main water tube into its water, rather than into the steam chest above, as in Matheson's arrangement.

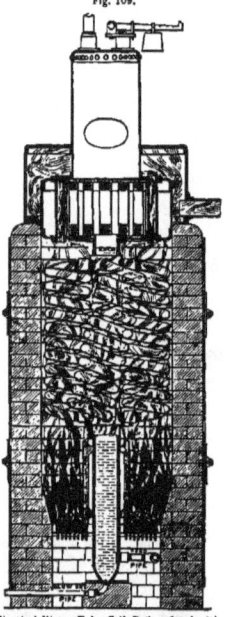

Hughes's Vertical Water Tube Coil Boiler, fitted with a central water tube, superheater, and steam chest. Patented in the year 1861.

As far back as the year 1853, Mr. Galloway introduced the flame and water tube boiler, as shown by Figs. 29 and 30, on page 16, and in this year, 1861, he again came forward with four arrangements, each claiming a separate advantage. The first is shown by Fig. 112,

which is a cylindrical shell fitted internally with three water pockets or chambers, with the top and bottom pockets connected by a central pipe. The midway pocket is connected to the two others by vertical pipes, and thus the heat finds a ready circulation; also the

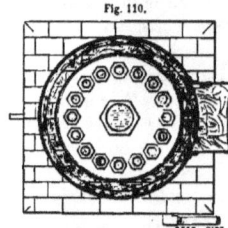

Sectional plan of the superheater of Hughes's boiler, shown by Fig. 109.

Sectional plan of the Water Tube Coil Tubes of Hughes's boiler, shown by Fig. 109.

tubes in each set are increased in diameter from the lower set.

The action of the flame in that boiler is this. The products of combustion rising from the furnaces impinge against the under side of the lower water chamber, and are guided through the space left between its circumference and the inner casing of the boiler; they then circulate around the first set of tubes, and pass through the space left between the conical pipe and the opening in the centre of the next or second water chamber; they then circulate around the midway set of tubes, and pass between the circumference of the upper water chamber and the interior of the inner casing; after which they circulate around the top set of tubes, and thence pass through the funnel to the chimney.

The second example is shown by Fig. 113,

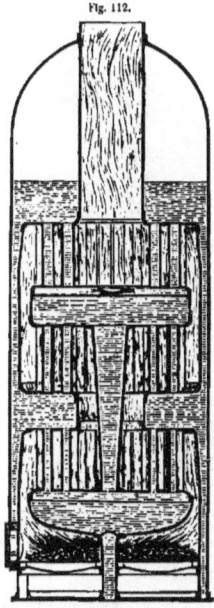

Fig. 112.

Galloway's Vertical Cylindrical Boiler, fitted with water pockets, water tubes, and central water pipe. Patented in the year 1861.

and in this arrangement the fire grate is supported on rollers, and is turned slowly round by any convenient mechanism. The pipe in the centre of the grate is open to the atmosphere, to supply air for assisting the combustion of the fuel, and this pipe and the outer

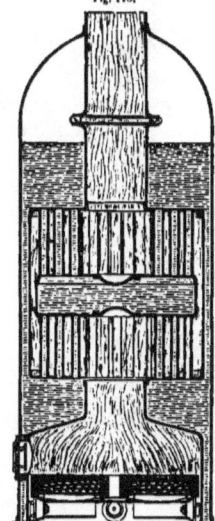

Fig. 113.

Galloway's Vertical Cylindrical Boiler, fitted with a midway water pocket, water tubes, and revolving fire grate. Patented in the year 1861.

circumference of the grate are furnished with a casing of fire bricks.

The products of combustion act first on the under side of the lower water chamber, and after passing through the opening in its centre, circulate around the first set of tubes, midway water chamber, and the second set of tubes;

from thence they proceed to the funnel and chimney. There are two man-holes in the midway chamber, by which access can be had to all parts of the boiler.

The third example is shown by Fig. 114,

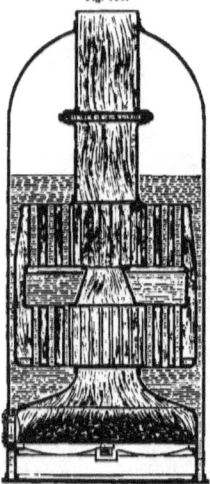

Fig. 114.

Galloway's Vertical Cylindrical Boiler, fitted with water pockets, water tubes, and a central flame pipe. Patented in the year 1861.

and in the place of the central water tube, as shown in Fig. 112, there is a central flame tube in this case, with a midway water chamber, as in the former example shown by Fig. 113.

The fourth example is the most complicated of the lot, as the sectional elevation shown by Fig. 115 indicates. In this arrangement the products of combustion rising from the two fire grates pass through the opening in the lower water chamber, and then through the radiating horizontal tubes, next between the midway water chamber and the inner casing of the boiler, and around the vertical tubes, some of which are conical, and thence to the funnel and chimney. The midway chamber is

Fig. 115.

Galloway's Vertical Cylindrical Boiler, fitted with a water pocket, conical water pot, horizontal flame tubes, and vertical water tubes. Patented in the year 1861.

connected to a conical water tube situated over the fire grate.

Mr. Galloway also proposed to divide the fire boxes of his boiler with water spaces, and two door openings, to be used in each example.

We have already shown, on page 15 of this work, that the cone-shaped flame space and central cone water space were invented in the year 1852; but in spite of that we show, by Fig. 116, that a Mr. Kinsey nine years after

Kinsey's Vertical Cylindrical Boiler, fitted with a central conical water pot, annular flame space, and water tubes. Patented in the year 1861.

patented an idea very similar in principle to the former. The central "pot" in this case has also the fire surrounding it, but the flame passes over the roof of the pot to the chimney: the top and bottom portions of the water in the shell being connected to the water in the pot by tubes. But Mr. Kinsey introduced, at the same time, a boiler with an annular "pot," as illustrated by Fig. 117, or a reverse position for the water spaces, as shown by Fig. 69, on page 30.

Our next illustrations, as shown in sectional elevations and plan by Figs. 118, 119, and 120,

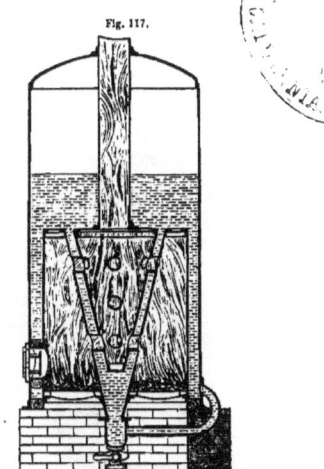

Kinsey's Vertical Cylindrical Boiler, fitted with an annular conical water pot. Patented in the year 1861.

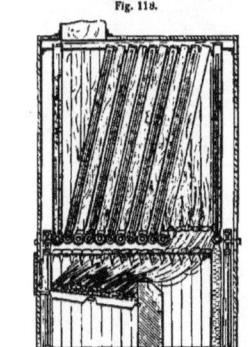

Williamson's Vertical Annular-Water, Steam-Tube Boiler, fitted with connecting horizontal branch tubes above the fire grate. Patented in the year 1861.

refer to an arrangement of a steam and water tube boiler invented by a Mr. Williamson, in which he put the steam tubes inside the water tube, and made the liquid surround the vapour.

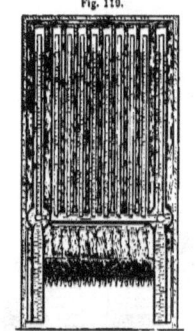

Sectional Vertical End Elevation of Williamson's Boiler, shown by Fig. 118.

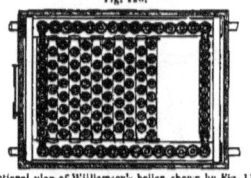

Sectional plan of Williamson's boiler, shown by Fig. 118.

The mechanical disposition is a series of double tubes angularly and vertically secured to horizontal main tubes, as shown; and the action of the generation of the steam is, that the flame from the fire grate surrounds the tubes above it, and from thence passes out at the chimney at the top.

The steam that is generated in the horizontal water tubes will find its way into the upper closed ends of the vertical tubes, which are only partly full of water, and the upper ends of which constitute so many steam chambers to the boiler; and in order to conduct the steam hence to wherever it is required, small internal steam pipes are introduced into each of the water tubes, and pass up to nearly the top of the same, where they are open, and communicate with the steam chambers. At the bottom these steam pipes are connected to branches formed on horizontal steam pipes, placed inside the water connecting pipes; and these are closed at one end, and are again connected by their open ends to the main steam pipes, fixed inside the main connecting tubes, and which pass out at the ends of the latter.

The steam pipes from the vertical side tubes are also connected to branches on the main steam pipe. The closed unconnected ends of all the vertical tubes are supported and stendied by means of stays or brackets, that do not prevent their expanding and contracting freely under the influence of heat; and the main side connecting tubes are held together in the position indicated by means of stays and straps, and are supported by saddles, fixed to the foundation, while the whole arrangement is surrounded by a suitable casing either of iron or brick.

Mr. Williamson adopted a twin angular arrangement of vertical tubes also, as shown by Fig. 121, in which it is apparent that the water tubes are secured at their lower ends to a series or set of boxes, and the steam tubes, within the water tubes, to another series of boxes below the others, but formed with them. The upper sets communicate with each other, as also do the lower, but each set is separate. The steam rising to the upper ends of the water tubes passes down the steam pipes to the boxes, and from thence to a pipe to the engine, as in the former example; the feed water being introduced in each case to the connecting pipes and boxes.

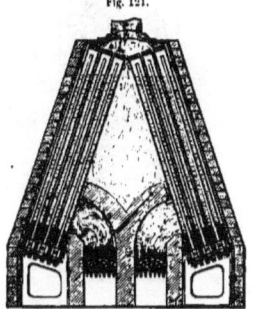

Williamson's Twin Vertical Annular-Water Steam-Tube Boiler, fitted with twin grates and brick arched furnaces. Patented in the year 1861.

Next was introduced by a Mr. Hewett a boiler evidently copied from a spirit still, where the flame acts on a series of pans situated directly over each other, and are connected by pipes, to allow a continuous flow.

In the arrangement of the boiler shown by Fig. 122, the flame passes from the fire box through a tube into two pockets or spaces, in which there are baffle-plates and tubes, each tube being covered by a ball that is removable for cleansing.

In the arrangement shown by Fig. 123, there is a water "pot" over the fire grate, and above the pot three pockets, fitted nearly as in the previous example, and in each case there

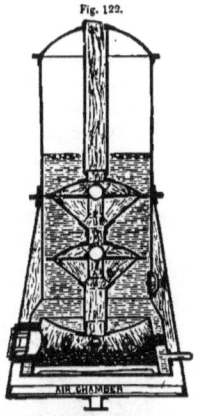

Hewett's Vertical Cylindrical Boiler, fitted with a central flame tube, two flame pockets, and an annular flame space. Patented in the year 1861.

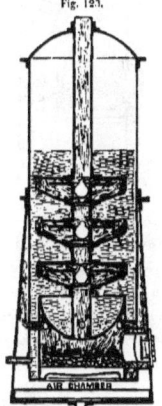

Hewett's Vertical Cylindrical Boiler, fitted with a water "pot" and three flame pockets. Patented in the year 1861.

is an annular conical flame space, connected to the fire box by tubes.

The chronology of boiler attainments in this year, 1861, was finished by a Mr. Selby, who perhaps gazed on the arrangement shown by Fig. 109 on page 44, and then thought out from that his boiler, as illustrated by Fig. 124,

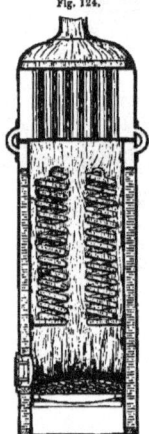

Fig. 124.

Selby's Vertical Cylindrical Boiler, fitted with water coils and a tubular superheater. Patented in the year 1861.

where it is seen that there are coils suspended in the combustion chamber, and a superheating chamber above, as in the former case. But in the place of a central tube there is an annular water space surrounding the fire box, that is in connection with the superheating chamber by curved tubes.

Another arrangement suggested by Mr. Selby is shown by Fig. 125, in which the coils are again introduced, but the superheater is displaced by a water box above the coils surrounded by a flame space.

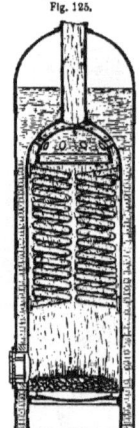

Fig. 125.

Selby's Vertical Cylindrical Boiler, fitted with water coils and a top water box. Patented in the year 1861.

Fig. 126.

Sectional plan of the Coils of Selby's boiler, shown by Figs. 124 and 125.

The illustration shown by Fig. 126, is a plan of Mr. Selby's boilers showing how the coils were arranged within the circle of the fire box, and a glance at Fig. 111, on page 44, illustrates how far he was anticipated.

The end of January, 1862, carried out with it the views of a Mr. Meriton on vertical boilers, as shown in sectional elevation by

Fig. 127, where the combustion chamber is fitted with water tubes and a water box, that can be used as a superheater, with the steam pipe direct to the engines, or as shown,

Fig. 127.

Merston's Vertical Cylindrical Boiler, fitted with water tubes and a water box in the combustion chamber. Patented in the year 1862.

A French engineer came next on the list, and made known his arrangement of boiler as illustrated in sectional elevation by Fig. 128, the pith of which is, that the fire box has a cylindrical water pot hung in it, and an annular water space surrounding the fire box, which space is surrounded by flame that first surrounds the pot, and then descends to the main flues at the sides above the fire grate. The pot has a cap on it that forms the bottom of the upper part of the boiler, and this part, and the water space below, are connected by pipes;

the left-hand upright pipe being prolonged to suit the idea of the inventor only. The feed pipe is prolonged from the cap down centrally near the bottom of the pot, so that the first water shall encounter the first fire.

Fig. 128.

Tolhausen's Vertical Cylindrical Boiler, fitted with a water "pot" and annular flame and water spaces. Patented in the year 1862.

The year 1862, although an exhibition year too, was not nearly as prolific in inventions of vertical boilers as the former year. We proceed to the end of the year's chronology, and describe next Mr. Merryweather's notion of a vertical boiler, as shown in sectional elevation by Fig. 129, and in plan by Fig. 130, which illustrates a shell containing an arrangement of hanging tubes suspended at their upper ends in the fire box. The fire grate, fed with fuel through the opening, is of the ordinary construction, and is enclosed in a casing in connection with the water and steam spaces. The roof of the fire box is a tube plate,

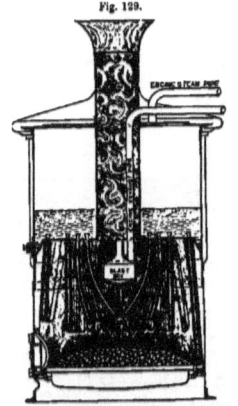

Fig. 129.

Merryweather's Vertical Cylindrical Boiler, fitted with straight and curved annular hanging water tubes. Patented in the year 1862.

Fig. 130.

Sectional plan of Merryweather's boiler, shown by Fig. 129.

from which descend into the box a series of straight and curved tubes, and those tubes contain other tubes, that are notched so as to be freely open at the bottom, and are formed with enlarged trumpet mouths at their upper ends, to encourage the steam and water ascending from the annular spaces, contained between the inner and outer tubes, in such manner as not to interfere with the downward current of the colder water, descending by the inner tube, and to replenish that water which has passed upward by evaporation between the inner and outer tubes.

That boiler, it will be noticed, is independently fixed; but Mr. Merryweather thought of another arrangement at the same time, which he enclosed in brickwork, as illustrated in sectional elevation by Fig. 131, in which

Fig. 131.

Merryweather's Vertical Cylindrical Boiler, fitted with vertical annular hanging water tubes, brickwork flues, and superheater. Patented in the year 1862.

the central hanging tubes are straight, and around the grate there are suspended longer tubes of the same class as the shorter tubes. The flame, after surrounding those tubes, passes out under the roof of the fire box through the horizontal tubes to the annular flues, above and below the water level in the boiler, and then act on the curved superheater, connected to the boiler by the pipe on the roof of the steam space, and from thence proceed through the main flue to the chimney.

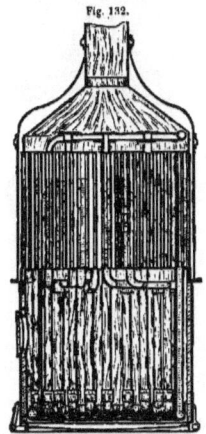

Fig. 132.

Roberts's Vertical Cylindrical Boiler, fitted with water tubes surrounding the fire box, and flame tubes above it. Patented in the year 1863.

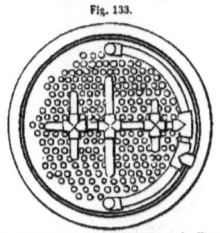

Fig. 133.

Sectional plan of Roberts's boiler, shown by Fig. 132.

About the middle of the ensuing year, 1863, Mr. Roberts patented an arrangement of tubes contained in a vertical cylindrical shell, as shown in sectional elevation and plan by Figs. 132 and 133, that illustrate cylindrical water and steam spaces, with tubes passing vertically through them; and a cylindrical fire box beneath the space, lined with fire bricks. At the bottom of them are the fire bars, and at the sides is a ring of vertical tubes that communicate at their upper ends with various parts of the roof of the fire box, some of those tubes at their upper ends being formed with elbow bends of different lengths in them, in order that they may be connected at varying distances from the centre. Those tubes at their bottom communicate with a ring tube, into which the feed water is introduced through a pipe passing through the outer casing of the fire box. The steam pipes from different parts of the top of the boiler are connected together, as is shown in the plan, and the steam is conducted by the branch steam pipe to the engine: the curved pipes being connected to safety valves.

Mr. Shand came next with his boiler, as illustrated by Fig. 134, on the next page, in sectional elevation. The arrangement is that the flame tubes and the combustion chamber are surrounded by water, and an annular steam space or pocket is constructed around the water space, for the purpose of reducing the height of the boiler; while another feature is—the shell is divided near the roof of the fire box, and connected by angle irons, bolts, and nuts.

Mr. Winan followed suit to a great extent, as his arrangement shows by Fig. 135, also on the next page, but he preferred to put the steam chest—or drum, as we have designated it—in the combustion chamber, and thus extended the height of his boiler.

54 LAND STATIONARY VERTICAL BOILERS.

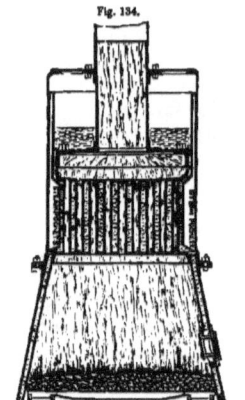

Fig. 134.

Shand's Vertical Cylindrical Boiler, fitted with flame tubes and an annular steam pocket. Patented in the year 1863.

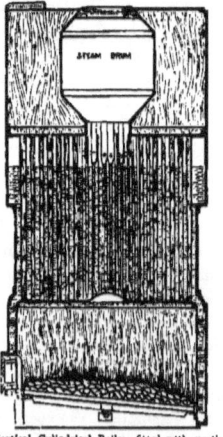

Fig. 135.

Winan's Vertical Cylindrical Boiler, fitted with vertical tubes, and a "steam drum." Patented in the year 1863.

A German engineer, named Meyn, put forward his ideas in England next, and claimed for novelty the arrangement shown by Figs. 136 and 137, in sectional elevation and plan.

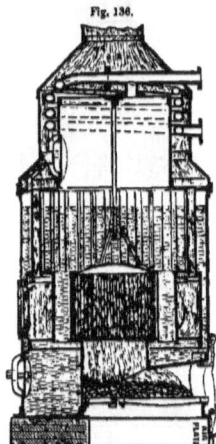

Fig. 136.

Meyn's Vertical Cylindrical Boiler, fitted with flat water tubes, round water tubes, and a coil superheater. Patented in the year 1863.

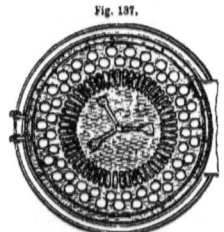

Fig. 137.

Sectional plan of Meyn's boiler, shown by Fig. 136.

The flame and gases from the fuel pass through the fire box, and surround the vertical flat tubes, and also act upon the stay plate, and then enter the side chambers, from which they pass up inside the round vertical tubes into a jacket surrounding the steam chest, and are finally carried off by the chimney.

The water is contained in the bottom spaces, the flat tubes, and in the annular space surrounding the flame tubes and chambers below.

The steam passes through the steam chest

Fig. 138.

Oakley's Vertical Cylindrical Boiler, fitted with a conical fire box, annular water spaces, horizontal and downtake flame tubes in connection with an annular flue. Patented in the year 1864.

into a coil of pipes carried round the steam chest, and becomes superheated therein on its way to the engine; and the upper end of the coil is fitted with a safety valve.

Early in the year 1864, a Mr. Oakley ushered in his ideas of the arrangement of a vertical boiler as illustrated in sectional elevation by Fig. 138, which shows that the fire box is shaped as a frustrum of a cone, and surrounded by an annular water space, which is

surrounded by an annular flame space, that communicates with the fire box by a ring of right angle tubes. Also there is a second annular water space surrounding the flame space for about midway of its depth. The flame, after descending into the flues below the outer water space, ascends through the annular brick flue to the chimney.

It will be remembered that, in the year

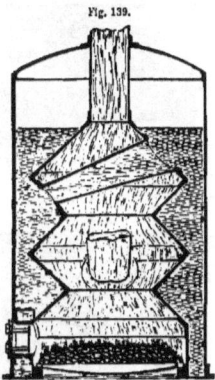

Fig. 139.

Winstanley's Vertical Cylindrical Boiler—the fire box being angularly formed and fitted with cross water tubes. Patented in the year 1864.

1848, a Mr. Millward constructed a cross tube boiler, and also Mr. Bowman in the year 1858, as shown on page 31, but this did not prevent a Mr. Winstanley, in the year 1864, from directing attention to his boiler, as illustrated in sectional elevation by Fig. 139, the arrangement of it being that the shell is cylindrical, and fitted with a fire box and combustion chamber formed with alternate angular recesses for the flame and water, and conical cross water tubes, connecting the water spaces surrounding the box and chamber.

The next arrangement of vertical boiler that we notice is shown in sectional elevation and plan by Figs. 140 and 141, being the

Fig. 140.

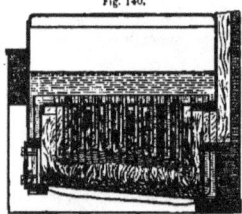

Marshall's Vertical Square Boiler, fitted with hanging water plate-division tubes in the fire box, and a water division-plate in the annular space. Patented in the year 1864.

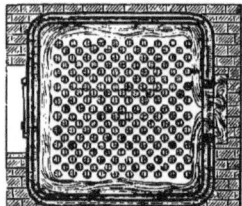

Sectional plan of Marshall's boiler, shown by Fig. 140.

invention of a Mr. Marshall, and in the place of the annular tubes he left out the inside tube, and put a plain flat piece of metal, curved at the bottom, in its place. The tubes are arranged in rows, as seen in the plan, and are suspended in the fire box from its crown plate, and project through it, as shown in the elevation.

The heat is presumed to rise on one side of the plate in the tube, and pass down at the other, and ascend again as before, and thus a continual circulation is maintained by the old and new heat. The water space surrounding the fire box is divided by a plate similar to the tubes, to accomplish a similar circulation of the heat in that space.

Fig. 142.

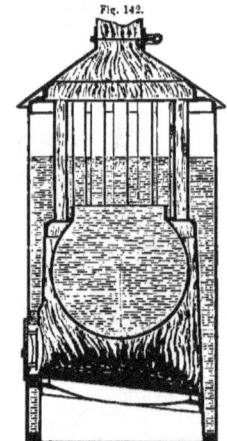

Thomson's Vertical Cylindrical Boiler, fitted with a water "pot" and vertical flame tubes above it. Patented in the year 1865.

Fig. 143.

Sectional plan of Thomson's boiler, shown by Fig. 142.

The year 1865 might have been famous

for a Mr. Thomson's boiler, as shown in sectional elevation and plan by Figs. 142 and 143, had not it been superseded as far back as the year 1851, as shown by Fig. 21 on page 12 of this work. The Stenson boiler has a water pot over the fire, and so has Thomson's, while in both examples the flame surrounds the pot, except at the opening, and passes through tubes to the chimney. Stenson's tubes are at the sides of the fire box as well as above it, while Thomson's are above it only.

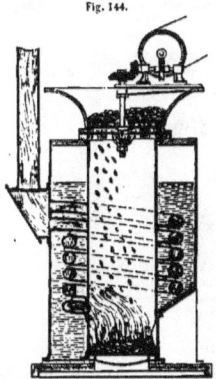

Fig. 144.

Durand's Vertical Cylindrical Boiler, fitted with a fire box—fed from the top—and two coils of flame pipes in the water space. Patented in the year 1865.

The French engineers were represented next, as to vertical boiler proficiency, by M. Durand, as illustrated by Fig. 144 in sectional elevation. The arrangement consists of a fire box extending through the boiler, and mounted with a circular coal box. The bottom of the box is perforated, and is secured on a plate with corresponding openings; the fuel is allowed to fall through those openings mechanically by the rotation of a perforated disc on the upper side of the bottom, and the motion for the disc is derived from bevel gearing, driven by worm and wheel motion.

The flame, on rising from the fuel, passes through two coils that surround the fire box centrally of the water space, and from thence to the chimney at the side.

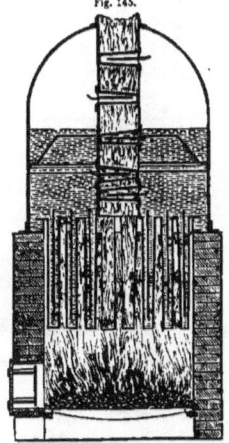

Fig. 145.

Smith's Vertical Cylindrical Boiler, fitted with hanging water plate division tubes in the fire box, and horizontal tubes in the combustion chamber. Patented in the year 1865.

On referring to page 56 of this work, an arrangement of suspended tubes with water division plates is illustrated by Figs. 140 and 141, and although the patent of that is plain enough, and Mr. Marshall evidently thought it his property, a Mr. Smith about twelve months after brought forward his patent, as illustrated in sectional elevation by Fig. 145, which shows a cylindrical boiler with tubes suspended from the bottom plate, and enclosed in a brick furnace. Each tube contains a plate that divides it, except for a short distance at the lower end, to which the plate does not extend.

The action of the flame is to surround those tubes, and from thence pass up through the vertical combustion chamber that connects the top and bottom of the boiler.

To more effectually heat the water and steam surrounding this chamber, in general it is fitted with horizontal tubes containing division plates also, and thus the surface is much increased. The motive of this arrangement of tubes and their plates, is to promote circulation of heat and surface, which motive is also in the former arrangement referred to.

A Mr. Wise came next with his tube boiler fitted with hanging angular water tubes, as shown in side and end sectional elevations by Fig. 146, but unfortunately he was anticipated to a great extent forty-one years before by Mr. Moore, as illustrated by Fig. 12 on page 5 of this work, who arranged his large pipes as rings, and connected them by vertical tubes, and Mr. Wise arranged his large pipes straight, and there the difference ends, because the hanging tubes were introduced in the year 1862, by Mr. Merryweather, and therefore form no part of Mr. Wise's arrangement. Those two examples, Figs. 12 and 146, clearly indicate how closely one inventor may copy his predecessor without knowing it.

But what can we say of Mr. Chaplin—who first introduced conical fire tubes, as shown

Fig. 146.

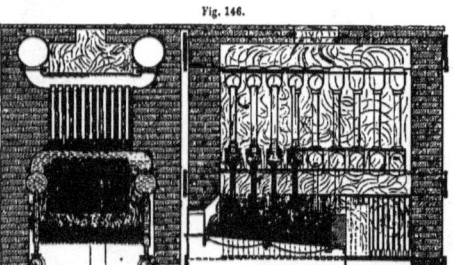

Wise's Vertical Tubular Boiler, fitted with hanging annular water tubes. Patented in the year 1865.

by Fig. 40 on page 21 of this work, when he stated in this year, 1865, that "it is preferred to dispense with appliances within the tubes for inducing circulation, such appliances being found practically unnecessary?" And he therefore introduced the more simple arrangement of hanging single tubes in a vertical boiler, as illustrated by Fig. 147, where the tubes are shown secured radiatively to a curved tube plate, that forms the top of the combustion chamber, and the flame, after passing amongst those tubes, proceeds through the uptake to the chimney, thereby upsetting all the prior

LAND STATIONARY VERTICAL BOILERS. 59

theories for inducing circulation of heat and water in tubes.

Tubular boilers being then the fashion, we next introduce Mr. Jordan's arrangement, as

Fig. 147.

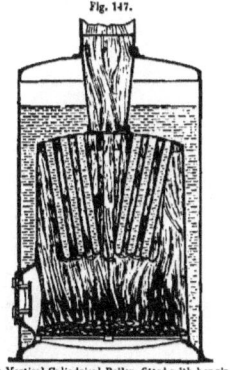

Chaplin's Vertical Cylindrical Boiler, fitted with hanging water single tubes. Patented in the year 1865.

Fig. 148.

Jordan's Vertical Tube Boiler, set in brickwork. Patented in the year 1865.

shown in sectional elevation by Fig. 148, and in sectional plan by Fig. 149. This con-

Fig. 149.

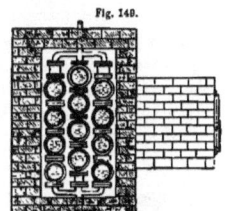

Sectional plan of Jordan's boiler, as shown by Fig. 148.

sists of a series of vertical pipes containing water and steam, that are connected at the top and bottom by branch pipes. The joints at the ends of the vertical pipes are made by covers, long bolts, and nuts, the removal of which admits access for cleansing and inspection, two functions imperative with boilers. The entire arrangement of pipes and their details are set in brickwork similar to Mr. Moore's, but horizontal bridges in this case are introduced to make the flame act with better effect on the boiler.

An American engineer named Davis next appeared with an overhead fuel feeding boiler, as shown in sectional elevation and plan by Figs. 150 and 151, on the next page; the arrangement of which is that the shell contains, first, an annular water space; secondly, an annular flame space; thirdly, a water space interspersed with flame tubes, closed at their upper ends; and lastly, a fire box fed at the top, the box also being a combustion chamber connecting the chimney. The fire grate is raised in the centre, and is "dished" at the sides with a flange to contain the fuel, which, when required to be "clinkered," or removed, the grate is lowered by the screwed spindle supporting it.

I 2

The connection of the inner and outer water spaces is by pipes directly over the fire grate roof plate, and to generate more steam from the flame in the outer space, a coil is therein arranged that is connected with the water and steam spaces.

Fig. 150.

Davis's Vertical Cylindrical Tubular Boiler, fitted with flame tubes closed at the top end, an annular flame space, and a central fire chamber fed at the top. Patented in the year 1865.

Fig. 151.

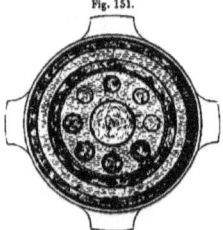

Sectional plan of Davis's boiler, shown by Fig. 150.

The inventor also provided steam jet pipes on each side of the grate, and in the chamber above, to promote a draught to assist combustion, and with that his arrangement was completed.

Mr. Davis having succeeded in obtaining English protection for his complication, another American named Wheeler obtained it for his simplicity, which constitutes that with a common vertical boiler, fitted with

Fig. 152.

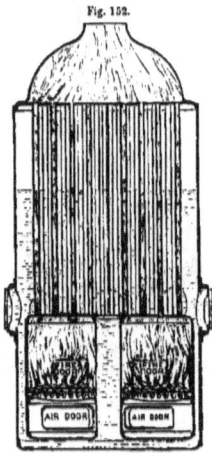

Wheeler's Vertical Cylindrical Boiler, fitted with numerous manholes around the crown of the fire box. Patented in the year 1865.

tubes connecting the fire box and chimney. The invention lies in the fact of proposing man-holes, in the shell, around and over the crown of the fire box, because all the remainder had been introduced many years before.

A Mr. Wilson came directly after Mr. Wheeler, believing that the circulation of the heat in the water could be induced by annular tubes—in spite of Mr. Chaplin's remark. He introduced no less than four arrangements of tubes for that purpose, the first being shown by Fig. 153, in sectional elevation. This is a common vertical cylindrical boiler, with flame tubes connecting the fire box and smoke box ; but with the addition of annular water tubes, extending from the fire box crown plate to a few inches above the water level in the boiler, and surrounding each flame tube for that length; so that each annular tube became a boiler of itself, and boiled over into the space around it.

The second arrangement, shown by Fig. 154, reversed matters, inasmuch that the water tube was the smaller, also enclosed in the flame tube, with a larger smoke box contained in the boiler. The water tubes also were curved at their ends, and were common to the space between the plates.

Then came the third example, illustrated by Fig. 155 in sectional elevation, showing that the fire box is conical, and surrounded by a casing open at the top in the steam space, and also at the bottom over the tube openings. Those tubes are curved to form right angle bends, and extend to the top of the fire box casing, and are bent over it, to better conduct any overflow of water to the space below.

The fourth example is shown by Fig. 156, and is very much in common with the previous example, excepting that the tubes

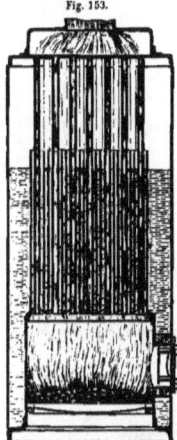

Fig. 153.

Wilson's Vertical Cylindrical Boiler, fitted with flame tubes, surrounded from the fire box to a little above the water level with water tubes. Patented in the year 1865.

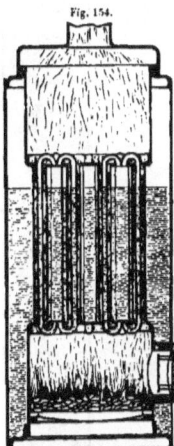

Fig. 154.

Wilson's Vertical Cylindrical Boiler, fitted with flame tubes, enclosing water tubes curved at their ends into the water space. Patented in the year 1865.

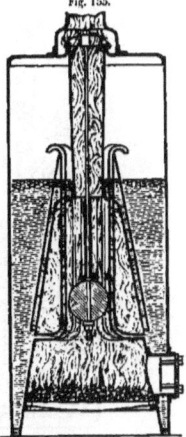

Fig. 155.

Wilson's Vertical Cylindrical Boiler, fitted with a conical fire box, containing curved water tubes and the box surrounded by a water division casing. Patented in the year 1865.

are syphon bent in the fire box and the casing is omitted.

Not to miss being recorded in the annals of the year 1865, a Mr. Barclay made known his pretensions as an inventor of vertical boiler improvements, close on the end of that year, according to the arrangement illustrated in sectional elevation and plan by

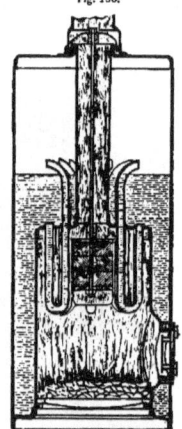

Fig. 156.

Wilson's Vertical Cylindrical Boiler, fitted with syphon water tubes in the fire box. Patented in the year 1865.

Figs. 157 and 158, in which it will be seen that the fire box contains a conical water pot fitted internally with another pot open at each end, and supported by brackets at the crown of the fire box. The fire box is surrounded by an annular water space, in which are fixed vertical pipes for the purpose—as the internal water pot—of circulating the heat absorbed from the flame; which, after acting in the fire box, passes out through the horizontal flues or tubes, near the crown, to the annular flame space surrounding the sides and top of

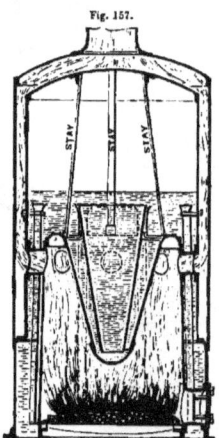

Fig. 157.

Barclay's Vertical Cylindrical Boiler, fitted with a conical water "pot," containing a divisional cone, and water pipes around the fire box. Patented in the year 1865.

Fig. 158.

Sectional plan of Barclay's boiler, shown by Fig. 157.

the boiler's shell. But Mr. Barclay, not being satisfied with that, added water tubes in the fire box to the arrangement, as illustrated in

LAND STATIONARY VERTICAL BOILERS.

sectional elevation and plan by Figs. 159 and 160, in which he was fully preceded, as shown by Fig. 105 on page 42.

Fig. 159.

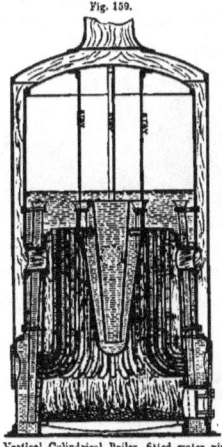

Barclay's Vertical Cylindrical Boiler, fitted water pipes in the fire box in addition to the arrangement shown by Fig. 157.

Fig. 160.

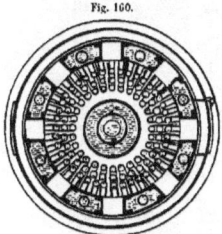

Sectional plan of Barclay's boiler, shown by Fig. 159.

Very early in the next year, 1866, a Mr. Adamson considered, and published it too, that a vertical boiler should have two fire boxes, connected by short tubes with the crowns connected to the smoke box by flame tubes, as illustrated in sectional elevation and plan by Figs. 161 and 162; and not being

Fig. 161.

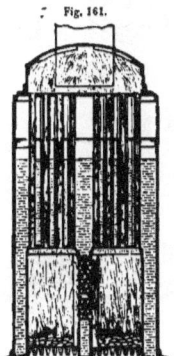

Adamson's Vertical Cylindrical Boiler, fitted with horizontal flame tubes, connecting twin fire boxes; and vertical flame tubes, connecting the fire and smoke boxes with a side chimney. Patented in the year 1866.

Fig. 162.

Sectional plan of Adamson's boiler, shown by Fig. 161.

content with that, Mr. Adamson proposed to double his scheme, by putting four fire boxes, common to one combustion chamber, in which is a central water tube; the top of the chamber being connected by flame tubes, as illustrated in elevation and plan by Figs. 163 and 164.

Following Adamson, an engineer named

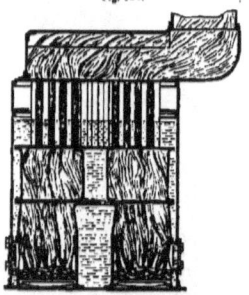

Fig. 163.

Adamson's Vertical Cylindrical Boiler, fitted with four fire boxes, common to one combustion chamber, which is connected by flame tubes to the smoke box. Patented in the year 1866.

Howard had the temerity to patent the idea of securing small thin circular vertical tubes, by long bolts and nuts, to large horizontal

tubes was permitted, by making the joints of their ends in the sockets of the horizontal pipes with india-rubber, or an equally elastic material. We need scarcely add that, in practice with high pressure steam, it would

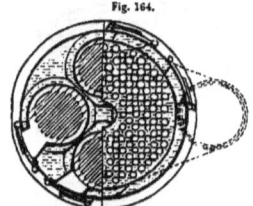

Fig. 164.

Sectional plan of Adamson's boiler, shown by Fig. 163.

have blown all the packing out, if the fire heat did not melt it.

Not having seen, evidently, Bowman's boiler, shown by Fig. 70, on page 31 of this work, a Mr. Woodward came forward directly after Mr. Howard, with his notion of

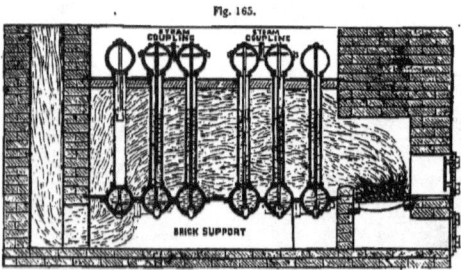

Fig. 165.

Howard's Vertical Tube Boiler, fitted to horizontal tubes. Patented in the year 1866.

thick tubes, and set them in brickwork, as illustrated in sectional elevation by Fig. 165. The contraction and expansion of the vertical

a vertical boiler, as illustrated in one view by Fig. 166.

The shell is cylindrical, as is the fire box, in

which are cross water tubes, secured at an angle, to better resist the flame. This boiler is conspicuous also by the absence of the fire grate; the inventor preferring the bottom of

Fig. 166.

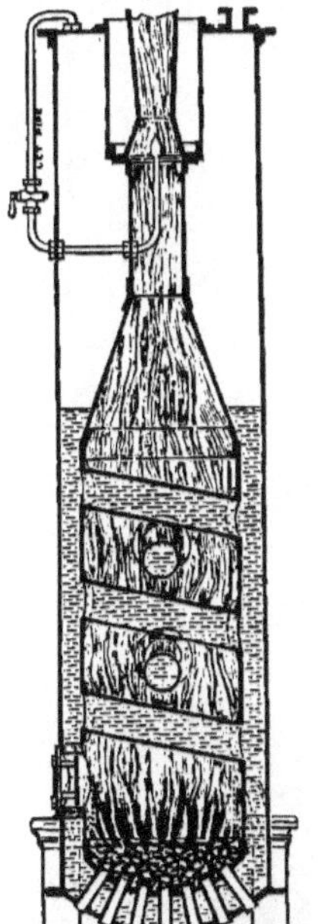

Woodward's Vertical Cylindrical Boiler, fitted with a fire box having cross water tubes in it, and the grate being the bottom of the fire box with air tubes in it. Patented in the year 1866.

the fire box fitted with air tubes in its place, and for that purpose mounted the shell on pillars, resting on brickwork.

We now refer to page 4 of this work, and direct attention to Fig. 10, which carries thoughts as far back as the year 1780, being 86 years from the year 1866, and we next direct notice to Figs. 167 and 168, as a contrast. The flame, in the former case, passed out of the chimney situated at the side of the arrangement; while in this case it escapes at the centre, with similar reverse positions for the fire grates also.

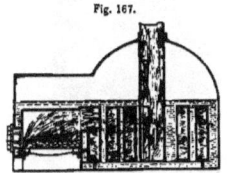

Fig. 167.

Schaubel's Vertical Boiler, with spiral internal water and flame spaces. Patented in the year 1866—In direct contradiction of that used in the year 1780.

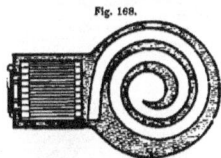

Fig. 168.

Sectional plan of Schaubel's boiler, shown by Fig. 167.

The next example of boiler worthy of our attention is illustrated by Fig. 169 in sectional elevation, and in sectional plan by Fig. 170, which was introduced by a Mr. Holt, an Austrian engineer in England, the arrangement being thus. In the central part of the boiler is a vertical cylindrical chamber containing water and steam, which is surrounded by horizontally corrugated water and steam chambers, and the space between those cham-

K

LAND STATIONARY VERTICAL BOILERS.

bers is the flame space or flue, leading from the fire box to the smoke box, the water and

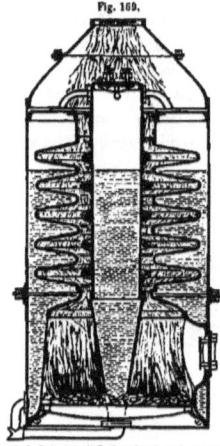

Fig. 169.

Holt's Vertical Cylindrical Boiler, fitted with a vertical water and steam central chamber, surrounded by a chamber horizontally corrugated. Patented in the year 1866.

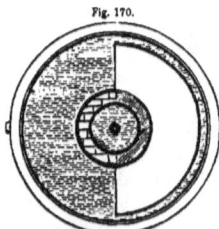

Fig. 170.

Sectional plan of Holt's boiler, shown by Fig. 169.

steam portions of the chambers being connected by pipes at the bottom and top.

If this example is compared with the example illustrated by Fig. 123, on page 49 of this work, the same contrast is very nearly similar as with Figs. 10 and 167.

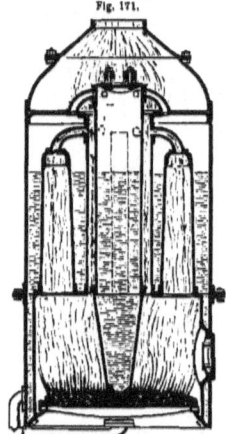

Fig. 171.

Holt's Vertical Cylindrical Boiler, fitted with a vertical water and steam chamber, surrounded by a chamber vertically corrugated, intermixed with flame tubes. Patented in the year 1866.

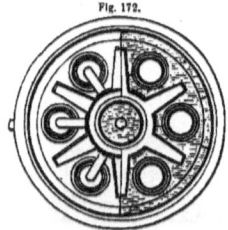

Fig. 172.

Sectional plan of Holt's boiler, shown by Fig. 171.

As we are in the spirit for contrasting arrangements, we refer next to page 18 of this work, and also to Figs. 171 and 172, when it will be apparent that the both inventors formed vertical corrugations in the chambers, but they were differently utilised for the same purpose.

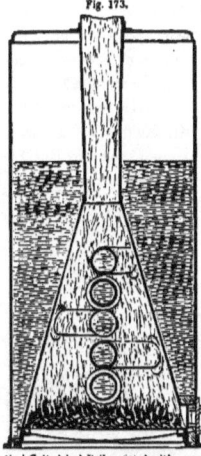

Fig. 173.

Green's Vertical Cylindrical Boiler, fitted with a conical fire box, having cross water tubes in it. Patented in the year 1866.

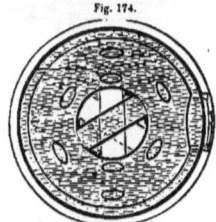

Fig. 174.

Sectional plan of Green's boiler, shown by Fig. 173.

The action of the flame in the latter example is that, after rising from the fire grate, it passes up through the large vertical and small curved tubes, and in the vertically corru-

gated flame spaces at the same time, to the chimney. The volumes of steam from the central and annular chambers are connected by similar pipes as for the flame's circuit.

Fig. 175.

Green's Vertical Cylindrical Boiler, fitted with cross water tubes in the fire box. Patented in the year 1866.

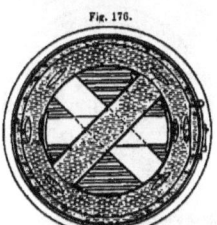

Fig. 176.

Sectional plan of Green's boiler, shown by Fig. 175.

Cross-tube vertical boilers were evidently much in favour about this time, from the fact that a Mr. Green introduced his ideas on the subject also, as illustrated in four views by Figs. 173 to 176, on the preceding page.

The first example shown in sectional elevation by Fig. 173, and in sectional plan by

Fig. 177.

Field's Vertical Cylindrical Boiler, fitted with a fire box containing hanging water tubes and a vertical water space. Patented in the year 1866.

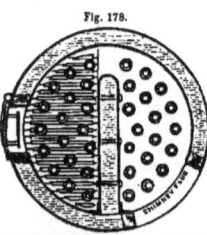

Fig. 178.

Sectional plan of Field's boiler, shown by Fig. 177.

Fig. 174, is a plain cylindrical shell fitted internally with a very conical fire box, in which are five cross water tubes secured at three different angles horizontally, and, in the second example, the same arrangement is repeated with the fire box less conical.

If Mr. Green's patent is valid, what has become of Mr. Bowman's, as shown on page 31 of this work, with the advantage of the eight years' interval between them; and of what use is legal protection?

Close upon Mr. Green came Mr. Wise, with his friend Mr. Field, who had seconded him in the arrangement shown by Fig. 146, on page 58 in this work, but in this case Mr. Field led

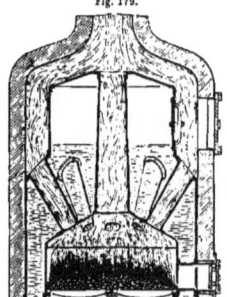

Fig. 179.

Dickins's Vertical Cylindrical Boiler, fitted with central and side angular flame tubes. Patented in the year 1866.

Mr. Wise to make known to the professional community the advantage of a cylindrical boiler as illustrated by Figs. 177 and 178; showing thereby, that if a vertical cylindrical fire box had a vertical water space in it sufficient to form a flame passage at one side, and that the fire spaces divided in the box were fitted with vertical hanging double tubes, and the flame passed out through a bottom flue, a good result should occur.

The next novelty is a very simple affair, as Fig. 179 illustrates, showing a fire box with an angular roof, connected by tubes to an annular flame space, and by a central tube with the chimney. The shell is surrounded by brickwork, and thus the flame acts inside and

Fig. 180.

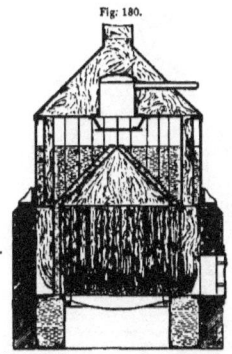

Miller's Vertical Boiler, consisting of a brick fire box, having a lower annular water space, with water tubes connecting upper water and steam spaces, containing flame tubes, and a water jacket over the fire box crown. Patented in the year 1866.

Fig. 181.

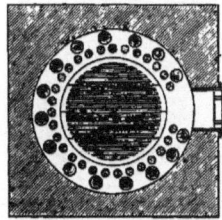

Sectional plan of Miller's boiler, shown by Fig. 180.

outside the boiler, but at the latter part mostly, to superheat the steam.

An American engineer next appeared on the scene with his boiler, as illustrated in sectional elevation and plan by Figs. 180 and 181, the arrangement of which is, that the bottom of the boiler is an annular water space, in the form of a ring, and in the central part is the fire grate, as seen in the plan. This water space is fitted on the top with vertical tubes that communicate at their tops with a water jacket of a conical shape, to correspond with the fire box roof that is arranged to project upwardly within the upper case to the water level. The water jacket and tubes contain the amount of water from which the steam is generated by the fire being made to act on them direct, to the exclusion of the remaining water in the boiler, which is contained within the space surrounding the water jacket, and within the water tubes that connect the lower water space.

Now the steam generated in the sheet of water has a ready and free escape—without passing through a superincumbent body of water into the steam space in the upper portion of the casing—and around and over a primary plate into the steam dome; while any water carried over the water jacket by the steam serves to keep up a circulation by its flow down the tubes.

The smoke and gaseous products pass off from the fire chamber, up the tubes in the water and steam casing, into a smoke box surrounding the steam dome.

Mr. Miller was not permitted long to be alone as the inventor of the "water jacket" arrangement, on account of a Mr. Fisken, early in the year 1867, appearing with his scheme, as illustrated in sectional elevations by Figs. 182 and 183, on the next page, which consists of a flat fire box, whose sides are at an acute angle, and on them are secured plates to form a water jacket, as in the former example.

The flame, after rising from the grate, is dispersed from right to left by the baffle plates, and from there ascends to the chimney, as shown in the side elevation.

Mr. Fisken also proposed to combine a series

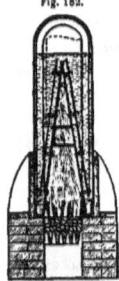

Fig. 182.

Fisken's Vertical Boiler, fitted with a flat angular fire box, and a water jacket on it. Patented in the year 1867.

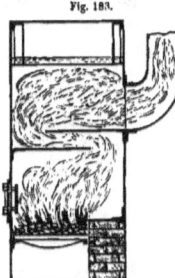

Fig. 183.

Side sectional elevation of Fisken's boiler, shown by Fig. 182.

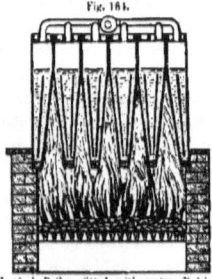

Fig. 184.

Fisken's Vertical Boiler, fitted with water divisional plates. Patented in the year 1867.

of angular flat boilers together, with their sharp ends next to the fire grate, as shown by Fig. 184—instead of upwards, as shown in Fig. 182—and fit them internally with water divisional plates to promote circulation. The upper parts of those boilers are vertical, and connected at the roof by pipes to a central main pipe that conveys the steam to the engine.

Fig. 185.

Lochhead's Vertical Cylindrical Boiler, fitted with horizontal water tubes and vertical flame tubes. Patented in the year 1867.

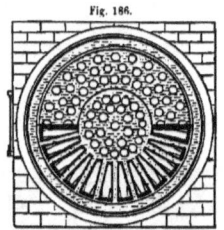

Fig. 186.

Sectional plan of Lochhead's boiler, shown by Fig. 185.

A far more sensible arrangement is represented in sectional elevation and plan by Figs. 185 and 186, sent from California by a Mr. Lochhead, which consists of a cylindrical boiler containing a fire box, having in its centre a water pot connected by flame tubes to the smoke box. The space around the pot is fitted with horizontal water tubes, that connect the inner and outer water spaces. The roof of the fire box has flame tubes on it,

Fig. 187.

Dunn's Vertical Cylindrical Boiler, fitted with a fire box containing a vertical circular water tube, connected at each end by flat water branches to the outer water space. Patented in the year 1867.

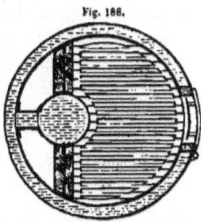

Fig. 188.

Sectional plan of Dunn's boiler, shown by Fig. 187.

also extending to the smoke box. The action of the flame in this boiler is that some of the flame enters the pot tubes direct, but the remainder circulates amongst the horizontal tubes, and after passes through the fire box tubes to the chimney, so that the water has

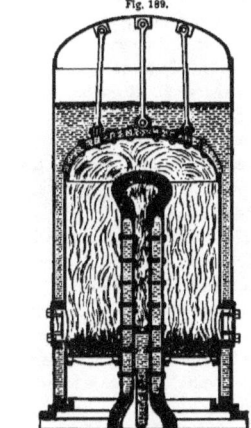

Fig. 189.

Dunn's Vertical Cylindrical Boiler, fitted with twin fire boxes, and central flame and water spaces. Patented in the year 1867.

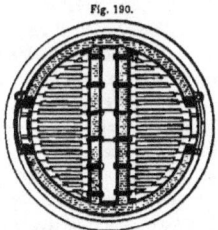

Fig. 190.

Sectional plan of Dunn's boiler, shown by Fig. 189.

the benefit of inner and outer flame surfaces for evaporation.

Next appeared Mr. Dunn again, with the same species of boiler as shown on page 25 of this work, but with certainly a much improved arrangement, as illustrated by Figs. 187 and 188, in sectional elevation and plan. The arrangement consists of a cylindrical shell, having a fire box in it, containing the fire grate, and a circular water tube, that is connected at the back by branches of a flat shape, as seen in the plan. Brick bridges are also formed with the portion that the tube rests on. The action of the flame is to envelope the circular tube and its branches, and then pass to the down flue below, opposite the fire grate outside the brickwork.

Mr. Dunn also introduced a twin arrangement of fire boxes, as shown in sectional elevation and plan by Figs. 189 and 190. In this case the flame ascends in each fire box, passes through the perforated brick bridge, on the top of the flat water spaces, and then descends through the central flame space to the chimney flue at the bottom.

Mr. Shand followed Mr. Dunn with an amalgamation of vertical flame tubes and flame and water tube chambers, common to one combustion chamber, which is situated in the shell of the boiler, and divides the roof of the fire box from the portion above, as shown in sectional elevation by Fig. 191. The action of the flame is for some of it to pass direct through the flame tubes, and the remainder to operate in the flame chambers, and pass out through the small tubes connecting the combustion chamber, that contains the water chambers, situated directly over the flame chambers: therefore Mr. Shand was evidently not content with his arrangement, as shown by Fig. 134, on page 54 of this work, and neither was Mr. Field content with his method, as shown by Fig. 177 on page 68, because he introduced, directly after Mr. Shand, the arrangement shown in side and end sectional

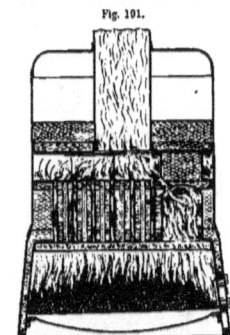

Fig. 191.

Shand's Vertical Cylindrical Boiler, fitted with vertical flame tubes, and three vertical flame and water chambers common to one combustion chamber. Patented in the year 1867.

elevations, illustrated by Fig. 192, which consists of pipes transversely arranged with hanging annular tubes attached to them over the fire grate, and the whole set in brickwork.

Again we direct attention to a previous example, as a comparison to the next we describe and illustrate. In the first place, refer to page 40 of this work, and notice Fig. 102, and now turn to Fig. 193, a Mr. Rogan's idea, on page 73, and it will be seen that, in each case, there is a central water casing, attached, at the sides, to the outer casing, by angular tubes; and the flame passes amongst those tubes to the chimney, but in the latter arrangement the top and bottom of the central casing is connected by tubes also. Besides that, Mr. Rogan proposed his boiler to be cylindrical throughout.

Mr. Holt, again from Austria, next appeared

with a very simple boiler indeed, and what is better, it can be cleaned and repaired easily, which is more than can be said of his arrangements that are shown on pages 26 and 27 of this work. The present example is shown in sectional elevation and plan by Figs. 194 and 195, on the next page; the arrangement being that the shell contains a fire box with an angular roof, and the top of the boiler has a roof parallel to the other, which are connected by sheet flame tubes, as shown in the plan.

The idea of combining a series of vertical separate shells, common to one chest or casing, in a circular arrangement, evidently proceeded from Mr. Dunn, as seen by Fig. 46 on page 24, but that did not stop Mr. Fisken, who began, as illustrated by Fig. 184, on page 70, to propose wedge-shaped shell boilers as a combination, and again another set, as now illustrated by Figs. 196 and 197, in sectional elevation and plan, on the next page. The form of each shell is wide at the parallel outer part, but narrow at the angular extremities, and being connected by straight tubes at the inner part, by curved tubes about midway across the angular side, and by flanges from end to end of the shell beyond, thus form a cylindrical combined boiler, with a circular steam ring chest at the top, and a similar feed water pipe at the bottom.

A Mr. Messenger, not caring, evidently, about what had been previously done in water

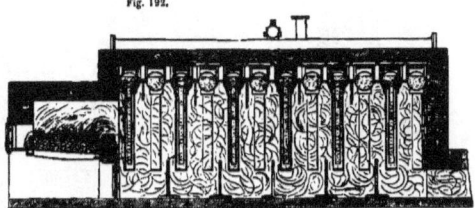

Fig. 192.

Field's Angular Tube Boiler, fitted with hanging angular tubes. Patented in the year 1867.

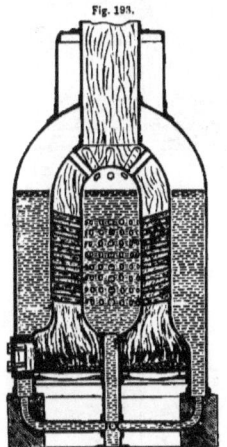

Fig. 193.

Regan's Vertical Cylindrical Boiler, fitted with a central water casing, connected at the top to the steam space, and at the sides and bottom to the water space by tubes. Patented in the year 1867.

pipes and coils for evaporation, introduced his arrangement of a vertical boiler, as shown

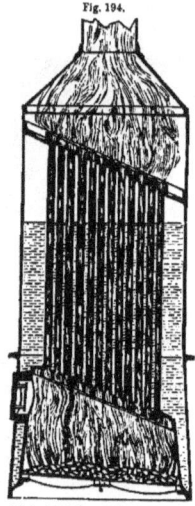

Fig. 194.

Holt's Vertical Cylindrical Boiler, fitted with angular tube plates and sheet flame tubes. Patented in the year 1867.

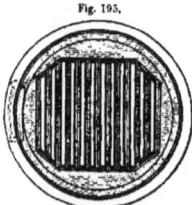

Fig. 195.

Sectional plan of Holt's boiler, shown by Fig. 194.

in sectional elevation by Fig. 198, by which is shown a vertical cylindrical boiler, with a fire box fitted with curved water pipes, brick disc,

baffle plate, and a coil in the chimney. The action of the flame is that the products of combustion rising from the furnace first exert their

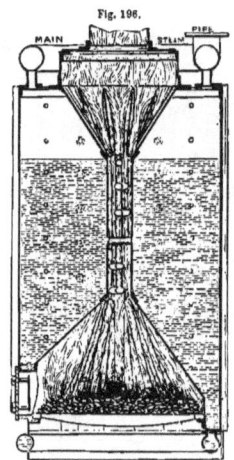

Fig. 196.

Fisken's Vertical Wedge-combined Boiler, secured together by tubes, bolts, and nuts. Patented in the year 1867.

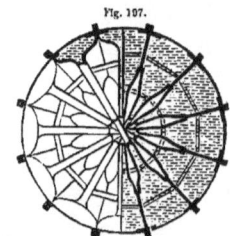

Fig. 197.

Sectional plan of Fisken's boiler, shown by Fig. 196.

influence in raising steam on the sides of the fire box below the water tubes. They are then baffled by the brick disc, and circulate around the lower ends of the water tubes, passing through the spaces between them. They then impinge against the sides of the fire box above the lower ends of the water tubes, and circulate around the upper part of those tubes,

enters the pipe at the roof of the boiler, then down it, and passes up again through the coil, from whence it is conducted into the boiler.

Fig. 198.

Fig. 199.

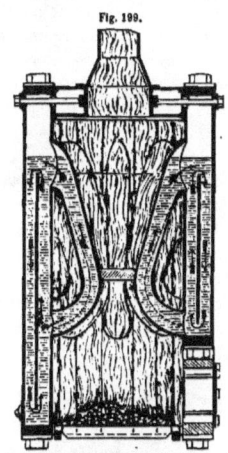

Messenger's Vertical Boiler, composed of curved and straight tubes bolted together. Patented in the year 1867.

Fig. 200.

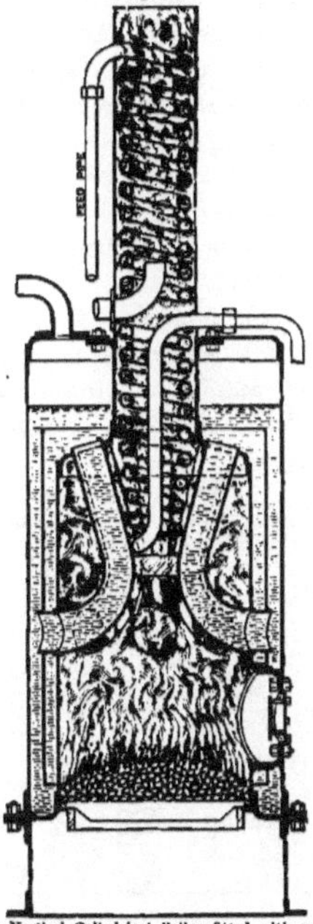

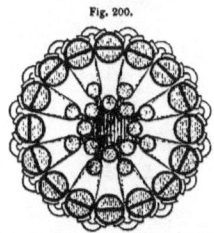

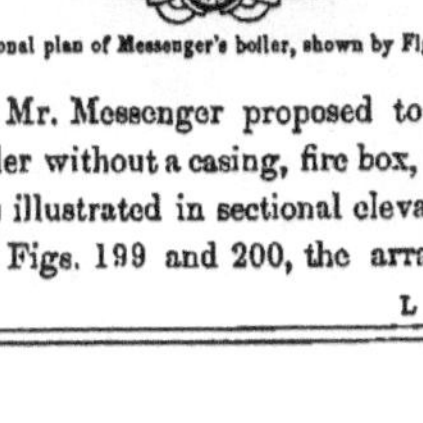

Messenger's Vertical Cylindrical Boiler, fitted with water tubes in the fire box, and a water coil in the chimney. Patented in the year 1869.

Sectional plan of Messenger's boiler, shown by Fig. 199.

passing a second time between them, and are then split up a third time in circulating from the outside into the inside, and passing up between the coiled pipe until they leave it at the top of the chimney. The feed water

Next Mr. Messenger proposed to make a tube boiler without a casing, fire box, or brickwork, as illustrated in sectional elevation and plan by Figs. 199 and 200, the arrangement being that the water tubes are bolted at the top to a ring that is circular inside and polygramic on the outside, the number of sides corresponding to the number of water tubes. One part of those tubes forms the external part of the boiler, and the other parts cause a circulation of the products of combustion by means of the baffle plate, similar to that already described. The larger tubes have a partition in them, which causes a circulation. The short curved pipes connecting the lower ends of the tubes are for the purpose of preserving the water level in them. The smaller water tubes are placed alternately in plan, so that the smaller branches are placed over the openings or spaces left by the larger branches, and this causes a further circulation of the products of combustion.

box that is surrounded by a water space, and that space surrounded by a flame space contained in another water space formed by the outer shell of the boiler.

The action of the flame is that, on rising from the grate, it surrounds the water pot, it then passes out amongst the vertical water tubes, and from those it descends in the annular space to the chimney at the side of the shell. It will be noticed that the "pot" is connected, by horizontal tubes, to the water space surrounding it; and that the top of that space is connected, by vertical tubes, to the main space above.

Mr. Howard, who came forward in the year 1866, with an arrangement shown by Fig. 165, on page 64 of this work, appeared again in this year, 1868, with a more com-

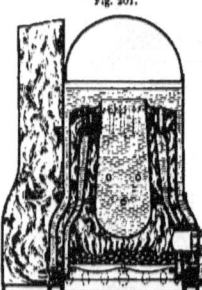

Fig. 201.

Allibon's Vertical Cylindrical Boiler, fitted with a water pot, surrounded with two annular flame spaces, and two water casings. Patented in the year 1867.

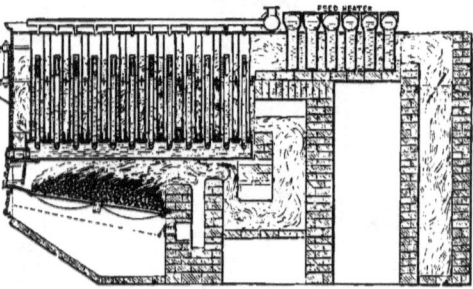

Fig. 201.

Howard's Vertical Tube Boiler, fitted with vertical annular circulating tubes, connected to horizontal water and steam pipes, with a vertical tube feed water heater behind. Patented in the year 1868.

The birth of the year 1868 gave birth, also, to a Mr. Allibon's idea of a vertical boiler, as shown in sectional elevation by Fig. 201, which illustrates a water pot hung in a fire

plicated apparatus, as shown in sectional elevation by Fig. 201, in which there are horizontal water pipes, secured and supported over the grate and brick bridges, and attached to those pipes are vertical tubes, containing smaller tubes, open at each end, of a shorter length. The upper extremities of the larger tubes are connected to horizontal steam pipes—in a line with the water pipes below—which discharge into a main steam pipe that is situated directly beyond the last vertical tube, and beyond that pipe is a vertical tube water feed heater, arranged in a very similar manner as the boiler proper.

A much more simple and cheaper arrangement was introduced next by a French engineer, named Bèzy, as illustrated in sectional elevation and plan by Figs. 202 and 203, showing, thereby, a cylindrical shell containing a large fire box, in which, over the fire door, are a series of tubes secured angularly, and amongst them the flame and heat from the grate passes, and generates steam. This boiler has the advantages of ready cleansing and access for repair.

Annular tubes, about this time, were much

Fig. 202.

Bèzy's Vertical Cylindrical Boiler, fitted with angular water tubes in the fire box. Patented in the year 1868.

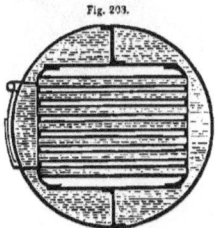

Fig. 203.

Sectional plan of Bèzy's boiler, shown by Fig. 202.

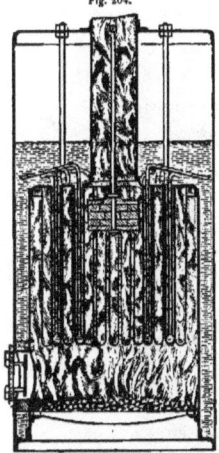

Fig. 204.

Moreland's Vertical Cylindrical Boiler, fitted with a fire box containing annular hanging tubes, and the inner tubes bent over the fire box crown. Patented in the year 1868.

in fashion, for be it remembered that, as their use was not generally understood, many would-be boiler inventors adopted them; and consequently a Mr. Moreland brought out his idea, as shown in sectional elevation by Fig. 204, that constituted a vertical cylindrical boiler, with hanging water tubes in the fire box, and in those tubes are smaller tubes, with their ends bent nearly at right angles over the fire box, and there the invention terminated.

there connected by cross branch pipes, and vertical circulating pipes, on the top of which is the main steam pipe, situated sufficiently high out of the effect of priming. The feed water enters the lower pipes at the back ends, and fuel is put in mainly exactly opposite, while a certain portion of fuel can be dropped

Fig. 205.

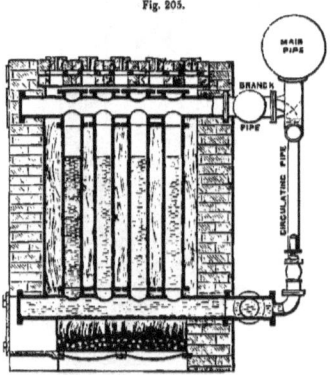

Fawcett's Vertical Tube Boiler, fitted to horizontal water and steam pipes. Patented in the year 1868.

Fig. 206.

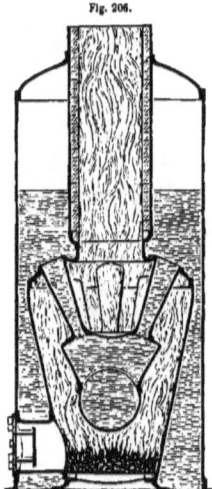

Morris's Vertical Cylindrical Boiler, fitted with a cross tube, water pot, and angular tubes in the fire box. Patented in the year 1868.

Although Mr. Howard had brought out his boiler two years previously, as shown by Fig. 165, on page 64 of this work, a Mr. Fawcett next appeared with his arrangement, as shown in sectional elevation by Fig. 205, which, like Howard's, has top and bottom pipes, horizontally situated and connected by vertical tubes, secured by flanges, instead of by packing joints. The horizontal pipes are prolonged at the back through the brickwork, and are

through the roof of the boiler space, by removing the fire clay plugs, as shown in section.

When Mr. Dickins introduced his boiler, illustrated by Fig. 179, on page 68, in this work, he perhaps thought he had discovered a good and simple arrangement, and one that needed no alteration; but not so Mr. Morris, two years after, because he borrowed Mr. Dickins's arrangement as flame tubes, and appropriated them for water tubes, in connection with a pot and a cross tube over the fire grate, as illustrated in sectional elevation by Fig. 206.

Next came Mr. Galloway, with a very peculiar arrangement of tubes, both for water and flame action, as shown in sectional elevation by Fig. 207.

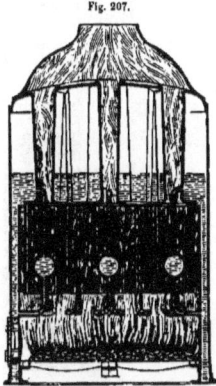

Fig. 207.

Galloway's Vertical Cylindrical Boiler, fitted with hanging water tubes, water space spheres, connecting pipes, and conical flame stay pipes. Patented in the year 1866.

The shell is cylindrical, with a curved top. The fire box is cylindrical also, with a flat crown, that is connected to the shell top by vertical conical flame tubes, and attached near to the lower end of those tubes are small water pipes, joined, about midway, to hollow spheres, and from thence extend below to the side of the fire box. Those spheres are surrounded by hanging tubes, suspended from the crown of the fire box, and thus the heating surface is said to be increased.

The action of the flame in this boiler is, that it circulates amongst and around the hanging tubes, spheres, and connecting tubes, and then passes up through the conical tubes to the chimney.

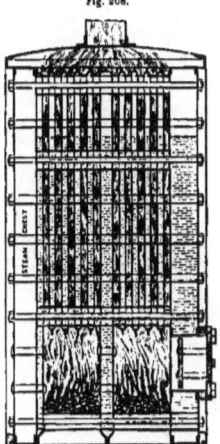

Fig. 208.

Wilkins's Vertical Square Shell Boiler, fitted with flat sheet flues and water spaces, and side main water and steam spaces. Patented in the year 1868.

It will be remembered that, on page 74 of this work, there is illustrated a flat sheet flued boiler by Fig. 194, and we next direct attention to another example by Fig. 208.

The boiler is composed of hollow sheets, formed by plates, bent and riveted together at the edges, and to maintain a uniform distance between the plates at the hollow part, hollow distance pieces are inserted. Each pair of plates are then similarly arranged vertically, and the whole series held firmly together by stay bolts, which also sustain the side main water and steam spaces. The top distance pieces conduct the steam from the sheets into the side steam space, and the water spaces are connected a little below the water level, directly over the fire grate, and at the base of the boiler, the hollow parts there forming the fire grate instead of the ordinary solid bars. The action of the flame is direct between the sheets to the chimney.

We need scarcely remark that the weak part of this arrangement is the jointings of the distance pieces, which need to be most carefully made; and if any single joint leaked, it would involve taking the boiler to pieces to repair it.

Mr. Shand, not being contented with his previous arrangement of boilers, thought fit to introduce another, as shown in sectional elevation and plan by Figs. 209 and 210, which consists of layers of cross tubes, each alternate

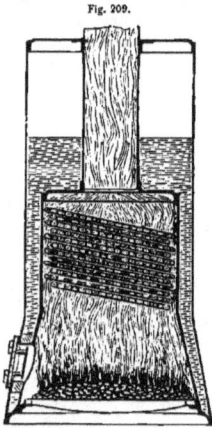

Fig. 209.

Shand's Vertical Cylindrical Boiler, fitted with layers of cross water tubes, angularly secured in the fire box. Patented in the year 1868.

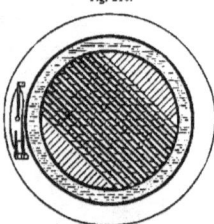

Fig. 210.

Sectional plan of Shand's boiler, shown by Fig. 209.

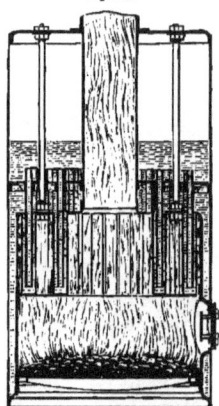

Fig. 211.

Smith's Vertical Cylindrical Boiler, fitted with short outer hanging tubes and long inner tubes, each length being secured to separate plates. Patented in the year 1868.

layer crossing the other, angularly secured in the fire box. But when we look at Green's boiler, shown on page 67, and Bezy's, shown on page 77 of this work, we cannot help thinking that Mr. Shand was fully anticipated, both in theory and practice, by two persons, while we give him the credit of blending their ideas into one arrangement.

We have already described and illustrated

Fig. 212.

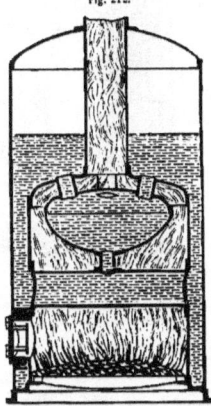

Arnold's Vertical Cylindrical Boiler, fitted with a cross tube, water pot, and connecting tubes in the fire box. Patented in the year 1868.

several examples of hanging tube boilers, but in each case the tubes were suspended from one plate. Now we refer to an arrangement on the preceding page, in which the outer tubes are suspended from the crown of the fire box, and the inner tubes from a separate plate, situated higher in the boiler, as illustrated in sectional elevation by Fig. 211. Of the advantage of this we decline to comment on.

If the principle of a certain arrangement can be patented, and the law says it can, how does Mr. Arnold's boiler, shown in sectional elevation by Fig. 212, affect Mr. Morris's? as illustrated by Fig. 206, on page 78 of this work, because they are substantially alike, except in proportion; and the law does not protect that more than it prevents one inventor from copying another, until ruinous litigation has been entered on, and then the largest purse has the best of it.

This year, 1868, was wound up, as far as regarded a Mr. Smart, in the most auspicious manner for that gentleman, because he invented and introduced three examples of syphon hanging tube boilers, as illustrated in sectional elevations and plans by Figs. 213 to 218. The first example, Figs. 213 and 214, is an arrangement of hanging tubes suspended from the fire box, containing tubes of a smaller diameter, that are hung from cross pipes over the crown of the fire box, and each of those pipes terminates in a down-take pipe in the water space surrounding the fire box.

The action of the flame is, of course, direct, but to preserve the lower ends of the tubes from the first intensity of the flame, a circular open baffle plate is secured directly under them.

Mr. Smart's second example, Figs. 215 and 216, is a fire box without any water space surrounding it; but it contains a circular set of hanging tubes secured to the crown, and within them are smaller tubes, suspended from a ring pipe situated over the crown.

His third example, Figs. 217 and 218, is nearly similar to the first, and the exception is, that the down-take syphon pipes are in a

LAND STATIONARY VERTICAL BOILERS.

central annular water space that surrounds the main flame tube or flue to the chimney, while, also, the water space surrounding the fire box is again omitted.

Mr. Smart's ideas were very ingenious, but, unfortunately, not altogether novel, on account of Mr. Smith's arrangement, as shown on page 80, by Fig. 211, in this work, where the

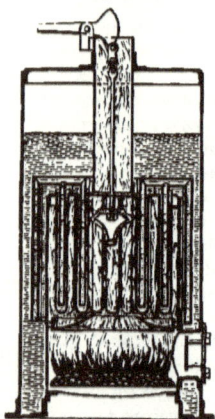

Fig. 213.

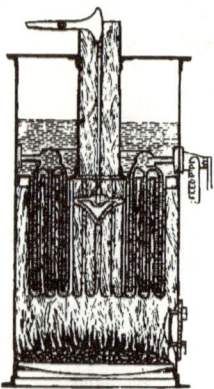

Fig. 215.

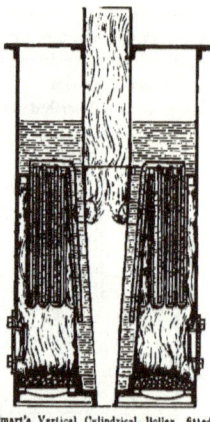

Fig. 217.

Smart's Vertical Cylindrical Boiler, fitted with syphon hanging tubes in the fire box and in the surrounding water space. Patented in the year 1868.

Smart's Vertical Cylindrical Boiler, fitted with hanging annular water tubes, the smaller tubes being suspended from a ring pipe over the crown of the fire box. Patented in the year 1868.

Smart's Vertical Cylindrical Boiler, fitted with syphon hanging tubes, and a water space in the fire box. Patented in the year 1868.

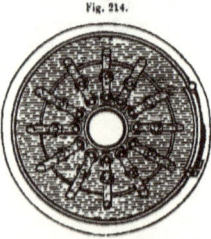

Fig. 214.

Fig. 216.

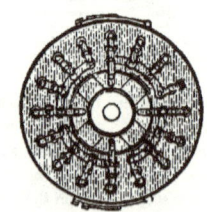

Fig. 218.

Sectional plan of Smart's boiler, as shown by Fig. 213.

Sectional plan of Smart's boiler, shown by Fig. 215.

Sectional plan of Smart's boiler, shown by Fig. 217.

longer and shorter tubes are fully illustrated, as they also are in a more syphon form by Fig. 204, on page 77 of this work.

The schemes and constructions of vertical boilers in the year 1869 next claim attention, and on the first day of that year Mr. Green, whose ideas have figured on page 67 of this work, again appeared with the arrangement shown by Fig. 219. In this case he ignored

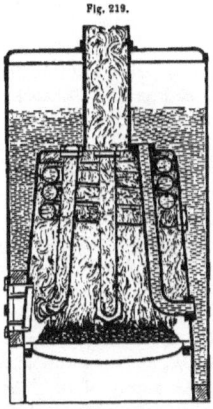

Fig. 219.

Green's Vertical Cylindrical Boiler, fitted with vertical water bend pipes and a water coil in the fire box. Patented in the year 1869.

cross tubes, but borrowed the vertical water pipes in the fire box, and for that sin introduced a water coil pipe surrounding them as a redemption.

Next came a Mr. Desvignes, who claimed personal honours for thinking that curved and straight syphon pipes hanging in a fire box were worthy of lawful protection, as illustrated by Fig. 220.

A Mr. Loader then appeared with a conical vertical boiler, as illustrated by Fig. 221, and

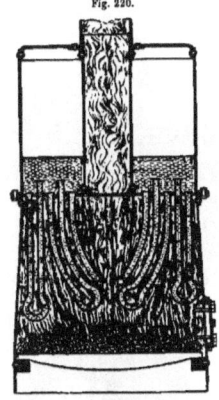

Fig. 220.

Desvignes' Vertical Cylindrical Boiler, fitted with straight and curved syphon tubes in the fire box. Patented in the year 1869.

Fig. 221.

Loader's Vertical Cylindrical Boiler, fitted with a conical water pot, containing a steam coil. Patented in the year 1869.

this boiler is really non-cleansing and unrepairable, while the arrangement is really a joke on science, on account of the steam coil

Fig. 222.

Barclay's Vertical Cylindrical Boiler, fitted with vertical water tubes in the fire box, and an annular water space surrounded with similar tubes, and a flame space. Patented in the year 1869.

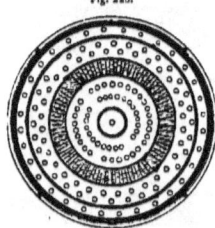

Fig. 223.

Sectional plan of Barclay's boiler, shown by Fig. 222.

being put in the water, thus subjecting it to a damp heat, whereas it should be to a dry heat.

On referring to page 63 of this work, the Figs. 159 and 160 illustrate a Mr. Barclay's arrangement for a vertical boiler, which was a modification of a prior idea shown on page 62, and thereby conveys the fact that the inventor felt rather dubious on both points of conclusion; but he became more so in this year, 1869, because he brought out various other modifications that are combined in the arrangement shown in sectional elevation and plan by Figs. 222 and 223.

In this case the vertical tubes with curved ends are again introduced in the fire box, but the water pot omitted, while the flame tubes are smaller and more numerous. The main feature, however, consisted in surrounding the water space with an annular flame space, in which are additional water tubes, connected at their upper ends to the water and steam chamber containing the circular steam pipe, perforated on its upper side, by which the steam is taken off to the engine. The space below the fire grate is shown closed in by the casing, and in this, an opening is made to admit the pipe, which is connected with a fan for producing a blast of air. The circular pipe is perforated, and through these perforations the compressed air escapes and mixes with the fuel on the grate. The pressure of the blast forces all the waste products of combustion to the upper part of the furnace, and the flue is provided with a damper, which is lifted to the requisite extent by the pressure of the blast, and the waste products of combustion are expelled therethrough into the chimney.

The outer tubes are enclosed by a casing, as shown, that is made double, and contains a space between the inner and outer plates, and which may be an air space or a water space, and when used as a water space it is connected directly by tubes or otherwise to the water spaces of the other parts of the boiler.

inventors of boiler novelties, and their attention was directed from tubes to surfaces of a more varied form.

Mr. Fletcher first appeared with an ar-

Fig. 224.

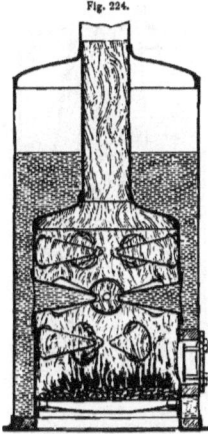

Fletcher's Vertical Cylindrical Boiler, fitted with conical water pockets in the fire box. Patented in the year 1860.

Fig. 225.

Sectional plan of Fletcher's boiler, shown by Fig. 224.

Fig. 226.

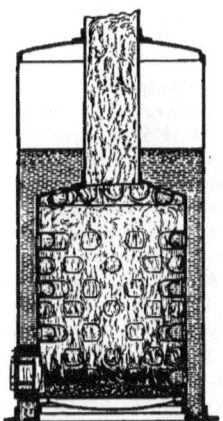

Barran's Vertical Cylindrical Boiler, fitted with "cups" in the fire box. Constructed in the year 1859.

Fig. 227.

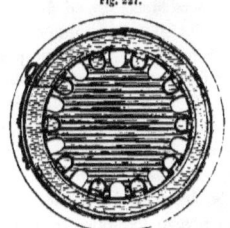

Sectional plan of Barran's boiler, shown by Fig. 226.

At this period of boiler chronology the tubular arrangements were withheld by the rangement shown in sectional elevation and plan by Figs. 224 and 225, in which the fire box is fitted internally with projecting water

cone spaces that are radially and horizontally secured. Those spaces were called "thimbles" by the inventor. But Mr. Fletcher had been forestalled by a Mr. Barran, ten years previously, who introduced the "cup" surface boiler, as shown in sectional elevation and plan by Figs 226 and 227, the arrangement of which is precisely as Fletcher's, with the addition of the cups at the crown. Barran had conceived a reverse position for the cups in the year 1855, as shown in sectional elevation and plan by Figs. 228 and 229. In that arrangement the fire box was fitted externally with cups, and suspended from the crown was a water pot containing flame tubes. The crown also was connected to the roof of the boiler by flame tubes that passed the products of combustion to the chimney.

Directly after Mr. Fletcher, an American engineer appeared with a novel means of increasing the evaporative power of boilers, as illustrated in sectional elevation and plan by Figs. 230 and 231; the arrangement of which is that the heating surface is formed of a cast metal dome-shaped fire box with projecting ribs extending into the fire or furnace chamber, and also with projecting ribs extending into the water space within the boiler. By this means the heat conducting power of metal is taken advantage of, and the ribs projecting into the fire are protected from injury by extending the ribs into the water within the boiler, and presenting a larger surface of ribs to the water than to the fire. The ribs within the water are surrounded by a shield, dividing the water between the ribs from the main body of water, and by this means ensuring a rapid circulation of the water between the fire box and the shield, and also increasing the absorption of heat by the water.

Tubular vertical boilers again became the

Fig. 228.

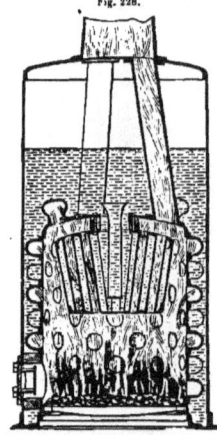

Barran's Vertical Cylindrical Boiler, fitted with a tubular water pot in the fire box and cups at the sides and crown, with flame tubes. Patented in the year 1855.

Fig. 229.

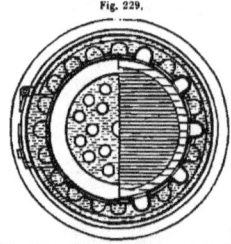

Sectional plan of Barran's boiler, shown by Fig. 228.

fashion, and Mr. Barker next appeared with the arrangement shown by Figs. 232 and 233 in sectional elevation and plans. The fire box is fitted with a ring of tubes that surrounds the fire grate, excepting at the door opening, and those tubes pass down below the the side, and crown of the fire box, passes through a flue that is bolted to the centre of the crown plate of the fire box. This flue is entirely below the water level, and branches off laterally to the chimney; a regulating

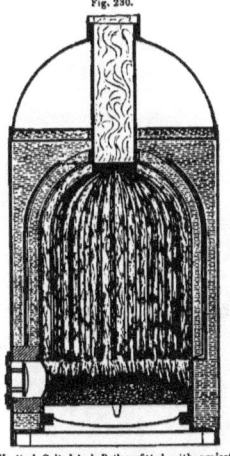

Fig. 230.

Miller's Vertical Cylindrical Boiler, fitted with projecting ribs inside and outside the fire box. Patented in the year 1869.

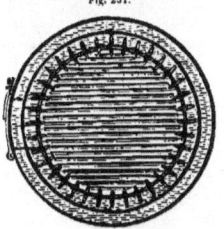

Fig. 231.

Sectional plan of Miller's boiler, shown by Fig. 230.

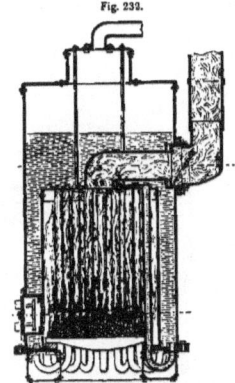

Fig. 232.

Barker's Vertical Cylindrical Boiler, fitted with vertical tubes having syphon ends below the fire grate. Patented in the year 1869.

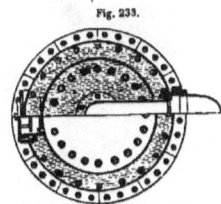

Fig. 233.

Sectional plan of Barker's boiler, shown by Fig. 232.

grate, and are bent around up into the water space that encloses the fire box plates.

The flame, after acting around the tubes, damper being employed for controlling the draught. The outer end of the flue is formed with a flange, which is bolted to an internal flange on an annular junction piece.

The fire door opening is composed of a

short tube provided with internal and external flanges at its inner and outer extremities respectively. These flanges are respectively bolted to rings riveted to the inner and outer shells for the purpose of presenting flat junction surfaces, against which the flanges of the tube are bolted.

Another syphon tube arrangement then came out by the endeavour of a French engineer, M. Thirion, who perhaps had seen a few plain tubular boilers, but not many of the syphon or circulating tube kind.

M. Thirion's ideas are illustrated by Fig. 234, in sectional elevation, which shows a plain cylindrical boiler, fitted with return or U shaped tubes in the fire box—but how the inside of the bend part of the tube is cleaned, we decline to explain, it being beyond our comprehension. Very similar also are Mr. Wilson's boiler tubes, as shown by Fig. 156, on page 62 of this work.

Mr. Kinsey appeared again, but this time with a more formidable arrangement than before, and not very unlike Mr. Ferinhough's boiler, as shown by Fig. 45, on page 23 of this work.

Mr. Kinsey's ideas are illustrated in sectional elevation by Fig. 235, and show that, in a series of boilers—fitted internally with circulating pipes—connected at the top and bottom by steam and water pipes, consists the arrangement. We may mention also that the circulating tubes were angularly slitted at the bottom ends, and that the top ends were fitted with ventilating cones.

Mr. Allibon, not being tired, came again

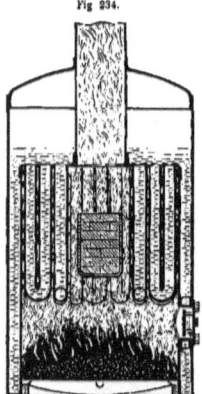

Fig. 234.

Thirion's Vertical Cylindrical Boiler, fitted with return syphon tubes in the fire box. Patented in the year 1869.

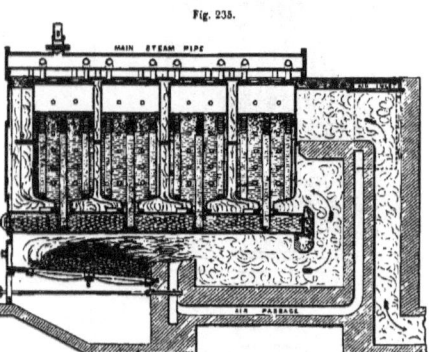

Fig. 235.

Kinsey's Vertical Cylindrical Boilers, fitted with circulating vertical water pipes, horizontal connecting water pipes, and horizontal main steam pipes. Patented in the year 1869.

with another boiler, and a very complicated one it is, too, as shown by sectional elevation and plan by Fig. 236. The mechanical arrangement is that in the fire box, suspended from its roof, is an annular water pot, with horizontal flame tubes at the top. The fire box is surrounded by an annular water space, that is connected at the top by

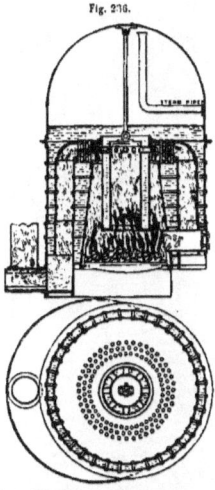

Fig. 236.

Allibon's Vertical Boiler, fitted with an annular water pot, and annular water and flame spaces and tubes. Patented in the year 1869.

vertical water tubes to the main water space. The annular water space is surrounded by a flame space, having a flue at its base, leading to the chimney at the side of the boiler.

The action of the flame is that, on rising from the grate, it enters the space in the pot as well as around it, and the two volumes mingle at the opening opposite the pot horizontal tubes. The flame passes amongst the vertical tubes, and then descends into the annular space, and finally out at the flue.

Although Mr. Allibon had borrowed some parts, and contrived the rest, to make a novel

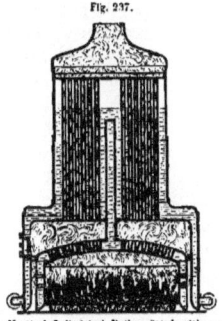

Fig. 237.

Salisbury's Vertical Cylindrical Boiler, fitted with a perforated water crown to the fire box, and above that a combustion chamber connected by flame tubes to the smoke box. Patented in the year 1869.

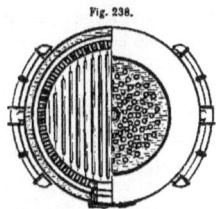

Fig. 238.

Sectional plan of Salisbury's boiler, shown by Fig. 237.

arrangement of boiler, as last illustrated, he was very nearly outwitted by an American engineer, named Silas Salisbury, who proposed the peculiar means for raising steam as shown in sectional elevation and plan by Figs. 237 and 238, consisting of a fire box surrounded—

excepting at the door—with a water space, which extends also over the crown; and this space, and the plates forming it, are pierced with flame tubes, to permit the flame to flow into an annular flame space, also into a combustion chamber over the crown. The roof of this chamber is a tube plate, containing the ends of a set of vertical flame tubes that extend to the base plate of the smoke box, which conducts the flame to the chimney over it. The water level in the tube chamber is about one-fifth of the total depth of the chamber from the top, and near that distance is the top of a central water pipe in connection with the water space over the fire box, and thus a circulation is permitted.

This arrangement, although at first sight appearing complicated, is really simple, and can be easily cleaned and repaired.

A Mr. Fraser next appeared with a vertical tubular arrangement, as shown in sectional elevation and plan by Figs. 239 and 240. Mr. Fraser of course knew that water tubes secured in the fire box, as he secured them, had been done eight years previously; but those tubes were the same diameter throughout. Now Mr. Fraser corrugated his tubes of unequal diameters, to make the flame act on them longer, and there his invention ended.

Mr. Chaplin concluded this year, 1869, of inventions, by a method of connecting the upper and lower water spaces of his previous boiler, as illustrated in sectional elevation by Fig. 241; the arrangement being a conical fire box with a water space around it, that is connected by outside pipes to the water chamber above; and flame tubes, of Mr. Chaplin's peculiar construction, connect the fire and smoke boxes.

The ensuing year, 1870, was not so famous

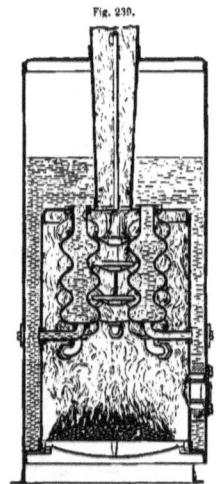

Fig. 239.

Fraser's Vertical Cylindrical Boiler, fitted with corrugated cylindrical vertical tubes. Patented in the year 1869.

Fig. 240.

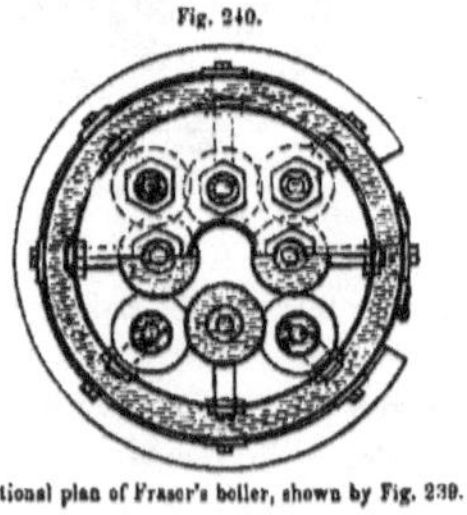

Sectional plan of Fraser's boiler, shown by Fig. 239.

for vertical boilers as the previous period; for the reason, perhaps, that inventors were rather exhausted, and it took them a full quarter of a year to recover; the first to return to the field being Mr. Green, who has already been noticed in this work.

Mr. Green's new arrangement is shown in side and end sectional elevations by Fig. 242, the arrangement being that four water pipes, situated directly over and beyond the fire grate, are placed horizontally and close to-

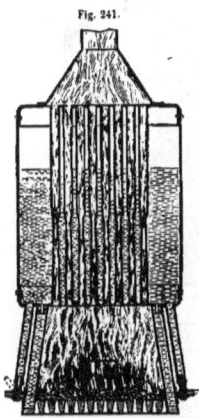

Fig. 241.

Chaplin's Vertical Cylindrical Boiler, fitted with water tubes, to connect the upper and lower water spaces. Patented in the year 1869.

gether upon the brickwork, so that no space is left between them; and by this means the flame and products of combustion are caused to act against their under sides, thereby imparting a large amount of heat to them before the products return over them on their passage to the flue or to the chimney. It will be observed, also, that nearly the whole of the exterior surfaces of the pipes are available for heating purposes, as they touch only one another at one portion of their circumference. Each pipe rests near its rear end upon an upright or support, that checks the passage of the flame and products, and tends to increase their efficiency.

The upper portions of the pipes have holes in them, over which tapered tubes are riveted or otherwise secured, the smallest ends of the

Fig. 242.

Green's Vertical Tube Boilers, fitted to horizontal pipes, situated directly over and beyond the fire grate. Patented in the year 1870.

tubes being downwards, to admit of a comparatively larger space being formed between each row, to enable a man to pass for cleaning the soot or other deposit from them.

These tubes are of a larger diameter at their upper parts, and are connected one to the other by short connecting pipes, so that the steam generated in one tube can mingle with the steam in the next one, and thus establish a large steam area proportionate to the size of boiler.

Each of the tubes has a cover or top plate, which is secured to a flange fitted in the inner edge thereof, so that the edges of the covers are kept flush, or nearly so, with the top edges of the tubes. Thus the tubes rest against each other, and prevent the products rising to the roof before they have given off nearly all their heating properties.

We next return to page 25 of this work, and notice that the flame descends in the examples of boilers there illustrated, similar to the illustration shown by Fig. 243, which is an American production, by a Mr. Montgomery, who, fourteen years after Mr. Dunn, came forward in the world of boiler producers

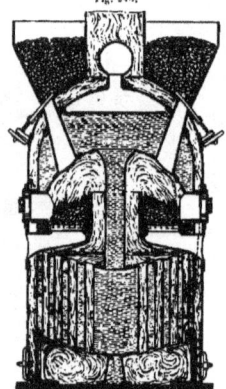

Fig. 243.

Montgomery's Vertical Cylindrical Boiler, fitted with an annular fire grate, about midway of the height of the shell, and a coal box above, with the flame acting for the generation of steam below the grate. Patented in the year 1870.

with his arrangement as a novelty; but not being content with borrowing the descent of the flame scheme, he also claimed the invention of the annular fire grate, revolving or not, as he chose, and an overhead feed-fuel apparatus, finishing with the idea that the flame should surround the water shell of the boiler.

Now, if Montgomery had seen the arrangement shown by Fig. 20, on page 11 of this work, he doubtless would not have committed the apparent mistake of patenting what had been invented thirty-four years previously.

Mr. Barker, who figures metaphorically, by

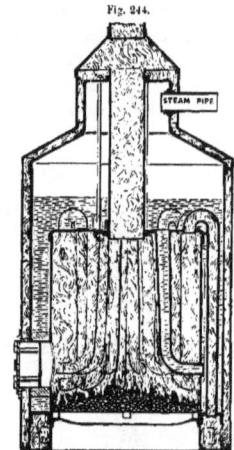

Fig. 244.

Barker's Vertical Cylindrical Boiler, fitted with water tubes in the fire box, and syphon flame tubes surrounding it. Patented in the year 1870.

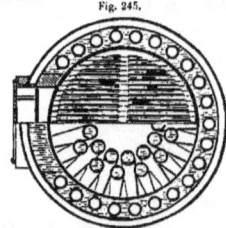

Fig. 245.

Sectional plan of Barker's boiler, shown by Fig. 244.

the illustration of his boiler, on page 87 of this work, is represented again in this year, 1870, in sectional elevation and plan, by Figs. 244

and 245, by which it is evident that he overthrew all his prior ideas; and, indeed, so much so, that he proposed the curved end water tubes to begin with, and in doing so, borrowed "Burch's" idea, as shown on page 42 of this work.

Mr. Barker extended his new scheme to flame tubes also, by connecting their ends in a syphon form, either to the crown of, or just below the fire box, and prolonged them to a

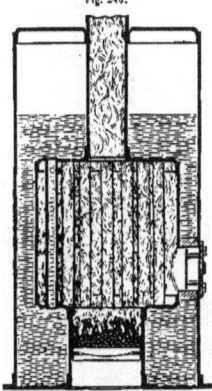

Fig. 246.

Wilkins's Vertical Cylindrical Boiler, fitted with a shallow fire box under a large combustion chamber containing water tubes, and a main flue overhead. Constructed in the year 1870.

ring chamber situated below the annular water space surrounding the fire box.

The flame from those tubes passes through openings in the chambers to the space surrounding the water shell of the boiler, and thus combines in the smoke box with the remainder of the flame that passes up through the central flue connected to the roof of the fire box. The return flame tube arrangement

was, however, invented in the year 1851, and the outer flame space in the year 1836, as shown on pages 12 and 11 of this work.

A Mr. Wilkins, about the middle of this year, constructed a vertical boiler, which we illustrate in sectional elevation by Fig. 246, the peculiarity being that the fire box is cylindrical, very shallow, and directly above it is a combustion chamber, of an enlarged

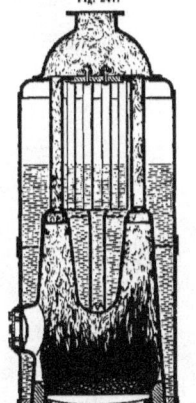

Fig. 247.

Riche's Vertical Cylindrical Boiler, fitted with a water pot in the fire box, and flame tubes connecting the smoke box. Constructed in the year 1870.

diameter, containing nearly a ring of water tubes, situated outside the diameter of the fire box. The fuel is put on the grate through a door opening directly over the roof of the fire box, so that is the reason why the water tubes are "nearly" a ring in arrangement. The main flue is directly over the fire grate in position, at the roof of the combustion chamber, and thereby the action of the flame in the combustion chamber depends on the area of the flue in proportion to the area of the fire grate, and the draught below it.

A Mr. Riche's example of boiler next appeared, as shown on the preceding page by Fig. 247, the arrangement being a "pot" containing a circulating pipe, and flame tubes passing through the water, and steam spaces above the fire box. This is a very good

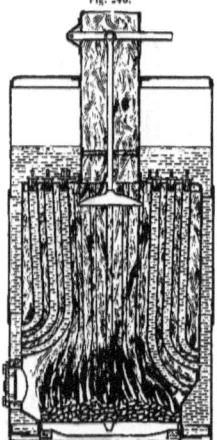

Fig. 248.

Paxman's Vertical Cylindrical Boiler, fitted with curved water tubes, having mushroom caps on their top ends. Patented in the year 1870.

arrangement, that may be termed a "compound" flame and water boiler, copied from two boilers of those types and blended into one.

The next example of boiler is as illustrated in sectional elevation by Fig. 248, the arrangement of which is that the fire box and shell are the ordinary shape, but the internal fittings form the novelty. Those fittings are

water tubes in the fire box, curved at their lower ends, and smaller than at the straight part; at the top of which, on each tube, is fixed a "mushroom cap," having three ribs, to permit the circulation of the water during the raising of steam.

Mr. Paxman also introduced at the same time another arrangement, as illustrated by Fig. 249. In this case the same class of tubes

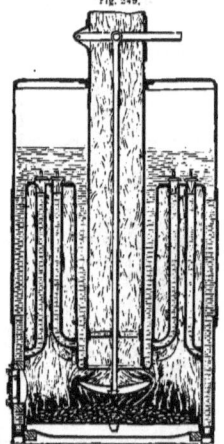

Fig. 249.

Paxman's Vertical Cylindrical Boiler, fitted with a water-spaced central flue in the fire box, and water tubes having mushroom caps on their top ends. Patented in the year 1871.

were secured in the fire box, but the main flue, instead of commencing at the roof of the fire box, descended into it with an annular water space, or what is more familiarly known as a "water pot with the flue passing through it." The tubes were secured to the water pot, and also to the side of the fire box, so that a circulation was permitted in both cases.

LAND STATIONARY VERTICAL BOILERS.

The novelty in those two arrangements is the caps on the top ends of the tubes, because the tubes are precisely as Burch's, as illustrated on page 42 of this work.

The intention of the use of these caps is to prevent the bubbles of heat and steam rising out of the tubes from passing up directly through the water immediately over the tubes, and thereby to cause the heat to disperse at right angles; or, as the inventor states, "tend to confine the current of water to the sides of the tubes, whilst allowing free egress 85 and 86 of this work we illustrated three examples of "cup" boilers—the cups being filled with water in two cases, and with flame in the other—and that it appeared that there was nothing left to improve in them as "cups." But not so Mr. Lees, the patentee of the example shown on the next page in sectional elevation and plan by Figs. 251 and 252, who evidently had seen the other examples, and thus improved on them by making the cups partially globular, and putting a circulating pipe in each.

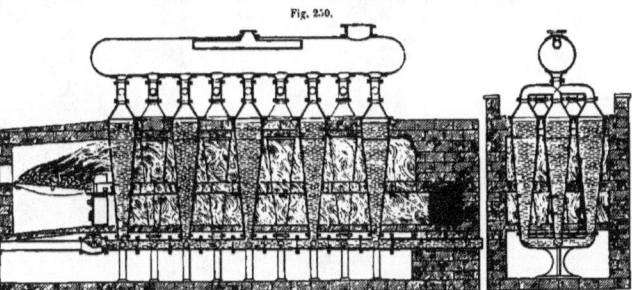

Fig. 250.

Kenyon's Vertical Conical Tube Boilers, fitted with a feed water pipe at their bottom ends, and a steam chamber at their top ends. Patented in the year 1870.

for the water, which, striking against the mushroom head, is directed downwards, or into its proper channel again, thereby reducing the tendency to priming."

Next appeared an arrangement of a series of vertical cone boilers, the invention of a Mr. Kenyon, as shown in side and end sectional elevations by Fig. 250, the main feature of which is that each boiler is separate from the other, and in the event of repair can be readily removed.

It is, of course, not forgotten that on pages

We next refer to page 35 of this work, and to the illustration, Fig. 85, on it, which boiler, excepting the horizontal flame tubes, is precisely arranged as the two present examples, shown in sectional elevation by Figs. 253 and 254, showing again how two people unknown to each other think alike at different periods.

Mr. Vaughan Pendred next came to the front line of inventors of vertical boilers in this year, 1871, with two arrangements of horizontal water tubes secured in the combustion chamber, the first example being illustrated in sectional elevation and plan by Figs. 255 and 256 on the next page.

This arrangement consists of a cylindrical chamber, fitted with horizontal water tubes, having at each end curved nozzles or bend pipes, open in opposite directions at each end,

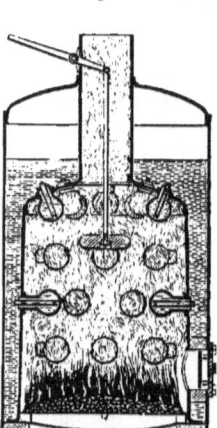

Fig. 251.

Lees' Vertical Cylindrical Boiler, fitted with "cups"—containing tubes—in the fire box. Patented in the year 1871.

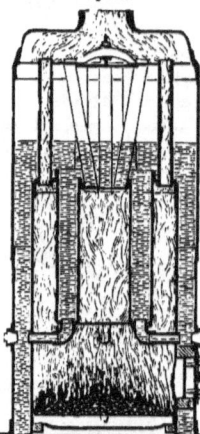

Fig. 253.

Jeffery's Vertical Cylindrical Boiler, fitted with central flame box, annular flame space, and vertical flame tubes. Patented in the year 1871.

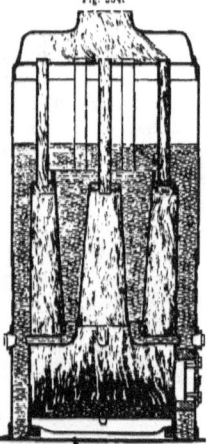

Fig. 254.

Jeffery's Vertical Cylindrical Boiler, fitted with central flame box, annular flame space, and vertical flame tubes. Patented in the year 1871.

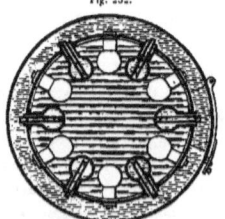

Fig. 252.

Sectional plan of Lees' boiler, shown by Fig. 251.

outer shell, enclosing a similar shaped fire box, above which is a square combustion to cause a free circulation of the water and heat through them.

Mr. Pendred's next arrangement is illustrated in sectional elevation by Fig. 257, and is undoubtedly the same as the former example, excepting the ends of the tubes, which in this case are connected in three sets, by vertical pipes at each end, open at the bottom on one side, and open at the top at the other; but how or where the advantage of that occurs is not apparent, inasmuch that free heat circulation, and separate tube arrangement, are similar in the principle of results.

LAND STATIONARY VERTICAL BOILERS.

Mr. Winstanley, whose original ideas on boilers are illustrated on page 55 of this work, appeared again this year, 1871, with a totally different view of the subject, as shown in sectional elevation by Fig. 258, the arrangement being a series of flame pockets over each other, connected by tubes, fitted in a cylindrical shell of ordinary form.

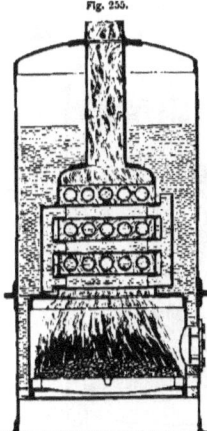

Fig. 255.

Pendred's Vertical Cylindrical Boiler, fitted with horizontal tubes in the combustion chamber, open at their ends. Patented in the year 1871.

Fig. 257.

Pendred's Vertical Cylindrical Boiler, fitted with horizontal tubes in the combustion chamber, connected at their ends by vertical pipes. Patented in the year 1871.

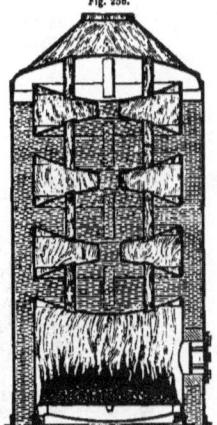

Fig. 258.

Winstanley's Vertical Cylindrical Boiler, fitted with flame pockets and tubes. Patented in the year 1871.

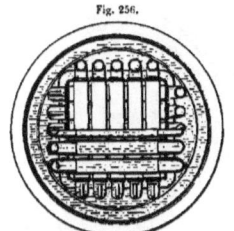

Fig. 256.

Sectional plan of Pendred's boiler, shown by Fig. 255.

The disadvantage of this boiler is that it cannot be cleaned easily—and as for repair, that means total destruction. Besides, if we refer to Figs. 122 and 123 in this work on page 49, a doubt is unquestionably there shown as to the validity of Mr. Winstanley's patent.

Another water tube boiler next made its appearance, introduced by a Mr. Ashton, who followed the arrangement by Mr. Galloway—shown by Fig. 29, on page 16 of this work—with the chimney at the side of the combustion chamber, as illustrated in sectional elevation and plan by Figs. 259 and 260.

The fire box is dome-shaped, with a central flame flue opening into a large combustion chamber, containing a series of tubes, that connect the lower and upper water spaces.

Fig. 259.

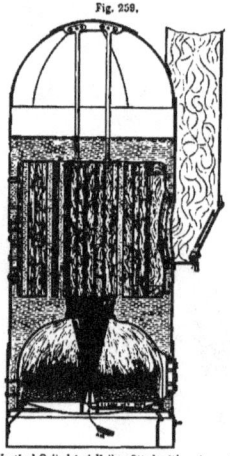

Ashton's Vertical Cylindrical Boiler, fitted with a large dome fire box, having a small flue in the centre, leading into a combustion chamber, containing a series of water tubes, with a side chimney. Patented in the year 1871.

Fig. 260.

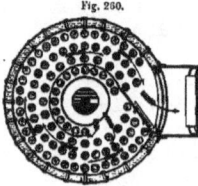

Sectional plan of Ashton's boiler, shown by Fig. 259.

The flame, after acting in the fire box, circulates amongst the tubes, and then passes up the chimney at the side of the chamber.

To accelerate the draught, as illustrated, a steam jet pipe is used, which Mr. Ashton states is for cleaning the flue and tubes also;

Fig. 261.

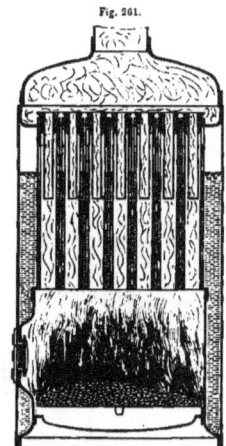

Oram's Vertical Cylindrical Boiler, fitted with water tubes enclosing flame tubes, having a short smaller tube in each to check the flame. Patented in the year 1871.

Fig. 262.

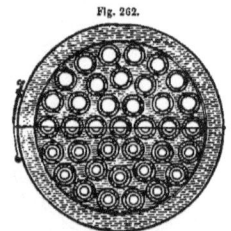

Sectional plan of Oram's boiler, shown by Fig. 261.

but how, he does not explain, and neither can we.

A Mr. Oram came directly after Mr. Ashton,

with an arrangement of flame tubes surrounded by water tubes, in which are slots at the two ends and below the water line, to allow the water and steam in the tubes to mix with the quantities of vapours and liquids surrounding them; and to prevent a too sudden exit of flame through the tubes, a short open tube of a smaller diameter is

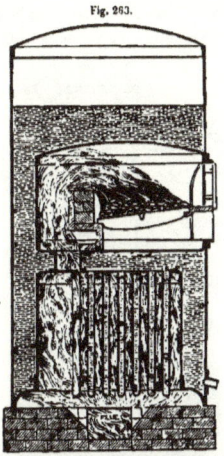

Fig. 263.

Laharpe's Vertical Cylindrical Boiler, fitted, central of its height, with a fire box, below which is a water space surrounding two flame flues that lead down to a combustion chamber containing hanging water tubes. Patented in the year 1871.

secured at the top of each tube, closing it by the method of securing. This arrangement is shown in sectional elevation and plan by Figs. 261 and 262.

The French engineers next were again represented, and, at this time, M. Laharpe had that honour; but he, doubtless, not knowing that Mr. Chaplin claimed the introduction

of using single tubes "hanging" from the tube plate—as shown on page 59 of this work—fell into the mistake of proposing the arrangement as shown by Fig. 263, which is a shell containing a fire box, midway of its height, with two flues in the bottom of the box, leading down into a combustion chamber containing the hanging tubes referred to.

M. Laharpe then, to be on the safe side as an inventor, proposed three more arrange-

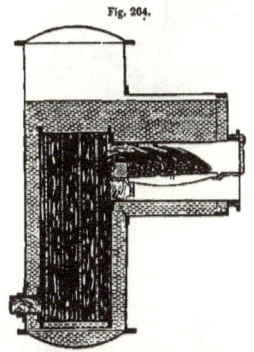

Fig. 264.

Laharpe's Vertical Cylindrical Boiler, fitted with a projecting midway fire box, at the inner end of which is a combustion chamber containing water tubes extending to the water space near the bottom of the shell. Patented in the year 1871.

ments, as illustrated by Figs. 264 to 266, which are sufficiently explained by the description under them.

The example of boiler we notice next is a novelty, because it has no free egress for the products of combustion, as illustrated in sectional elevation by Fig. 267; the arrangement being that the fire box is fitted with a flue, closed with a lever weighted safety valve, and a little above the water line the flue is fitted with downtake flame tubes that extend in the water space surrounding below the fire grate, and are there in connection with non-return valves that permit any flow from the tubes, but no return back through them.

According to Mr. Adams, the inventor,

Fig. 265.

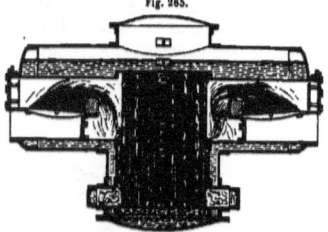

Laharpe's arrangement, as the preceding example, with twin fire boxes. Patented in the year 1871.

Fig. 266.

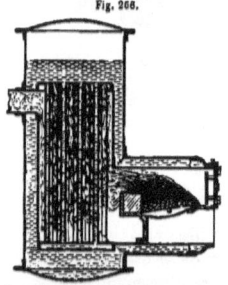

Laharpe's Vertical Cylindrical Boiler, fitted with a projecting fire box near the bottom of the shell, leading into a combustion chamber containing water tubes that are situated nearly midway of the shell. Patented in the year 1871.

the action of the flame in this boiler will be as follows:—" When the fire is first lighted in the fire box, the safety valve on the top of the chimney or flue is raised, to allow a free passage through the chimney. The air enters the ash pit in the ordinary manner, and steam is raised in this boiler precisely as in other boilers; and when the steam has attained a pressure sufficient to work a blowing engine, not shown, the furnace and ashpit doors are closed air-tight, and the safety valve is lowered to close the chimney, the fire being thus cut off from all communication with the external air. The products of combustion are then forced through the downtake tubes, open

Fig. 267.

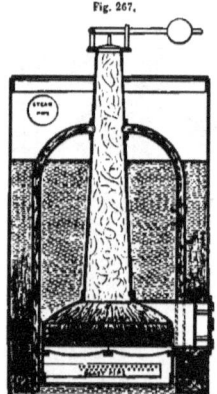

Adams' Vertical Cylindrical Boiler, in which the products of combustion are forced into the water in the boiler. Patented in the year 1871.

the valves, and thereby rush up into the water as illustrated."

We here add that natural laws are entirely at variance with the above theory, and therefore have no hesitation in stating that all the steam generated in that boiler would be required for the blowing engine; consequently, the motive power is entirely absorbed in the " working " of the boiler only.

The example next noticed is illustrated by Fig. 268, showing thereby that a Mr. Martin, although in Australia, sent his notion of what a vertical boiler should be, to England; the notion being that baffle or circulating plates should encircle the flame and water spaces, and thus disturb the free circulation of the heat in the boiler and outside of it.

Next followed in the train of boiler inventors a Mr. Clark, with the arrangement as illustrated in sectional elevation and plan by

Fig. 268.

Martin's Vertical Cylindrical Boiler, fitted with circulating plates in the water and flame spaces. Patented in the year 1871.

Figs. 269 and 270, the main feature of which is that the fire box and the central water tube are corrugated, to cause the flame to act on those surfaces with more effect than if they were plain.

Mr. Clark no doubt thought himself the originator of his scheme, and the law protected him in that belief; but he perhaps had not seen the illustrations shown by Figs. 271 and 272,

on the next page while the law omitted to remind him of it, although an interval of eighteen years had elapsed, and the original patent was of course very well ventilated.

Fig. 269.

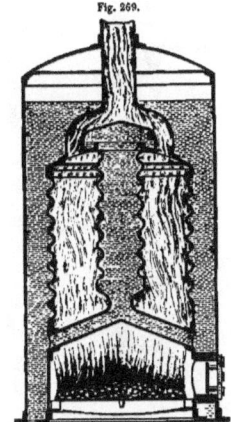

Clark's Vertical Cylindrical Boiler, fitted with a corrugated fire box containing a central corrugated water tube, connected to the sides by tubes over the fire grate. Patented in the year 1871.

Fig. 270.

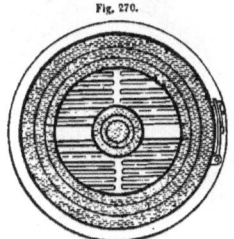

Sectional plan of Clark's boiler, shown by Fig. 269.

Mr. Dunn, who claims the first honours in the matter, it will be seen, proposed precisely what Clark did in the way of corrugating: and more than that, Dunn proposed also the boiler shown by Figs. 273 and 274, at the same time.

Fig. 271.

Dunn's Vertical Cylindrical Boiler, fitted with a corrugated fire box and a side chimney. Patented in the year 1853.

Fig. 272.

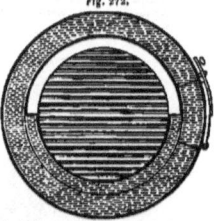

Sectional plan of Dunn's boiler, shown by Fig. 271.

Fig. 273.

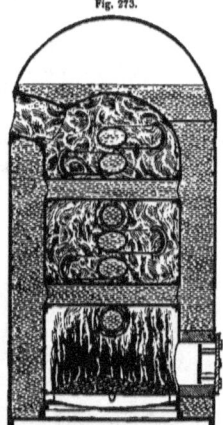

Dunn's Vertical Cylindrical Boiler, fitted with a dome fire box, having cross water tubes in it and a side chimney. Patented in the year 1853.

Fig. 274.

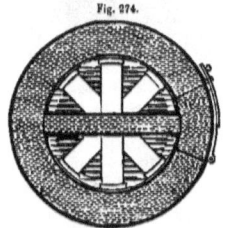

Sectional plan of Dunn's boiler, shown by Fig. 273.

CHAPTER II.

LAND STATIONARY HORIZONTAL BOILERS.

THE horizontal or longitudinal boiler was introduced by Watt, in the year 1788, being twenty-five years after the first cast iron globular boiler was introduced, as shown by Fig. 1, on the first page of this work.

Fig. 275.

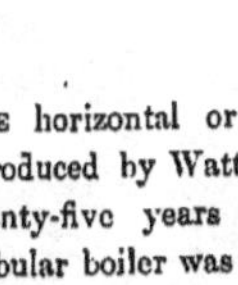

Transverse Sectional Elevation of Watt's "Wagon" boiler, with straight sides and bottom. Erected in the year 1788.

Fig. 277.

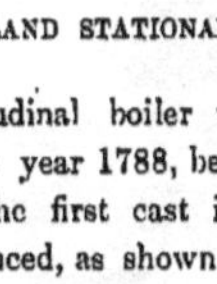

Transverse Sectional Elevation of Watt's "Wagon" Boiler, with recessed sides and curved bottom. Erected in the year 1793.

Fig. 276.

Transverse Sectional Elevation of Watt's "Wagon" Boiler, with curved sides and bottom. Erected in the year 1789.

Fig. 278.

Transverse Sectional Elevation of Watt's "Wagon" Boiler, with a central flue in it. Erected in the year 1793.

Watt's first horizontal boiler is illustrated in transverse sectional elevation, by Fig. 275, and was known then as the "wagon" boiler, from its cumbersome form resembling the covering of that article. The sides and bottom of that boiler being the weakest parts, Watt curved them next, as shown by Fig. 276, and to make the flame impinge more on the sides, he recessed them, as shown by Fig. 277. And after that, to increase the heating surface, he put a flue through the shell, as shown by Fig. 278.

There were rivals in the field of boiler arrangements at this date, but Watt held the palm in his hand for his wagon boiler for

Fig. 279.

Sectional elevation and plan of Brindley's Granite Boiler, fitted with a copper flame tube. Erected at Camborne, Cornwall, in the year 1800—being the first flame tube boiler.

some time, and it was only until the year 1800 any attempt was made to wrest it from him; when Mr. Brindley, the then well known Cornish engineer, introduced the arrangement as shown by Fig. 279, which illustrates a granite boiler containing a copper return flame tube, and according to Nicholson's Journal, "the boiler was twenty feet long, nine feet wide, and eight feet six inches deep, and the three copper tubes were each twenty inches in diameter."

LAND STATIONARY HORIZONTAL BOILERS.

Watt, however, not to be beaten, made a wooden boiler with a copper return flame tube, as illustrated by Fig. 280, which could only be worked at 2¼lbs. on the square inch.

There was no need, however, for this departure from the use of metal for the shells of boilers; because, at about this period, 1800, wrought iron plates came into boiler use

Fig. 280.

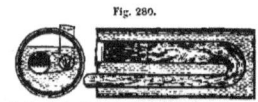

Sectional elevation and plan of Watt's Wooden Boiler, fitted with a copper flame tube. Constructed in the year 1800.

Fig. 281.

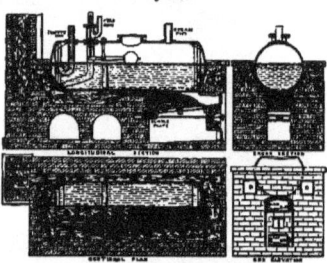

General arrangement of the setting of a Cylindrical Horizontal Boiler, with curved ends—the flame first passing under the shell; secondly, at one side and front end; and thirdly, back at the other side and up the chimney. Erected in the year 1800.

generally, and the common practice amongst the few makers of boilers then, was to make the shell of wrought iron, where exposed to the fire, and the remainder of cast iron, and thus the wagon form of boiler gave way to the shape and setting as illustrated by Fig. 281. The ends of this boiler, it will be noticed, are curved, or arcs of a circle, from which came the egg-end boiler, as shown in dotted lines. To be mentioned also as a feature in this arrangement, is the stop plate —at one side of the feed pipe in the shell—to prevent any deposit from passing back to the portion of the boiler over the fire.

Fig. 282.

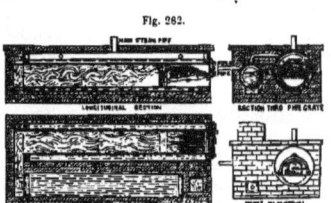

General arrangement of Evans's Cylindrical Horizontal Boilers, set in brickwork, the flame passing through the tube of the larger shell and under the smaller boiler. Erected in the year 1804.

Fig. 283.

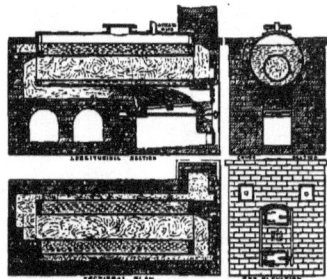

General arrangement of the setting of a Single Tube Horizontal Cylindrical Boiler; the flame first passing under the shell; secondly, back through the tube; thirdly, return at the one side flue; and fourthly, back through the other flue to the chimney. Erected in the year 1810.

Four years after—or in the year 1804— Mr. Evans, an American engineer, invented the arrangement of cylindrical boilers as shown, set in brickwork, by Fig. 282. The smaller boiler is a plain shell with straight or flat ends, and the larger of the same shape, with a tube directly through it, for the flame to operate in first, and then to pass under the small boiler to the chimney.

Trevithick and Vivian, the well known Cornish engineers, were then in full power with their locomotive engines and boilers, and Evans perhaps worked with them for land

Fig. 284.

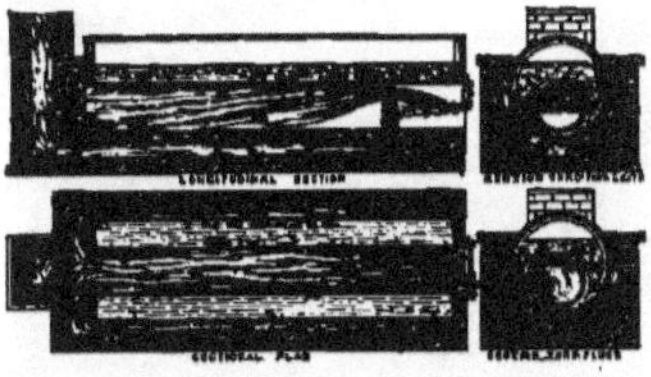

General arrangement of the setting of a Horizontal Cylindrical Boiler, with an internal tube, known now as the "Cornish" boiler; the flame first passing through the tube; secondly, split and return at the sides to the downtake flue; and thirdly, pass under the shell to the chimney. Erected in the year 1810.

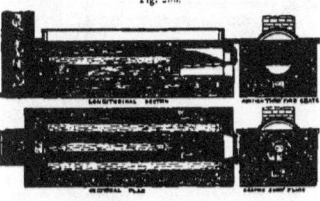

Fig. 285.

General arrangement of the setting of a "Cornish" Boiler, with a horizontal water tube in the boiler-flue, the flame passing as in Fig. 284. Erected in the year 1811.

purposes also, and thus the boiler and setting, as illustrated by Fig. 283, came forth in the year 1810, as also did the "Cornish" boiler and setting shown by Fig. 284, after which appeared the same boiler with a central water tube in the flame tube, as illustrated with the setting by Fig. 285.

But the men of the West were not allowed to carry their sway altogether any longer; because in this year, 1811, the Midland engineers brought out the "Butterly" boiler, as shown by Fig. 286, which was first manufactured at the works now bearing that name as the prefix. This boiler, however, differs but very little from the "Cornish" arrangement, the difference being that the front end

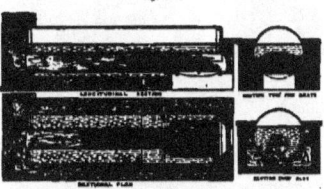

Fig. 286.

General arrangement of a Horizontal Cylindrical Boiler, known as the "Butterly" boiler, with half portion removed for the grate, the remainder being as the Cornish boiler, shown by Fig. 284. Erected in the year 1811.

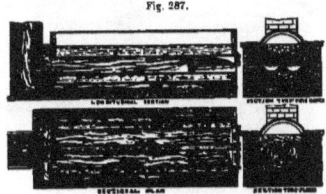

Fig. 287.

General arrangement of a Horizontal Cylindrical Boiler, with two internal tubes, known as the "Lancashire" boiler; set as the boiler shown by Fig. 284. Erected in the year 1811.

of the boiler is "cut away" for the length of the fire grate under the crown of the furnace, which crown is curved with a larger curve than the diameter of the tube beyond.

Next appeared the now known "Lancashire" boiler, with its two tubes instead of one, but set precisely as the Cornish prior example, and of course, unless in large diameters, cannot be better, as the illustration, Fig. 287, truthfully shows.

The Lancashire boiler makers next produced, in the year 1814, the egg-end boiler with two tubes in the shell, each tube passing

Fig. 288.

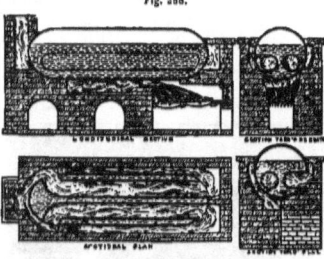

General arrangement of the "Lancashire" Boiler, with egg-ends; the flame first passing under the shell; secondly, back through the side flame flues; and thirdly, return through the tubes to the chimney. Erected in the year 1814.

Fig. 289.

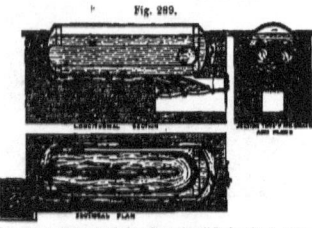

General arrangement of the "Lancashire" Boiler, fitted with a return tube; the flame first passing under the shell; secondly, up to and through the return tube; thirdly, back at the nearest side, then around the front end; and fourthly, at the other side to the chimney. Erected in the year 1819.

through the shell at the sides, at the back end, as set and shown in four views by Fig. 288. This forms a strong shell and tubes, and is said to be equal in generating steam with the best of any other arrangement of the same class.

But the Lancashire people were not exhausted in the way of boiler arrangements, as the next example and setting, shown by Fig. 289, illustrates. In this case the original shaped end is adopted, but the tubes are U shaped in the shell at the front end, and pass direct out at the other. The flame in

Fig. 290.

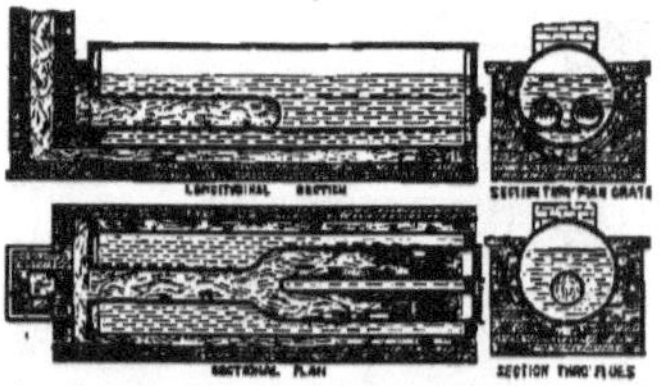

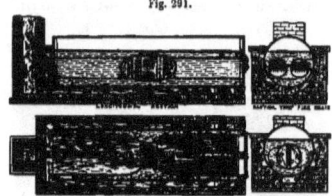

General arrangement of the Horizontal Cylindrical Boiler, now known as the "Breeches" Tube Boiler, and "set" precisely as that shown by Fig. 284. Erected in the year 1820.

Fig. 291.

General arrangement of a Horizontal Boiler, fitted with twin "breeches" tubes for the fire boxes, and flues inside the boiler, and also the intermediate portion, or combustion chamber, fitted with vertical water tubes to increase the circulation of the heat. Erected in the year 1822.

this setting travels a longer circuit than in any previous arrangement.

In the year 1820, the Cornish and the Lancashire boilers were blended by the Midland engineers into one arrangement as illustrated in setting by Fig. 290, which is termed the "Breeches" Tube Boiler, and from that issued the twin breeches tube boiler, with vertical water tubes in the single portions, as shown by Fig. 291.

Very soon after this date, 1822, the French engineers bestirred themselves, and produced, three years after, their "Elephant" Boiler,

Fig. 292.

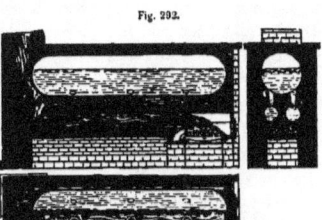

General arrangement and setting of the three Horizontal Boilers; connected by four vertical tubes, known as the French "Elephant" Boiler. The flame first passes under the two small boilers; secondly, returns at one side; and thirdly, back the other side under the large boiler, and up the chimney. Erected in the year 1825.

Fig. 293.

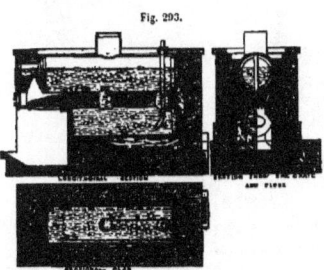

General arrangement of the top and bottom Horizontal Boilers; connected by vertical tubes, known as the "Double" Boilers. The flame passes first under the top boiler; then, secondly, descends and surrounds the bottom boiler; and finally passes through the flue to the chimney. Erected in the year 1825.

as they termed it—but why "Elephant," we cannot comprehend; because it partakes no form of that animal, except perhaps the vertical tubes, which might be said to be the legs, and the small tubes, in the end view, the feet. This arrangement and its setting is shown in the three sectional views by Fig. 292.

Next was introduced in England the arrangement shown by Fig. 293, which is two boilers—one over the other—connected by two vertical tubes, the passage of the flame being under each boiler.

The French engineers came forward again in the year 1830, and introduced a little improvement in their "Elephant" boiler, which

Fig. 294.

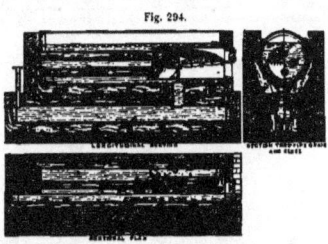

General arrangement of the Horizontal Boiler, fitted with a cylindrical fire box, four tubes beyond it, and a bottom water tube under the boiler. The flame first passes through the tubes; secondly, returns under the shell of the boiler; and thirdly, passes back around the water tube. Erected in France in the year 1830.

improvement consisted of a cylindrical shell, fitted with a cylindrical fire box, and four flame tubes beyond it. The shell is connected under the bridge, to a long water tube situated below the boiler — which situation, by the way, is not conducive for evaporation; but the improvement lay in the flame tubes, which "split" up the flame in the largest body of water, as illustrated with the setting by Fig. 294.

In the year 1836, Mr. Holmes introduced his wagon-shaped, flued boiler, as shown in sectional elevations by Fig. 295; in which the flame passes through a series of flues that are

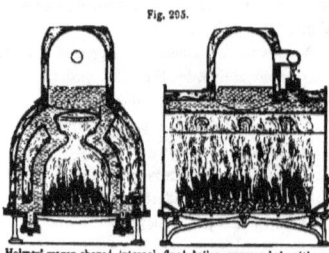

Fig. 295.

Holmes' wagon-shaped internal flued boiler, surrounded with a flame casing at the sides. Patented in the year 1836.

formed by water spaces, and finally around the shell of the boiler and up the chimney.

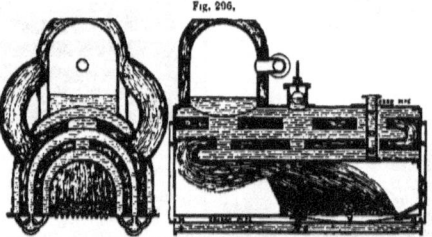

Fig. 296.

Holmes' Horizontal Saddle Boiler, formed of a series of saddle-shaped shells, containing water, and the flame passing between them. Patented in the year 1836.

Another and better arrangement is shown by Fig. 296: where there are shown a series of saddle-shaped shells, connected by large tubes on the tops, containing water, and the flame passing between them.

It is worth remarking that the flame ascends and descends in its circuit in the first arrangement, but traverses forward and backward in the second arrangement.

From this date, 1836, horizontal land stationary boilers were permitted to be "left alone" by the inventors wise in those matters, and it was not until the year 1849 that any example, worthy of notice here, appeared; which example is illustrated by Fig. 297, the arrangement being that the shell is cylindrical, with a flame tube in it, and the tube is divided in the fire box by a water space perforated with flame tubes, and beyond this there are vertical and horizontal water tubes, so arranged that the products of combustion can act on them to generate steam.

Matters pertinent to the present subject, as patents were then withheld for some time, and indeed sufficiently to allow inventors to look around them and think about what a boiler for generating steam should be, the first result of the cessation being an invention by a Mr. Lawes, who really deserves some credit for it, as Fig. 298 illustrates; which is an end and side sectional elevation, showing a cylindrical boiler, fitted with an internal fire tube of such a diameter as to leave a very small water space around it. At the sides of the shell are two horizontal tubes, and above them, centrally, a third tube; the two side tubes being connected by six short tubes, and the top tube by a channel piece.

The main improvement in this boiler, is the adoption of a small amount of water by Fig. 299, in which was fitted a cylindrical fire box and horizontal flame tubes—of a very small diameter, comparatively—extending to the end of the shell. This boiler is the first

Fig. 297.

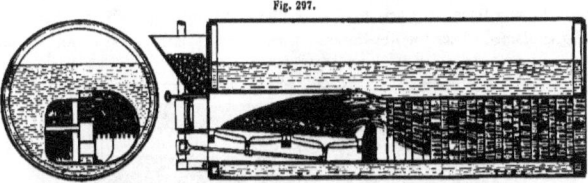

Leigh's Horizontal Cylindrical Boiler, fitted with a flame tube, containing vertical and horizontal water tubes in it beyond the bridge, and horizontal flame tubes above the fire grate, which is composed of hand-rocking fire bars. Patented in the year 1849.

Fig. 298.

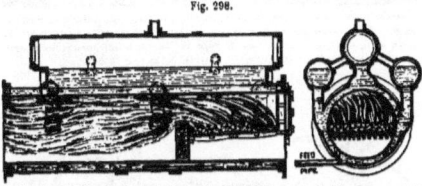

Lawes's Horizontal Cylindrical Boiler, in which the flame tube is entirely surrounded by water, and the water space is in connection with two tubes, situated at the sides, and a third tube, centrally supported above. Patented in the year 1852.

Fig. 299.

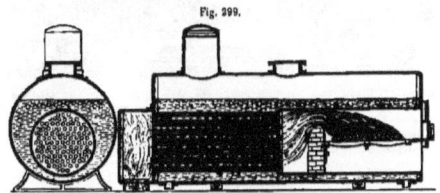

Schofield's Horizontal Cylindrical Boiler, fitted with a cylindrical fire box, and horizontal flame tubes beyond it, leading to the combustion chamber. Patented in the year 1852.

around the fire tube, so as to increase the rapidity of evaporation.

In this same year, 1852, a Mr. Schofield introduced the arrangement of boiler, as shown multitubular boiler introduced for general land stationary purposes.

The advantage of tubes for those purposes having been proved, other tubular arrangements came into notice, one of which is illustrated in end and side sectional elevations by Fig. 300; showing one large shell of Watt's original type, fitted by three tubes at each side, with separate water chambers, having flame tubes through them—the action of the flame being explained under the illustration.

tion to, is very peculiar; not on account of vertical and horizontal tubes being fitted in the fire tube, because that was proposed in the year 1849, as shown by Fig. 297, on page 109; but the peculiarity in this example, as illustrated by Fig. 301, is that there is a water boiler surrounded by an air chamber,

Fig. 300.

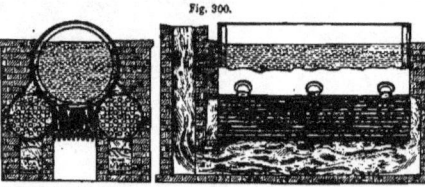

Hopkins's Cylindrical Horizontal Boiler, fitted with two flame tube water chambers, one on each side, arranged and set in brickwork; so that the flame, after acting under the main shell, splits, and returns back through the tubes; thirdly, passes under each chamber; and finally up the chimney. Patented in the year 1852.

Fig. 301.

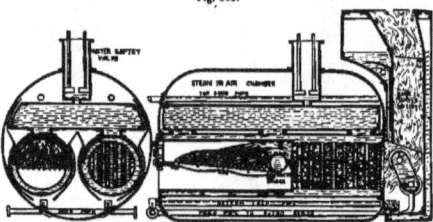

Ramsell's Horizontal Twin Fire Tube Boiler; each tube being fitted, with a water bridge, and cross tubes beyond; and above the fire tubes is a water chamber, extending transversely and longitudinally over them; the whole being enclosed in an air and steam chamber. The uptake is fitted with water spaces, termed a supplementary boiler. Patented in the year 1852.

This arrangement is a direct contradiction of the fact that the generation of steam is always quicker through shallow than through deeper water, over the heating surfaces; because the less amount of water the heat has to pass through, the least resistance offered.

The next arrangement that we direct atten-

which have no direct communication with each other.

The air chamber is of sufficient capacity and strength for the required steam and the hot air; but the internal boiler is used merely as the generator, being filled with water heated by the furnace, at the end of which is the water bridge, and beyond that the usual boiler flue, intersected with water tubes, alternating in series, the one series placed at right angles to the next series, and so on throughout.

The upper part of the boiler is connected with the lower part by means of large water tubes or socket joints made of wrought iron; this boiler, which is supplied with a separate feed pipe.

The steam generated in the main boiler is permitted to rise first in the steam chest in the supplementary boiler, and is from there conveyed by a pipe into the air chamber, and from it passes to the engine.

We must mention also that Mr. Ramsell

Fig. 302.

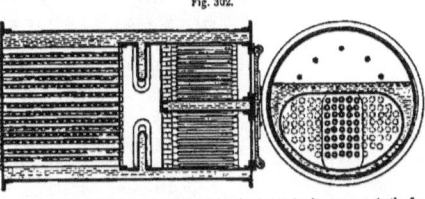

Johnson's Horizontal Cylindrical Boiler, fitted with a longitudinal water space in the fire box, and transverse water spaces in the combustion chamber, with flame tubes beyond. Patented in the year 1852.

Fig. 303.

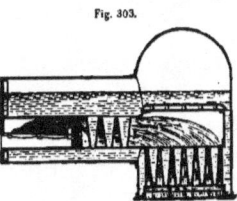

Galloway's Horizontal Boiler, fitted with vertical water tubes in the fire box and the combustion chamber. Patented in the year 1853.

Fig. 304.

Galloway's Horizontal Boiler, fitted with vertical water tubes in the fire box and the combustion chamber. Patented in the year 1853.

and a lesser tube communicates with a safety hot water valve properly weighted.

At the up take or end of the flue, is placed a supplementary boiler, consisting of a series of flat tubes or sheet water spaces, forming alternately a series of water spaces and smoke flues; the water spaces having their front and back ends opening into the water casing of proposed a tubular water bridge in the place of the ordinary brick bridge.

Following Mr. Ramsell, a Mr. Johnson laid his claim for legal protection as a national matter, for the arrangement shown in end sectional elevation and sectional plan by Fig. 302; which consists of a fire box, flat at the roof, fitted longitudinally with a water space, and transversely beyond, with two water spaces, and a series of longitudinal flame tubes.

Mr. Galloway commenced the year 1853 with introducing three examples of horizontal boilers, fitted with water tubes, as shown in side sectional elevation, by Figs. 303 to 305, inclusive.

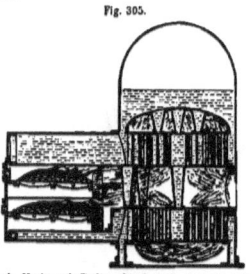

Fig. 305.

Galloway's Horizontal Boiler, fitted with two fire boxes, and vertical water tubes in the combustion chamber, to cause a duplex action for the flame. Patented in the year 1853.

grate, and the third example that the flame proceeds from two fire boxes situated over each other, and then passes up and down in duplex action to the combustion chamber, which is divided by a water space for the purpose.

The next example is the production of a Mr. Murgatroyd, as shown in side and sectional elevations by Fig. 306; where he proposes to concave the roof of the fire box, and stay it also, which really amounts to making weak that portion first, and strengthening it afterwards.

Another example of a stationary tubular boiler is illustrated by Fig. 307, the idea of a Mr. Pearce; that consists of a cylindrical shell with flat ends, containing three tubes for fire boxes at the central line, and underneath them a series of small tubes, through which the flame passes to the main flue leading to the chimney.

A corn factor next appeared on the scene,

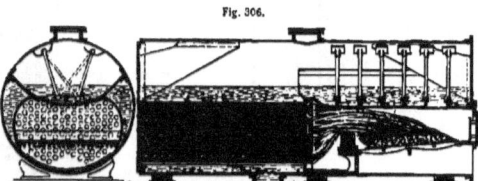

Fig. 306.

Murgatroyd's Horizontal Cylindrical Boiler, fitted with a fire box, having a concave crown, strengthened by side plates, and stays between them. Patented in the year 1853.

The first example illustrates that the flame passes over the bridge amongst vertical conical water tubes, and then down amongst tubes of a similar form.

The second example shows that the flame passes up and down after leaving the fire of the name of Evans, the younger; and his scheme was for wagon boilers, as illustrated by Fig. 308, which is described underneath.

Mr. Dunn next appeared again, with an arrangement, as shown by Fig. 309, in sectional elevations, of a series of egg-end boilers placed side by side, within a few inches of each other; and supported at the ends by an iron frame, which rests upon the side walls. This iron frame forms a saddle to each, by which means the adjoining cylinders, or the brick work, is kept undisturbed by the

Fig. 307.

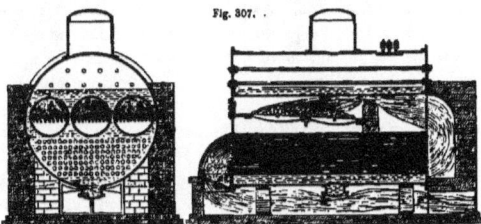

Fearce's Horizontal Cylindrical Boiler, fitted with three fire boxes, situated centrally of the shell, which contains also a set of flame tubes under the fire boxes. Patented in the year 1853.

Fig. 308.

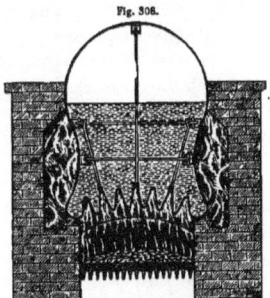

Evans's "Wagon" Boiler, with a corrugated bottom. Patented in the year 1853.

Fig. 309.

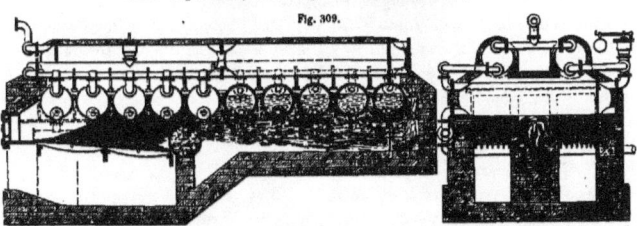

Dunn's arrangement of Horizontal Cylindrical egg-end Boilers and Steam Chambers combined, and a feed water cylinder as a fire bridge. Patented in the year 1853.

removal and replacing of one or more of the cylinders. Between each cylinder a column is carried up from the frame or saddle, upon which are carried beams for supporting the steam chambers — which are shown placed longitudinally across the top of the lower chambers or boilers—and also for the purpose of supporting the covering of the boiler.

The steam reservoir is formed by uniting two of the duplicate cylinders by their flanges, end to end, using as many cylinders as are necessary for the steam room required.

It will also be seen that there are two ranges of pipes, and which pass alongside one end of the cylinders, one range being a little above and the other a little below them; to which they are connected by angle or elbow pipes, being thus arranged, that the cylinders may be withdrawn from their places, without disturbing the main pipes. The elbows being removable from the short branch at the end of the cylinders, give also facility for uniting any two or more cylinders together, end to end. Another range of pipes may be employed at the other ends of the cylinders, if thought desirable, for the passage of the steam to the reservoir cylinders as shown, upon either of which as many safety valves may be fixed as may be deemed requisite.

The bridge of the furnaces is formed by a cylinder, through which should pass the water for feeding all the other cylinders, by means of the lower range of pipes, which are connected with the bridge cylinder, which will remain full of water; the other cylinders— with the exception of the reservoir cylinders which are not supplied with water — being maintained at about the water line.

By this arrangement it will be seen also that two furnaces are used, one on each side, leaving the division walls to form a flue between the furnaces which meet at the ends, so that the gas flame and smoke may mix and ignite, and then pass along the middle flue towards the front or furnace end, where arriving, it may pass by a down flue to the chimney.

As the boilers are not close together, stopping pieces are inserted between them, so that the intense heat from the furnaces will be confined to the lower parts of the cylinders.

To prevent the heat from passing out at the top, a covering of brick work or other material is used, supported by the beams; while radiation from the steam cylinders may be prevented to a great extent, by a covering of felt, as commonly employed for such purposes.

Mr. Dunn at the same time proposed another arrangement of boilers, as illustrated in sectional elevations by Fig. 310. In this case, cylindrical flat end shells containing flame tubes are supported on brick work at their ends, which forms a flue, and by which their ends are connected; brick partitions are placed in the flue, by which the draft is directed through each of the tubular portions in succession, until it has passed through the last, which, in this arrangement, is that boiler directly over the furnace. The spaces between the boilers are stopped by a layer of bricks, so as to prevent the heat ascending between them.

When either boiler requires detaching, it is to be lifted perpendicular from the brick seating, and replaced by another duplicate of it; while, to further facilitate the removal of parts, a valve should be placed at each junction of the main steam and water pipe with each of the portions, so that the communication between each part and the others can be cut off by the valves, which will enable the parts to be connected and disconnected with-

back end, was the best form of shell to enclose a cylindrical fire box, having beyond it syphon return tubes that also surrounded the fire box.

The action of the flame is that it first passes from the bridge through the five small syphon tubes, to the smoke box at the front end;

Fig. 310.

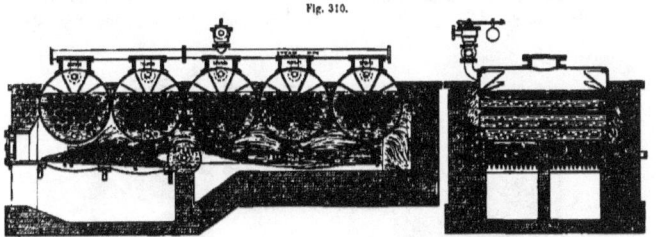

Dunn's arrangement of Horizontal Cylindrical flat-end Boilers, fitted internally with flame tubes. Patented in the year 1853.

Fig. 311.

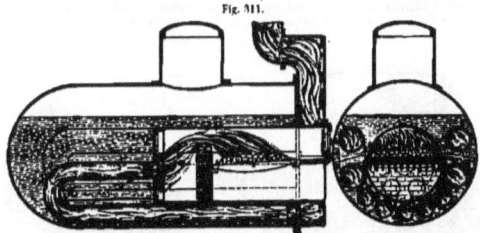

Culpin's Horizontal Boiler, with flat and semi-spherical ends, fitted internally with a cylindrical fire box, and beyond and around it, syphon return flame tubes, that lead into a smoke box, surrounding the front end of the fire box. Patented in the year 1853.

out greatly interrupting the working of the other parts.

A Mr. Culpin's arrangement of horizontal boiler is now referred to as being worthy of notice here, as illustrated in sectional elevations by Fig. 311. Mr. Culpin considered that a flat front end boiler, having a curved

secondly, back, and return on each side of the fire box, through the two large syphon tubes which are divided by stop-flame-plates in the smoke box, to cause the return circuit for the flame, before it passes from the top large tubes to the chimney.

Passing from arrangements of return action

for the flame, we direct attention now to arrangements for a more direct circuit, which was introduced by a Mr. Fearnley, as shown in two sectional views by Fig. 312.

In this example the roof of the fire box is formed by water and flame spaces, and over them is a large combustion chamber, at the end of which is a series of flame tubes that convey the flame to the chimney.

tubes can more easily receive the heat than if the combustion chamber were omitted—in fact, the combustion chamber in all cases is the main portion of any boiler.

A Mr. Bellhouse followed Mr. Fearnley, with the idea that the best known means of generating steam was to connect two flat-end cylindrical boilers with water tubes, as shown in sectional elevations by Fig. 313, and

Fig. 312.

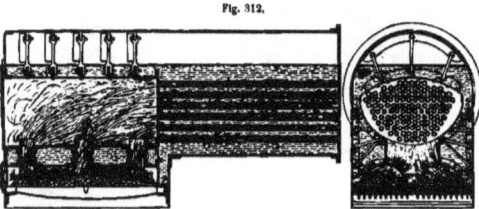

Fearnley's Horizontal Cylindrical Boiler, fitted with water and flame spaces over the fire grate; also a combustion chamber, with flame tubes connecting the smoke box, situated at the back end. Patented in the year 1853.

Fig. 313.

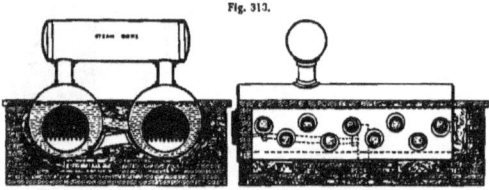

Bellhouse's Horizontal Cylindrical Twin Boilers, connected by water tubes. Patented in the year 1853.

This boiler is founded on correct principles; because, in the first place, the heat from the fuel is partially extracted by the water in the water spaces; and, secondly, the remainder of the heat is permitted to have time to "mingle," "mix," and "ignite" in the combustion chamber before entering the small flame tubes; and then the water around the

arrange the setting so that the flame acted on each boiler and tubes alike.

Mr. Kendrick's arrangements are next worthy of attention; his first example is shown by Fig. 314. In this case he deeply corrugates the fire box horizontally, and the flue beyond vertically, sufficiently to make alternate flat flame and water spaces. In the second example, shown by Fig. 315, he vertically corrugates the fire tube only, and at such distances to introduce brick bridges between them.

Directly after Mr. Kendrick, Mr. Horton bridge, beyond which is the flame tube, containing a longitudinal water space, perforated with flame tubes. The action of the flame is that, after leaving the bridge, it passes along one side of the water space and through the

Fig. 314.

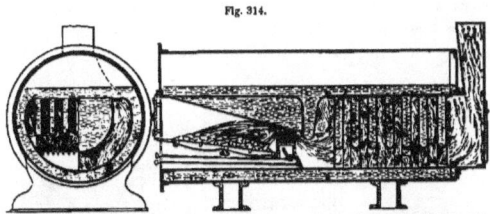

Kendrick's Horizontal Cylindrical Boiler, with the fire box and flame tube corrugated horizontally and vertically, to form alternate water and flame spaces. Patented in the year 1853.

Fig. 315.

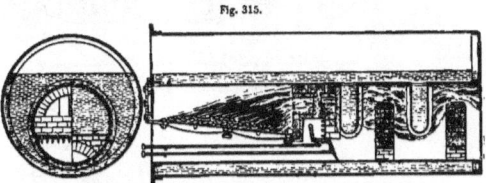

Kendrick's Horizontal Cylindrical Boiler, with the flame tube vertically corrugated, to make deep flat water spaces, with brick bridges between them. Patented in the year 1853.

Fig. 316.

Horton's Horizontal Cylindrical Boiler, fitted with three fire boxes, and a fire tube, containing a longitudinal water space, perforated with flame tubes. Patented in the year 1853.

appeared with his boiler, as shown by Fig. 316; the arrangement being that, in the place of one ordinary fire grate and fire box, there are three, common to one fire flame tubes to the other side, and from these back and forward through the outer flues to the chimney.

The French engineers were next represented by M. Erard, who believed he had accomplished a great deal, every way, from his arrangement, as shown by Fig. 317; which consists of a cylindrical shell, containing a cylindrical fire box and two fire tubes, passing through which are a series of water tubes, in connection at the back end with the water space below, and in connection at the front end with the steam chests above. There is also a combustion chamber enclosed in the shell—that is, common to the fire tubes—situated directly over them, and in communication with the chimney above.

The action of the flame in this boiler is, that it passes from the fire box direct amongst the water tubes, and returns back in the combustion chamber to the chimney, situated between the steam chests. Our opinion of this arrangement is, that the inventor involves it in such complication, that the practical difficulties overbalance the theoretical advantages, and thereby his scheme is that much reduced.

Mr. Barran next appeared with his scheme, as illustrated by Fig. 318; in which he claimed that, by arranging a cylindrical shell internally with a combustion tube, containing two water chambers, with flame tubes through them, and flame "cups" on the roof of the main tube, an economical evaporative result would be attained; and to ensure a circulation of heat, he put two tubes over and under each water chamber.

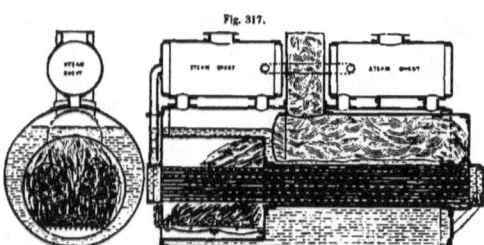

Fig. 317.

Erard's Horizontal Cylindrical Boiler, fitted with a Cylindrical fire box and two flame chambers, containing water tubes; above which is a cylindrical combustion chamber, in connection with the chimney. Patented in the year 1853.

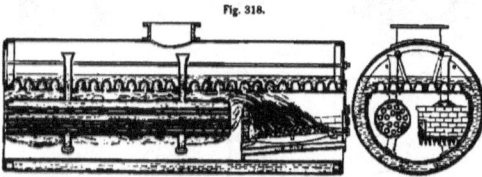

Fig. 318.

Barran's Horizontal Cylindrical Boiler, containing a combustion tube; in which are two water tubes, filled with flame tubes, and on the roof of the combustion tube are flame "cups." Patented in the year 1853.

That a certain advantage would result by this is obvious; but at what cost of repair and cleansing—two evils which Mr. Barran seemed to have overlooked in this case.

A French engineer followed in the rear of

Mr. Green came forward next, with two arrangements, as shown by Figs. 321 and 322; in which he proposed, that by perforating the shell with flame tubes, a great result must occur in the way of generating steam.

Fig. 319.

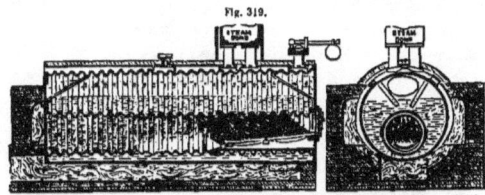

Rémond's Horizontal Cylindrical Boiler; the shell being transversely corrugated for nearly the entire diameter, and the combustion tube entirely corrugated. Patented in the year 1853.

Fig. 320.

Rémond's Horizontal Cylindrical Boiler, fitted with three transversely corrugated fire boxes, in connection with a corrugated double forward and backward tube. Patented in the year 1853.

Fig. 321.

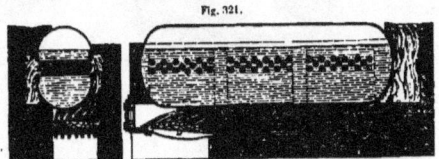

Green's Horizontal curved ends Boiler, fitted with transverse flame tubes through the shell. Patented in the year 1853.

Mr. Barran, named Rémond, with two arrangements, as shown by Figs. 319 and 320; from which it is apparent that he wished to increase the surface in the smallest capacity by corrugating instead of cupping.

The next year, 1854, was not nearly so prolific in boiler invention as the year preceding, and we attribute the cessation to the fact that the subject had been so hacked, that the interest became reduced in proportion.

A Mr. Weatherly appeared in this year, with his notion of what a horizontal land boiler should be, as illustrated by Fig. 323. The shell of the boiler is cylindrical, with flat ends, and contains a large flue tube, as for Cornish boilers: within this flue tube are fitted a series of pipes over the fire grate, that communicate with an expansion disc, beyond the bridge, that is connected by a series of pipes to a second disc, and it by pipes is con-remained in the discs, they would soon burn and leak.

Mr. Holt came forward also with two boiler arrangements, as shown by Figs. 324 and 325.

In the first example the shell is cylindrical, with flat ends, and the main flue tube contains across it a water space, perforated with flame tubes; on each side of which are brick bridges, to cause the flame to descend and ascend through the tubes.

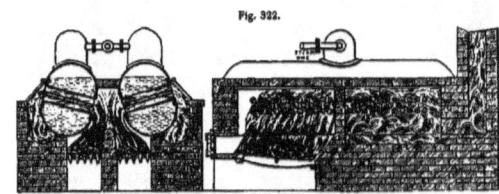

Fig. 322.

Green's Twin arrangement of the Boiler, shown by Fig. 321.

Fig. 323.

Weatherly's Horizontal Cylindrical Boiler, fitted with a fire tube, containing, beyond the bridge, a water tube, having flame tubes through it, with expansion discs and guide wheels. Patented in the year 1854.

nected to a third disc, that is situated at the back end of the tube; these three discs are supported on wheels, which permit any movement from expansion and contraction of the tubes. The other pipes, shown in the drawing, relate to connections and feed water.

Our opinion of this boiler is, that it requires much more than ordinary skill to make it, and to repair it; and that, if any deposit

The second example is merely the addition to the first of a deeper fire box, and over the grate a perforated water space. Both of those boilers are creditable productions, but require great draught.

Mr. Henley began the year 1855 with the idea, that round or circular water tubes obstructed the draught; and to get rid of that obstruction, he proposed, and patented

LAND STATIONARY HORIZONTAL BOILERS.

it too, that vertical water tubes in the main flue of horizontal cylindrical boilers should be Rhombus-shaped, or what is more commonly weakness, and cheap manufacture for costly production, in carrying out his idea.

We now begin with remarks on the boiler

Fig. 324.

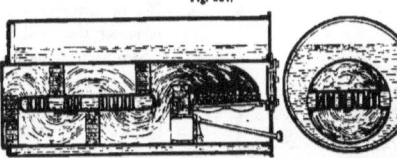

Holt's Horizontal Cylindrical Boiler, in which the fire tube is fitted with a transverse water space, with small vertical flame tubes, and intervening brick bridges. Patented in the year 1854.

Fig. 325.   Fig. 326.

   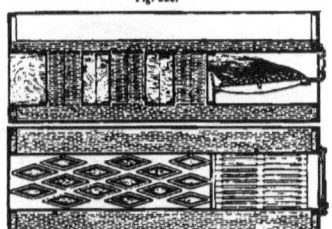

Holt's Boiler, as shown by Fig. 324; fitted with a deeper fire box, and a perforated water space over the fire grate. Patented in the year 1854.

Henley's Horizontal Cylindrical Boiler; fitted with Rhombus-shaped tubes in the main-flue tube. Patented in the year 1855.

Fig. 327.

Henley's Boiler; fitted similar as in Fig. 326, but with the fire grate under the boiler. Patented in the year 1855.

known as diamond shape, in section, as shown arranged by Figs. 326 and 327. Mr. Henley, however, sacrificed comparative strength for events that occurred in the year 1855, and direct attention to a Mr. Jeffrey's very novel idea, as illustrated by Fig. 328; which consists of the use of two fire boxes and grates, over each other, for the purpose — according to Mr. Jeffreys — of consuming the smoke from the higher fire; but how the smoke in the lower fire is consumed, or if at all, Mr. Jeffreys does not explain, and we confess ourselves in the same dilemma.

A Mr. Lee Stevens being second in the list, we direct attention to his invention, shown by Fig. 329; the arrangement being that the fire grate is under the shell, and the flame, after acting there first, passes back through a large flame tube into a cylindrical chamber, and from thence through small flame tubes, to the chimney. The advantage in this arrangement is, that the flame has time and space given to it for mixing, and thus combustion is better carried out.

The next example is by a Mr. Cowburn, as shown in sectional elevations by Fig. 330;

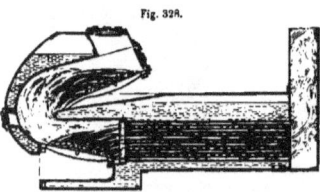

Fig. 328.

Jeffrey's Horizontal Cylindrical Boiler, having two fire grates, one above the other, and stoked in opposite directions. Patented in the year 1855.

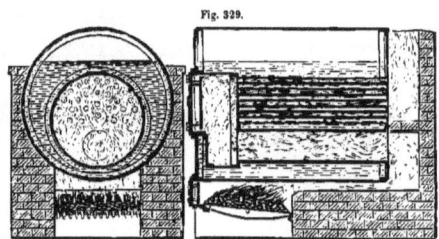

Fig. 329.

Stevens's Horizontal Cylindrical Boiler, fitted internally with tube and box combustion chambers and flame tubes. Patented in the year 1855.

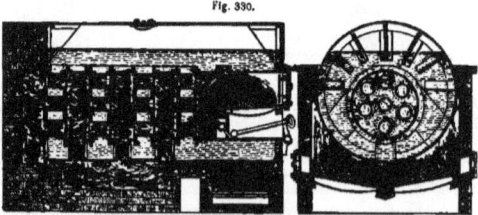

Fig. 330.

Cowburn's Horizontal Cylindrical Boiler, fitted with flame chambers and tubes in the tube flue, and three fire grates, one grate being in the tube flue, and one grate on each side, meeting under the boiler. Patented in the year 1855.

the arrangement of which is, that the flame tube beyond the bridge is formed into chambers, perforated by tubes, through which the flame passes. There are three grates also, one in the tube, as usual, and two under the boiler shell, so as to distribute the flame more than one grate would do, according to the idea of the inventor.

This boiler is rather complicated, and the chambers difficult to clean, either inside or out, and repair equally difficult, if not impossible.

Mr. Dunn commenced the year 1856, and this time with his brain and hands full; for he brought out no less than forty arrangements of retort boilers—as he termed them—of which we illustrate six examples, being the "pick" of the lot.

Fig. 331 is a transverse section of an arrangement, consisting of five retort boilers, placed horizontally; the steam generated in these retort boilers is conveyed into the steam chest, placed above them: there are also eight cylindrical water heaters placed in two vertical rows. All these retort boilers and water heaters are supported at their ends by brickwork, and the products of combustion rising from the fire grate pass under the retort boilers, then along the flues and mixing chamber formed by the two rows of vertical water heaters and the brickwork, and thence to the flue which is in communication with the chimney, this passage of the products of combustion being illustrated.

The supply of water is pumped into the uppermost water heater, through the pipe, from whence it flows into the others through the lower pipe, and into the retort boilers through the pipe under them. The small box on the first heater contains a valve opening inwards, to allow the water to flow into the pipe from the water heaters, and to prevent the steam and water from the boilers flowing into the water heaters.

The passages in the brickwork are for admitting air to the flue, in order to increase the combustion of the smoke.

Fig. 332 represents a modification of this steam generator; the retort boilers and the water heaters being surrounded by a double casing containing water. The products of combustion, after passing under the retort boilers and around the water heaters, are conveyed over the retort boilers to re-heat or dry the steam, and thence pass through the flue to the chimney.

Fig. 331.

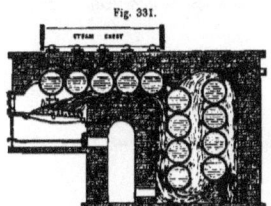

Dunn's Horizontal Retort Boiler, stacked in brickwork; shown in transverse sectional elevation. Patented in the year 1855.

Fig. 332.

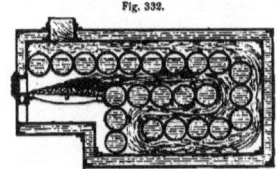

Dunn's Horizontal Retort Boiler, stacked in a water casing; shown in transverse sectional elevation. Patented in the year 1855.

Fig. 333 represents another modification of the boiler shown by Fig. 331. In this instance the cylindrical water heaters are placed

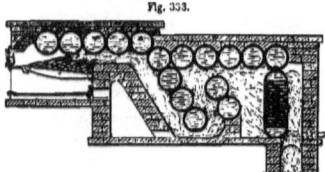

Fig. 333.

Dunn's Horizontal Retort Boiler, stacked in brickwork; shown in transverse sectional elevation. Patented in the year 1855.

diagonally and horizontally, and the products of combustion, after passing through the flues formed by the sides of the water heaters, are conveyed through the tubes in the multitubular water heater before they pass off to the flue. That water heater is a flat vessel of wrought or cast iron, the sides of which are stayed by the tubes.

Fig. 334 represents a section of a boiler of large dimensions. In this arrangement there are two sets of retort boilers, placed on each side of the central boiler, into which the steam from all the retort boilers is collected by side pipes. Each set of retort boilers is heated by a fire grate. The products of combustion pass over the bridges along the flues, and then through the tubes in the multitubular water heaters, where they unite before entering the flue.

The steam generator, shown by Fig. 335, consists of a row of retort boilers, and another

Fig. 334.

Dunn's Horizontal Retort Boiler, arranged in two sets; shown in transverse sectional elevation. Patented in the year 1855.

Fig. 335.

Dunn's Horizontal Retort Boiler, stacked in brickwork; shown in transverse sectional elevation. Patented in the year 1855.

row of cylindrical water heaters below them, connected together by junction pipes. The products of combustion from the fire grate, pass around and between the retort boilers and water heaters, thereby communicating the heat to the water contained in them in their passage to the flue.

Fig. 336 is a longitudinal section of a boiler composed of three cylinders, made of any suitable length, supported by brickwork, or partly by the brickwork and partly by the multitubular water heating chambers.

The products of combustion from the fire

grate pass over and through the tubes in the bridge formed by the first water heating chamber, then over and through the tubes of the second chamber, and then through the

Fig. 336.

Dunn's Horizontal Retort Boiler and Tube Chambers, stacked in brickwork; shown in longitudinal sectional elevation. Patented in the year 1855.

Fig. 337.

Dunn's Horizontal Retort Boiler, stacked in brickwork; shown in sectional elevation. Patented in the year 1855.

and the fire grate is the water heating chamber, furnished with tubes, to allow part of the products of combustion to rise up and impinge against the under surface of the cylinders. The steam from these cylinders is conveyed by pipes to a steam chamber, not shown, and the water is conveyed from the chamber to the upper cylinders by the curved pipes.

Mr. Pearce again showed the world his knowledge of boilers, by the illustration, Fig. 338; which shows a boiler with an undulated bottom, to better receive the action of the flame—but whether that is worthy of a patent, seems doubtful.

Mr. Holt appeared next, with two examples of boilers, as illustrated by Figs. 339 and 340 on the next page. The first example shows that the flue tube is fitted with cross water spaces, connected longitudinally by tubes, and there are brick bridges, above and below the cross tubes, to cause the flame to undulate during its transverse in the flue tube.

The second example shows cross tubes only,

Fig. 338.

Pearce's Horizontal Boiler, with a corrugated bottom, for the flame to impign against. Patented in the year 1856.

tubes of the third chamber, from whence they pass to the flue.

Fig. 337 represents a steam generator, composed of four cylinders, between which

with brick bridges in the flue tube, to cause the flame's action as before. Mr. Holt had been anticipated, however, in the use of cross tubes, as shown by Fig. 297, on page 109.

A very much better boiler than Mr. Holt's, was introduced by Mr. Cater, early in the year 1857, as shown by Fig. 341, which is a cylindrical return tubular arrangement. The products of combustion pass from the flue second set of tubular flues, of lesser diameter than the former set, conducts the smoke and gases back through the boiler to a smoke box at the rear end of the boiler, and from thence the gases escape into the chimney.

Fig. 339.

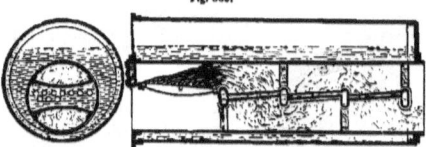

Holt's Horizontal Cylindrical Boiler, fitted with brick bridges and transverse and longitudinal water tubes in the flue tube. Patented in the year 1856.

Fig. 340.

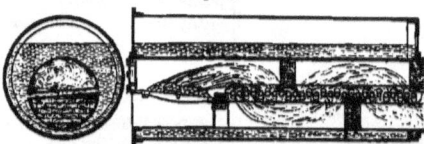

Holt's Horizontal Cylindrical Boiler, fitted with cross water tubes and brick bridges in the flue tube. Patented in the year 1856.

Fig. 341.

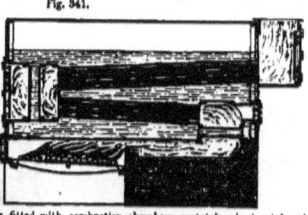

Cater's Horizontal Cylindrical Boiler, fitted with combustion chambers, containing horizontal and vertical water tubes, and the chambers connected by small flame tubes. Patented in the year 1857.

up into a chamber made in the bottom, and near to the back end of the boiler, and from there lead the lower set of tubular flues, which conduct the heated gases to another chamber at the front end of the boiler. A From this description it will be seen, that this boiler consists of a combination of the cylindrical and tubular boiler, and that the flame and heat is taken from the bottom and conducted along tubular flues, during which time the water surrounding the tubular flues will abstract a portion of the caloric contained in the gases, and while the latter are returning along the second set of tubular flues, which are of lesser diameter than the former, a considerable portion of the caloric still remaining will be given off to the water in the boiler surrounding the flues. By thus causing the gases to pass in a circuitous or serpentine direction through sets of tubular flues, novelties in horizontal boilers. The first arrangement is shown by Fig. 342; which it will be seen is an elliptical boiler containing an ordinary flue tube: but the novelty is first in the rectangular water bridge, and second, in a cylindrical conical water chamber in the flue beyond the bridge, the chamber being connected to the steam space above by a branch pipe at the back end. But Taylor improved on that, as illustrated by Fig. 343. In this

Fig. 342.

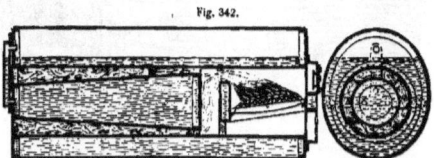

Taylor's Horizontal Elliptical Boiler; in which the flue tube is fitted with a cylindrical conical water chamber. Patented in the year 1857.

Fig. 343.

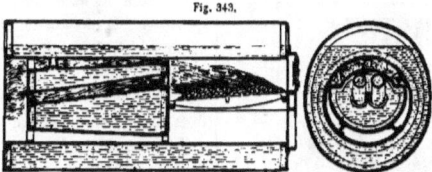

Taylor's Horizontal Elliptical Boiler; in which the flue tube is fitted with a water chamber, containing two water tubes. Patented in the year 1857.

varying in size, and whereby a large amount of heating surface can be obtained, a greater amount of heat is given off from the gaseous products of combustion than in other general steam boilers. It will be seen, on referring to the drawing, that doors are fitted to the ends of the boiler, for the purpose of enabling the attendant to get at the tubular flues when required.

A Mr. Taylor next appeared, with certain case he still put the chamber in the flue tube; but he put two flame tubes in the chamber also. Not being yet contented, Taylor introduced flame tubes in the water chamber in the flue tube, with a water tube on the fire grate, as illustrated by Fig. 344. In our opinion, Taylor omitted the proper means for the circulation of the heat in the water in each arrangement.

Mr. Green having remained content for

some time on boiler arrangements, came forward again early in the year 1858, as to the advantage of two common boilers in connection with steam and water tubes; in which the flame superheated the steam, and the steam heated the water, as arranged in Fig. 345. For our part, we fail in seeing the advantage of this arrangement for raising steam: because, for that purpose, the heat should always affect the water first.

A Mr. Adshead put forth next his opinion on the best means for retarding the flame in the tubes, which was by water bridges directly behind the back end of the fire bars, as shown by Fig. 346.

A very much more effectual means for flame action is shown by Fig. 347, being a Mr. Price's arrangement; which is a cylindrical boiler with two flue tubes, in which are six vertical water tubes, and between

Fig. 344.

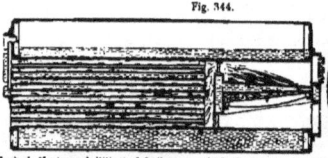

Taylor's Horizontal Elliptical Boiler; in which the flue tube is fitted, with a water tube on the grate, in connection with a rectangular water bridge, beyond which is a cylindrical water chamber, with flame tubes passing through it. Patented in the year 1857.

Fig. 345.

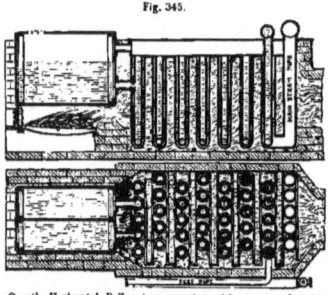

Green's Horizontal Boiler, in connection with steam and water tubes in the combustion chamber; the flame being intended to superheat the steam, and the steam to heat the water in the pipes. Patented in the year 1858.

them and beyond, at each end, are angular flame tubes on each side, in connection with the lower side of the shell of the boiler, underneath which is the fire grate and the combustion chamber.

The action of the flame is, that on rising from the grate, a portion of it passes along under the boiler, and the remainder up through the angular tubes into the flue tubes, where it meets with the other portion of the flame that is returning back through the flue tubes, and both volumes then proceed to the chimney.

Mr. Price also proposed the arrangement of cross flame tubes passing through the shell, as illustrated by Fig. 348.

Following Mr. Price with the same idea, that the flame in the flue tubes should be obstructed in its passage, to increase evaporation of water, came a Mr. Hopkinson with his arrangement, as shown by Fig. 349; in which the increase referred to is supposed to be accomplished by securing spiral water tubes in the flue tube—but how the inventor cleanses them inside, or proposes to do it, is not known.

water spaces as the best means, occupying the main portion of the shell, as illustrated in sectional elevations by Fig. 350 on the next page; the arrangement of which is, that there are two fire boxes that lead into a combus-

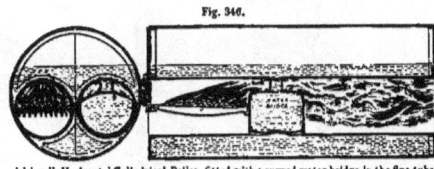

Fig. 346.

Adshead's Horizontal Cylindrical Boiler, fitted with a curved water bridge in the flue tube. Patented in the year 1858.

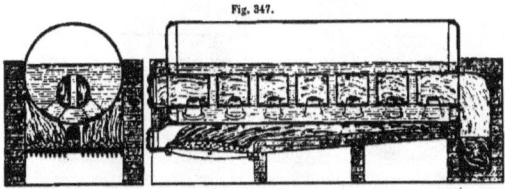

Fig. 347.

Price's Horizontal Cylindrical Boiler, in which the flue tube has vertical water tubes in it, and is connected below to the shell of the boiler, longitudinally, by angular flame tubes. Patented in the year 1858.

Fig. 348.

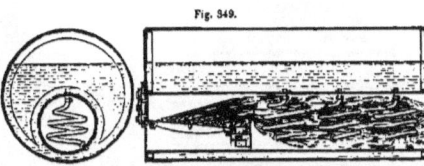

Fig. 349.

Hopkinson's Horizontal Cylindrical Boiler; having a flue tube fitted with spiral water coils in it. Patented in the year 1858.

Price's Horizontal Cylindrical Boiler, fitted with angular cross flame tubes through the shell. Patented in the year 1858.

Next came a Mr. Hunt, whose ideas were also relating to the obstruction of the flame in the flue tube. In his case, he ignored tubes for the purpose, but considered flues and

tion chamber, formed of water and flame spaces, but the flame passed alternately through small openings, and thus impinged more on the sides than if a direct zigzag passage was only provided for.

A Mr. Tapp, in the next year, 1859, introduced his arrangement for obstructing the flame, as illustrated by Fig. 351; which consists of a flue tube formed with a curved top, and the bottom rising in the centre, so as to form a water space throughout the length. The top of this space is connected to the top of the tube, by angular water tubes, and thus the obstruction of the flame is attained.

Next appeared the invention in boilers, by a Mr. Harman, as shown in sectional elevations, by Fig. 352. The brickwork in which the boiler is supported is built in the usual manner: the boiler has two internal furnace or flue tubes, the wrought iron plates of which are riveted to external circular stays of T-iron. The contiguous plates of the fur-

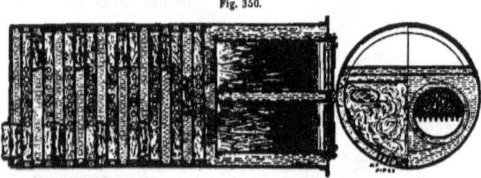

Fig. 350.

Hunt's Horizontal Cylindrical Boiler, fitted with two cylindrical fire boxes, that lead into a large combustion chamber—occupying more than half the shell—formed with flat flame and water spaces. Patented in the year 1859.

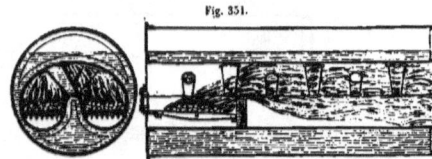

Fig. 351.

Tapp's Horizontal Cylindrical Boiler, containing a curved top flue tube, the bottom of which is a longitudinal water space, connected by angular water tubes to the top. Patented in the year 1859.

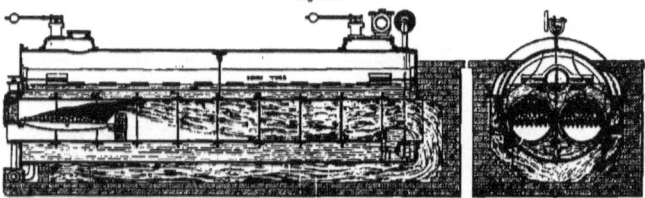

Fig. 352.

Harman's Horizontal Cylindrical Boiler, fitted with twin flue tubes, two steam chests, twin superheaters, scum pipe as a stay tube and working gear, and a screw propeller in the boiler to circulate the water. Patented in the year 1859.

nace tubes are arranged so as to leave a small intervening space, which is filled up with caulking. The front extremity of each of the flue tubes projects out beyond the end plate of the boiler, so that the ends of the fire bars are in a line with the end plate. This arrangement ensures the heating of the flue box crown plates to the end of the boiler. The projecting portion of each flue tube is secured to the boiler by means of external rings of angle iron, the weight of the tube being, however, mainly supported by the end plate of the boiler through which the tube passes.

The front ends of the fire bars are carried on the inner end of a narrow dead plate, which fills up the space between the ends of the bars and the furnace doors.

The flame and heated products of combustion arising from the burning fuel pass along the furnace tubes, out at the backward ends, and along the flue beneath the boiler to the front part, thence upwards and along the lateral flues to the chimney.

The backward end of each furnace tube projects outwards a short distance beyond the end plate of the boiler, to which it is attached by means of angle iron, in a manner corresponding to the arrangement of the front end.

The end plates of the boiler are stayed internally by means of the longitudinal scum tube, which is arranged above the furnace tubes, the lower portion of its periphery dipping a little below the proper water level.

Connected with the boiler is an arrangement for superheating the steam before it reaches the steam pipe which conveys it to the engine. The saturated steam, as it rises from the water, passes into the front end steam chest, which communicates with the superheating chambers that are formed of iron plates riveted to the sides of the boiler—they are flat at the bottom—which partly rests on the brickwork, so as to partially support the weight of the boiler. In this manner also the bottoms of the chambers form the roofs of the lateral flues, from which they receive their heat. The backward ends of the superheating chambers communicate with the back end steam chest, which forms a segmental belt, partially encircling the boiler in manner similar to the other steam chest.

In order to obtain a continuous circulation of the water, and so ensure the colder portion being effectively exposed to the more highly heated parts, the water is put in motion by means of a small screw propeller, which is kept rotating while the boiler is at work. The arrangement for effecting this object is on the back end of the boiler, where is fixed a hollow standard, the upper part of which is forked and carries the bearings of a short horizontal spindle. To this spindle is keyed the pulley, which is driven by means of an endless belt from the engine. The spindle has also fixed on it the bevel wheel, which gears with the pinion on the upper end of the vertical shaft, which extends down through the standards to a footstep bearing in the projecting part of the standard, which rests on the bottom of the boiler. At the lower part of the shaft is keyed a bevel wheel, which gives motion to the pinion, fixed on a short shaft carrying the screw propeller. Upon the wheel being put in motion, the water is caused to circulate in a continuous current, which passes over the more highly heated portion of the furnace tubes, so as to equalize the temperature in all parts of the boiler, and thus avoid the unequal tension to which boilers are ordinarily subject.

On referring back to page 128, there will be seen a boiler containing a flue tube, in which is a water chamber, fitted with flame tubes, shown by Fig. 344; and we next direct attention to the fact, that only two years after, that arrangement was re-patented by a Mr. Musgrave, as illustrated by Fig. 353, as far as concerns the water tube—our patent law being so elastic to permit that occurrence.

Mr. Horton's belief in elliptical flue tubes, fitted with vertical and horizontal water tubes, prompted him to patent the arrangement illustrated by Fig. 354, which is rather ingenious; but it shows no deep thought on the subject of instantaneous evaporation.

A Mr. Hill commenced the ensuing year, 1860, with a method for admitting a certain amount of contraction and expansion of the flue tubes, as shown by Fig. 355, which was accomplished by connecting the plates forming the tubes with rings, in section, as the primitive rails for experimental or temporary railways.

Fig. 353.

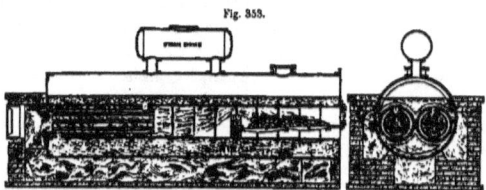

Musgrave's Horizontal Cylindrical Boiler, fitted with twin tubes, in which are cylindrical water chambers, containing also flame tubes; so that the flame passes around the chambers, and through them at the same time. Patented in the year 1859.

Fig. 354.

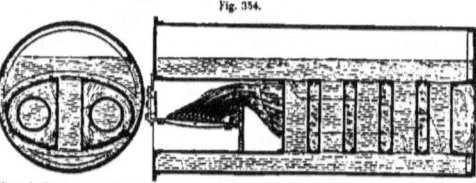

Horton's Horizontal Cylindrical Boiler, fitted with an elliptical flue tube, containing five vertical, and two horizontal water tubes. Patented in the year 1859.

Mr. Galloway next appeared, with two complex arrangements of vertical tubes, for the obstruction of the flame. One example is shown by Fig. 356; the arrangement being that there are two fire grates, separated by the water passage connecting the shallow water chamber, directly over the grates, with the sides of the front portion of the boiler; there are small tubes, by which a communication is established between the water in the upper part of the boiler and that in the shallow chamber; there are also large conical tubes for the same purpose. The main vertical flue is brought down to the top of the shallow chamber, and is provided with a water casing, in which are formed side openings, for the passage of the flame. The main flue is formed so as to pass through the horizontal the perfect circulation of the water in the boiler and the shallow chamber.

The flame from the fire grates, after passing round the shallow chamber and in between the small tubes and conical tubes, escapes through the side openings into the main flue, where it imparts heat to the conical tubes beyond.

The other example is illustrated by Fig. 357.

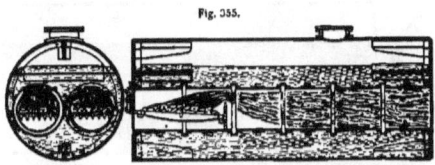

Fig. 355.

Hill's Horizontal Cylindrical Boiler, fitted with twin tubes, manufactured with short plates and tram-rail section rings. Patented in the year 1860.

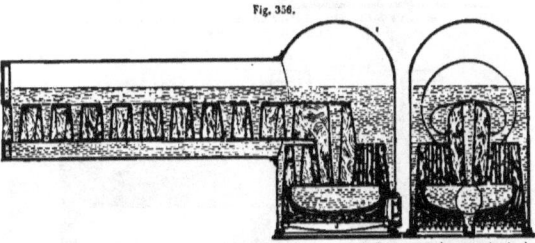

Fig. 356.

Galloway's Horizontal Boiler; fitted with a shallow water chamber over the fire grate, and water tubes in the combustion chamber overhead, in connection with a series of vertical water tubes in the flue tube beyond. Patented in the year 1860.

cylindrical portion of the boiler, where it is of a flat elliptical shape, as shown in the transverse section. In this flue are fixed a number of conical water tubes.

A similar conical water tube connects the centre of the shallow chamber with the upper part of the boiler, and this tube, together with the side passages and the small tubes, effect

In this arrangement the shallow chamber and fire box are rectangular in plan, instead of circular, as in the previous example, and the whole of one side of the chamber is open, and communicates with the body of the boiler.

The fire grates are separated by the water passages, connecting the chamber with the sides of the boiler. The top of the chamber is connected to the upper part of the boiler by a number of small water tubes, as also by the conical tubes, by which means a circulation of the water in the chamber is effected.

The flat elliptical flue, in which are fixed the conical water tubes, is contracted in plan, where it joins the space over the chamber, in order that the heat may be more effectually

Mr. Roddewig begin to show to the world his achievements in boiler improvements, as illustrated by Fig. 358, by which it will be seen that the arrangement consists of a horizontal cylindrical boiler, that is fitted with two plates, at a certain distance apart in the shell, extending for nearly its length, so as to form a central chamber containing water and steam,

Fig. 357.

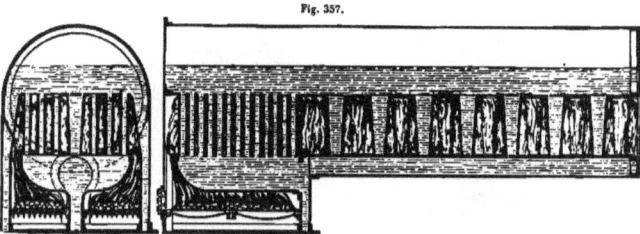

Galloway's Horizontal Boiler, fitted with a shallow water chamber and tubes over the fire grate, and water tubes in the flue tube. Patented in the year 1860.

Fig. 358.

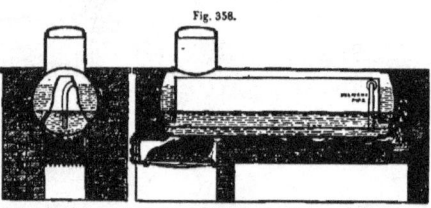

Roddewig's Horizontal Cylindrical Boiler, fitted internally with two longitudinal division plates, to form a steam and water space, between them. Patented in the year 1861.

retained among the small tubes and conical tubes. The flames from the furnaces pass round three sides of the chamber, and after giving off the greater portion of their heat to the tubes, pass into the flue tube, where their heat is still further absorbed by the conical tubes.

The year 1861 next began, and so did a

and the level of the water in that chamber depends on the amount of water discharged through the delivery pipe: it always being supposed that the amount of feed water forced into the two side chambers does not cause any overflow into the central chamber.

Then came a Mr. Harlow, with two arrangements for obstructing the flame, under

LAND STATIONARY HORIZONTAL BOILERS.

the shell and in the flue tube, as shown by Figs. 359 and 360. The arrangement of the first example is, that at the front end of the shell is a water box, containing flame tubes, the box being connected by a branch pipe on its top, to the shell, and at the bottom by a branch pipe to a horizontal water tube, connected at the other end to the shell.

The arrangement of the second example is rather more complicated: because two flue tubes are secured in the shell, each containing the aforesaid tubular water box, and beyond it cross water tubes, as in former examples illustrated in this work.

Mr. Harlow seemed to have forgotten, that to obstruct the flame directly at the back end of the fire grate is to check the draught most injuriously.

A very peculiar boiler next appeared, being the invention of a Mr. Fanshawe, as illustrated in end and side sectional elevations, by Fig. 361 on the next page.

The boiler is composed of a series of narrow compartments, which are the water spaces. These compartments are composed of side plates, and narrow rings of trough iron, which pass all round the compartments, and are secured to side plates by means of rivets. The side plates have helical openings cut through them at the lower part, which openings are covered by a helice formed of trough iron, which is riveted to the side plates, so that when two compartments are brought together, the two contiguous helices will form a curved convolute, or serpentine flue, or passage for the gases from the furnace or firebox, to the central flue; a tubular tie rod, provided with screw nuts at one or both ends, is passed from end to end of the boiler, so that by screwing up the nuts, all the compartments

Fig. 359.

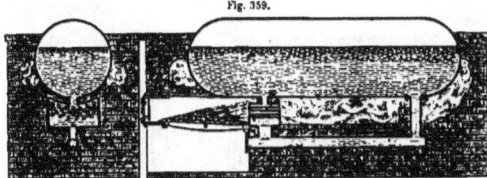

Harlow's Horizontal Cylindrical egg-ends Boiler, fitted with a water bridge, containing flame tubes, also in connection with a water tube underneath the shell. Patented in the year 1861.

Fig. 360.

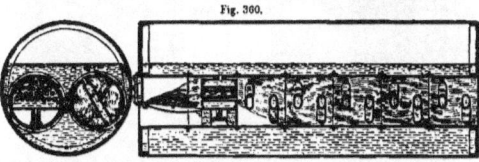

Harlow's Horizontal Cylindrical Boiler, fitted with two flue tubes, containing angular water tubes and a flame tube water bridge. Patented in the year 1861.

may be held tightly together. This tubular tie rod is extended outside the boiler at both ends, one end entering the base of the boiler, the other end passing downwards and under the rear end of the furnace, enters the boiler, so that a constant circulation of water is kept up in the tie rod. The central flue is formed by a series of hollow rings, through which the tie rod passes, the rings being perforated all round, so as to allow the heated gases to enter from the helical flues; which heated gases by forming large circular openings in the sides of the latter. The intervening spaces between the compartments are closed by means of wrought-iron rings, formed of trough iron, which are secured to the sides of the water spaces by means of bolts. The upper parts of the compartments are further held together and strengthened by the longitudinal tie rod, which extends from end to end of the boiler, and is secured upon the end plates by screw nuts. A considerable space above the

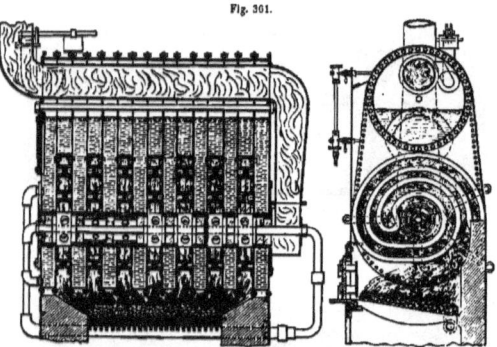

Fig. 301.

Fanshawe's Horizontal Spiral Boiler, with central flame flues. Patented in the year 1861.

passing along the central flue through the return flue, which is placed in the steam space at the upper part of the boiler into the chimney, superheat the steam before it is allowed to pass to the cylinder of the engine.

The upper parts of the compartments, together with the intervening spaces between the several compartments, is made to form one large water space from end to end of the boiler, and common to all the compartments, water level is left for steam, as shown; but in addition to this, Mr. Fanshawe proposed to place a steam chamber above the boiler. When it is desired to obtain access to the interior of the boiler, the manhole cover on one of the end plates may be removed; a man can then get inside, and can remove the nuts of any of the compartments, so that the latter may be taken out, if required, after removing the tie rods.

The next year, 1862, although an "Exhibition" period of international renown, was peculiarly silent on boiler improvements of the class under notice: so much so, indeed, that we have only one example on record, worthy of introduction here, which is illustrated in end and side sectional elevations by Fig. 362. The arrangement is that the shell is cylindrical, with one fire box, in which are two fire grates. The fire box is joined to an oval flue tube, in which are a series of vertical tubes or water passages, open at each end to the interior of the boiler. The upper grate bars are wrought-iron tubing, open at one end to the boiler above the flue, and at the other to a water space, which opens at each end to the boiler, and also forms the dead plate. The lower grate bars are the ordinary kind, with an ashes box underneath, and there is a brick bridge at the back, as usual.

The upper furnace is supplied with coal, and the flame and smoke or gases pass down through the grate bars to the flue. The lower furnace is fed with coke and the incandescent fuel falling from the upper surface, the flame and intense heat from which, arising and passing towards the flue, come in contact with and ignite the unconsumed gases and carbonaceous matter passing from the upper furnace. Thus states the inventor as to the "working" of the grates; but we decline to endorse the statement, although we give it place here as a novelty.

Mr. Inglis commenced the Annals of boiler productions in the year 1863, by the arrangement shown by Fig. 363, on the next page, which consists of a cylindrical shell in connection with two water chambers, situated below it at each end, that are connected by a series of water tubes, the chambers and those tubes being set in brickwork, as shown, with the flame passing amongst the tubes, and then descending to the flue below at the back end.

Next came Mr. Galloway again, with two very simple arrangements of "retort" boilers, as illustrated by Figs. 364 and 365.

The first example is an arrangement of six retorts situated over each other, each pair being connected by eight vertical water tubes, and above the top retorts, or boilers, is a steam chamber dome, secured across them.

The second example is shown in transverse sectional elevation only, which illustrates four

Fig. 362.

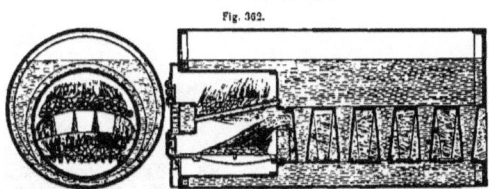

Eastwood's Horizontal Cylindrical Boiler, fitted with a cylindrical fire box, containing two fire grates over each other; the upper grate bars being water tubes, and the lower bars the ordinary kind. Beyond the fire box is an oval flue tube, containing conical water tubes. Patented in the year 1862.

T

pairs of retorts, connected alternately by angular water tubes.

A Mr. Stewart came forward next with two arrangements to obstruct the flame in the tube, which are illustrated by Figs. 366 and 367, and explained underneath each view.

Fig. 363.

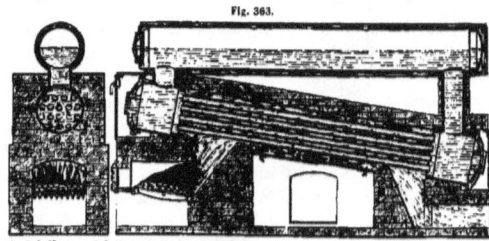

Inglis's Horizontal Cylindrical Boiler, in connection with two lower water chambers, connected by water tubes. Patented in the year 1863.

Fig. 364.

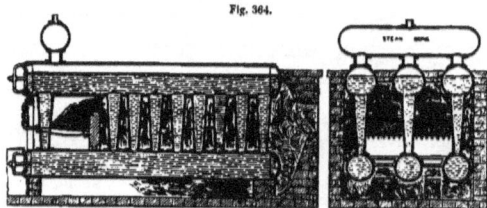

Galloway's Horizontal "Retort" Boiler, connected by vertical water tubes. Patented in the year 1863.

Fig. 365.

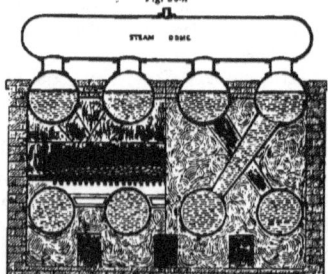

Galloway's Horizontal "Retort" Boiler, connected by angular water tubes. Patented in the year 1863.

There was nothing more done worthy of record on boilers from this example until the year 1865, and the first inventor of that year was a Mr. Amos, who arranged two cylindrical shells, one over the other, as illustrated by Fig. 368. The lower shell contains a cylindrical fire box and short flame tubes, connecting a combustion chamber. And the upper shell contains long flame tubes, connecting the smoke box, the shells being connected by two flange tubes.

A miller named Lake appeared next, with an arrangement of a saddle-shaped fire box and cylindrical shell, containing four large flame tubes, fitted with damper plugs at their back ends, as illustrated by Fig. 369.

A Mr. Smith's arrangement of vertical in them; in some cases the tubes were open through the flue, and in other cases hanging only and closed at the bottom end.

We now direct attention to three examples

Fig. 366.

Stewart's Horizontal Cylindrical Boiler, containing a flue tube, curved at the bottom, and deeply corrugated at the top, to cause an impinging of the flame during its traverse. Patented in the year 1863.

Fig. 367.

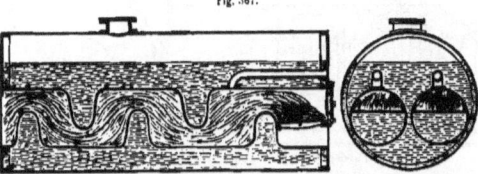

Stewart's Horizontal Cylindrical Boiler, containing a flue tube, deeply corrugated at the top and bottom, to cause an impinging of the flame during its traverse. Patented in the year 1863.

Fig. 369.

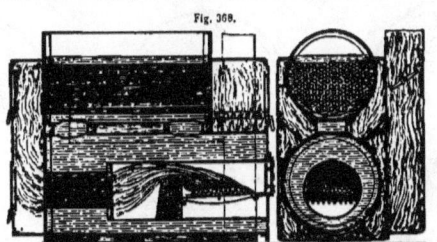

Amos's Horizontal Cylindrical Combined Boiler, fitted with a cylindrical fire box, and short flame tubes in the lower shell, and in the upper shell long flame tubes. Patented in the year 1865.

tubes in the flue tube was next introduced, as shown by Fig. 370. The tubes were of various shapes, fitted with circulating plates of boilers, in which the steam chambers are below the level of the water in the shell— which arrangement is in direct opposition to the practical result that saturated steam should be dried by a dry heat. The inventor, a Mr. Townsend, claimed also that he obviated priming by the positions of the steam pipes, as illustrated by Figs. 371 to 373.

A Mr. Wilson followed Townsend, with an arrangement of pipes, as illustrated by Fig. 374.

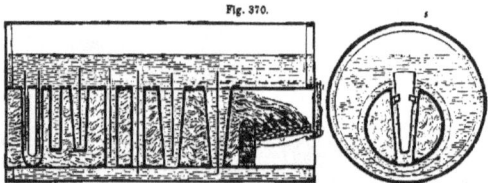

Fig. 369.

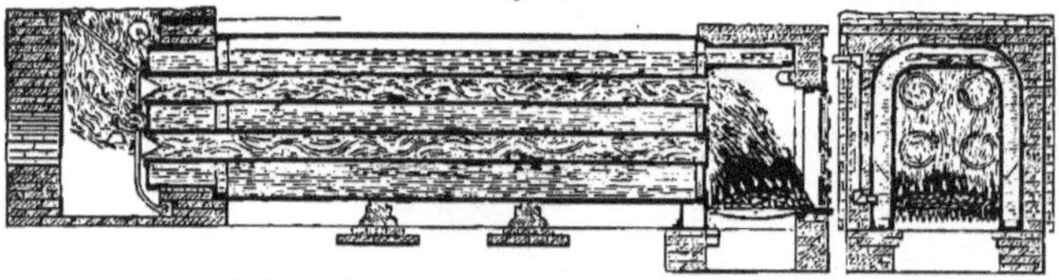

Lake's Horizontal Cylindrical Boiler, containing four large flame tubes, and fitted at their ends with stop plugs and lever damper gear. Patented in the year 1865.

Fig. 370.

Smith's Horizontal Cylindrical Boiler, in which the flue tube is fitted with vertical water tubes, of various shapes, having also circulating plates in them. Patented in the year 1865.

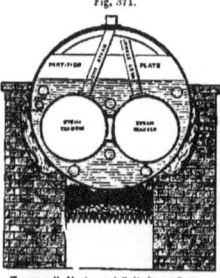

Fig. 371.

Townsend's Horizontal Cylindrical Boiler, containing two steam chambers below the water level. Patented in the year 1865.

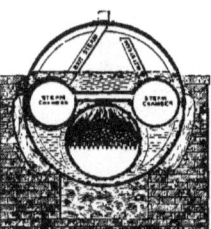

Fig. 372.

Townsend's arrangement of steam chambers and flue tube below the water level. Patented in the year 1865.

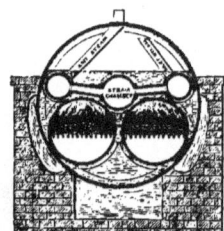

Fig. 373.

Townsend's arrangement, with two flues, and three steam chambers below the water level. Patented in the year 1865.

The year 1866 next began, and with it the ideas of a Mr. Woodward on Boilers, or rather the abolition of fire bars, which he accomplished by perforating the bottom of the fire box, as shown by Fig. 375.

Fig. 374.

Wilson's Horizontal Cylindrical Boiler, in which the flue tube is fitted with cross water tubes, extending above the water level in the shell. Patented in the year 1865.

The grate extends nearly to the back end of the boiler, and in the back end is placed a fire-brick partition, having spaces or perforations, through which the flames pass into the combustion chamber, of fire-brick, formed outside the boiler. From this chamber the flames and hot gases pass through a number of fire tubes, which conduct the same through the water space of the boiler, and into a front smoke box, from whence the flames pass back again through larger fire tubes in the steam space of the boiler, and through a flue, formed over the combustion chamber, into the chimney.

For producing an effective draught, a pipe of smaller diameter than the chimney is fixed in the latter, and the exhaust steam from the

Fig. 375.

Woodward's Horizontal Cylindrical Boiler, with the fire box perforated at the bottom, instead of using fire bars. Patented in the year 1866.

Fig. 376.

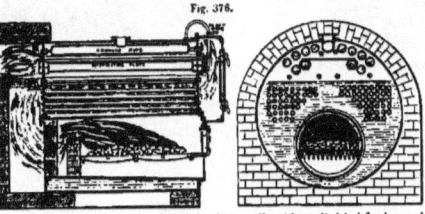

Thomson's Horizontal Cylindrical Boiler, fitted internally with a cylindrical fire box, and above it, two sets of tubes that are arranged to cause a twin return action for the flame. Patented in the year 1806.

A peculiar arrangement of flame tubes in a cylindrical shell was next introduced by a Mr. Thomson, as illustrated by Fig. 376. steam engine is allowed to pass from a blast pipe up the same. In order further to increase the draught, a blast pipe conducts steam from the boiler to a number of small nozzles on the pipe, situated inside the smoke box, in front of each of the fire tubes, from which nozzles jets of steam consequently pass into the tubes, and thus accelerate the passage of the flames or gases through the same.

The next example we refer to shows great ingenuity on the part of the inventor, Mr. Holt, to elude the patent right of inventor to rise into the spaces at the same time : a fact that we doubt, for the reason that there is nothing to cause any rise of the flame artificially, while naturally it will pass out at the nearest and largest opening.

Next came a Mr. Field with a cylindrical boiler, fitted with hanging tubes, secured to its under side, as shown by Fig. 378, also

Fig. 377

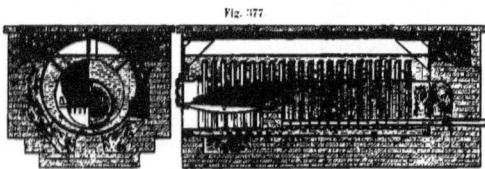

Holt's Horizontal Cylindrical Boiler, in which the flame tube is eccentrically corrugated, to form flat flame and water spaces. Patented in the year 1866.

Fig 378.

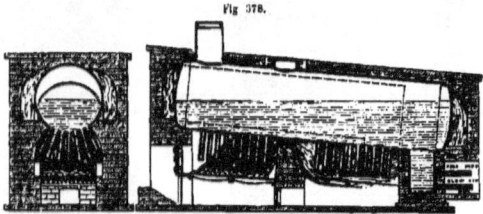

Field's Horizontal Cylindrical Boiler, with hanging water tubes at the under side of the shell. Patented in the year 1866.

Mr. Kendrick, whose arrangement is shown by Fig. 314, on page 117 of this work.

Mr. Holt's arrangement is shown by Fig. 377, which is a cylindrical boiler, fitted with a flue tube, constructed with vertical water and flame spaces that are eccentrically formed into a crescent shape, with circular openings throughout the flue tube for the passage of the flame, which is stated by the illustrating how small an idea our Patent law can be applied to protect.

A Mr. Daglish also, about this period, thought he would test the protecting powers of the Patent law, and finding it favourable, brought forward the arrangement shown by Fig. 379, which is a return-tube boiler, with vertical water tubes in the combustion chamber, which latter addition is very advantageous.

The Patent law must have been very elastic at this period in the minds of boiler inventors, for here is another illustration of its stretching powers, as illustrated by Fig. 380, which is a cylindrical boiler with two flame tubes, having tubular water bridges in them at the back ends; and on turning back to page 135 in this work, we see by Fig. 360 that the same bridge in principle and very nearly in construction was patented five years previously, but was proposed to be at the end of the fire grate then.

Next the United States of America were represented by a Mr. Miller, as illustrated on the next page by Fig. 381, the arrangement of which is—two horizontal cylindrical boilers are set in brickwork, and instead of the fire box being of brick, as generally, it is formed at the sides with vertical pipes, that are connected at each end to horizontal pipes: the top pipes being connected to the steam spaces of the boilers. The vertical pipes have circulating plates in them to cause the heat in the water to have more effect. There is nothing in this arrangement novel, because the pipes referred to are really Dunn's "retort boilers" in a peculiar position.

The French engineers considered that it was time to be represented as boiler inventors in England again, and accordingly M. Chevalier heralded his invention, as shown by Fig. 382, which illustrates that an elliptical shell, containing an elliptical fire box, a cylindrical combustion chamber, and return tubes above, that are bent on to the top side of the chamber to allow for free expansion and contraction, was deemed worthy of protection by our Patent law. And an equal amount of protection was granted to a Mr. James for his invention, introduced early

Fig. 379.

Daglish's Horizontal Cylindrical Boiler, fitted with under and over brick bridges, and a water bridge, having vertical water tubes on it in the combustion chamber. Patented in the year 1866.

Fig. 380.

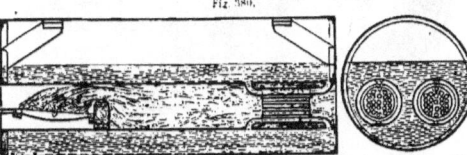

Twibill's Horizontal Cylindrical Boiler, containing two flue tubes, in which are tubular water bridges at the back ends. Patented in the year 1866.

in the year 1867, as illustrated by Fig. 383; which is a Cornish boiler, containing in the flue tube a water chamber, with a flame tube through it, the chamber being connected to the flue by short tubes or branches. But on referring to page 118 of this work, we see by Fig. 318 that the law had already granted one protection for that arrangement, and was therefore kind enough to honour Mr. James also.

After that came a Mr. Beeley, with the arrangement contained in a cylindrical shell, as shown by Fig. 384. The main part of the patent refers to bending the plates at right angles to make the joints, instead of by laps, while the remainder is explained under the illustration.

A similarly situated explanation is sufficient also to explain the next example, illustrated by Fig. 385.

Mr. Dunn's ideas being so fertile, he brought forward again thirty examples of boilers, out of which we have selected six, of the land horizontal class, the first of which is shown by Fig. 386. This boiler is made out

Fig. 381.

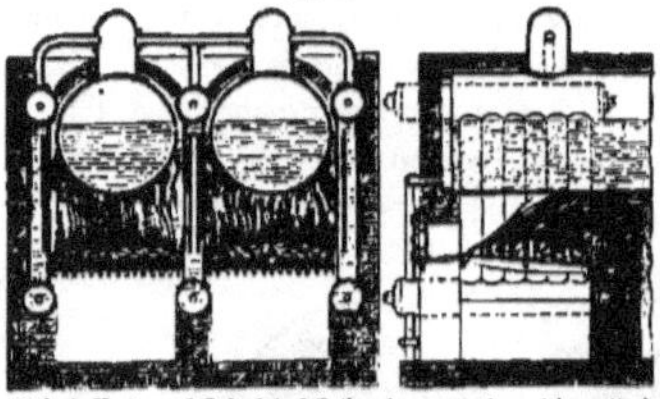

Miller's Horizontal Cylindrical Boiler, in connection with vertical and horizontal pipes, that form the walls of the fire box. Patented in the year 1846.

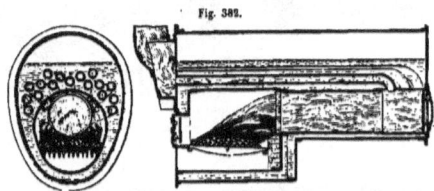

Fig. 382.

Chevalier's Horizontal Boiler, fitted with an elliptical fire box, cylindrical combustion chamber, and return flame tubes, "bent on to the combustion chamber." Patented in the year 1866.

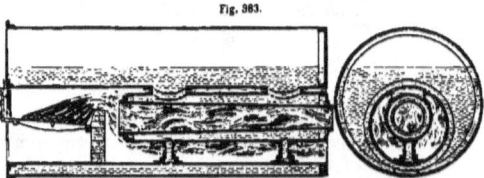

Fig. 383.

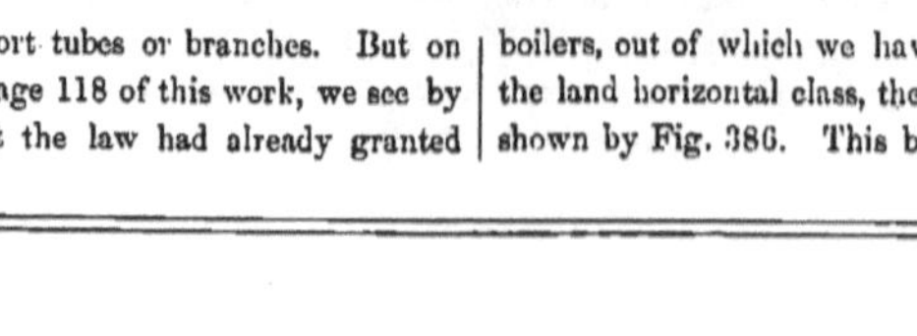

James's Horizontal Cylindrical Boiler, fitted with a flue tube, in which is connected a water chamber, having a flame flue through it. Patented in the year 1867.

of an ordinary double-flued boiler cut in two, and joined together again with the parts one above another, and with the flues connected, so as to leave a water space between the flues fixed a channel piece, to collect the sediment and to distribute the feed water.

The second example is illustrated by Fig. 387; and is a cylindrical Cornish boiler, bent,

Fig. 384.

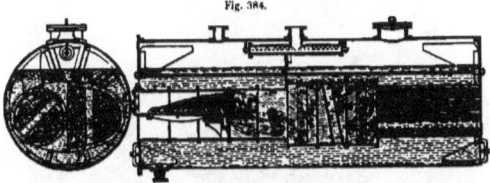

Beeley's Horizontal Cylindrical Boiler, fitted with two cylindrical fire boxes, with angular water tubes beyond the bridge, an elliptical combustion chamber, with vertical water tubes, and a series of horizontal flame tubes of unequal diameter beyond the chamber. Patented in the year 1867.

Fig. 385.

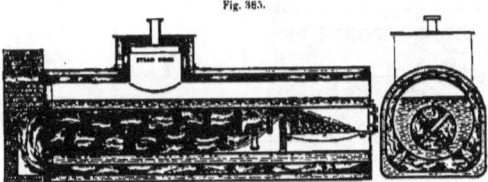

Storey's Horizontal Cylindrical Boiler, surrounded by a flame space, contained in a water casing. Patented in the year 1867.

Fig. 386.

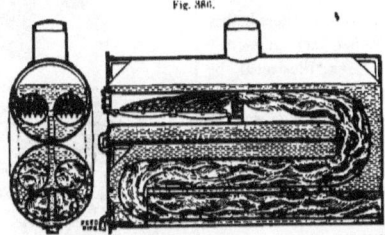

Dunn's Horizontal Cylindrical Boiler, fitted with twin return flue tubes, and an elliptical lower flue. Patented in the year 1867.

and the ends of the boiler; the two flues unite in a double flue, which is provided with a water space. At the bottom of the boiler is as shown, so that the main hollow portion of the boiler is filled with water, but the remainder with steam and water: and, to heat the feed water before it enters the boiler, a return longitudinal coil pipe is fitted in the longer portion of the flue tube, and in the downtake part, a triple return pipe is fitted.

and escape to the chimney by the flue; in this boiler the fire-brick bridge rests upon a hollow, perforated cast-iron chamber, within which is a sliding wedge block; this block is

Fig. 387.

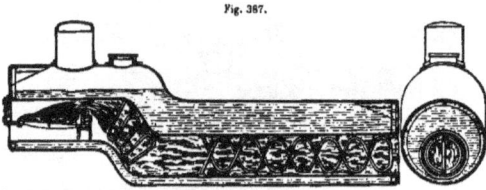

Dunn's Horizontal Cylindrical Boiler, having a flue tube, containing longitudinally serpentine water coils as feed pipes. Patented in the year 1867.

Fig. 388.

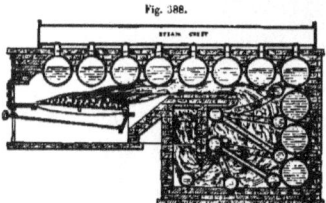

Dunn's arrangement of retort boilers and pipes, shown in transverse sectional elevation. Patented in the year 1867.

The third example relates to retort boilers —for which Mr. Dunn has been previously noticed in this work — as illustrated by Fig. 388; and are arranged very much as before, but with the addition of small retorts, connected by cross pipes, in the main combustion chamber.

But the fourth example is a much more exaggerated form to fill the boiler with water, as the Fig. 389 illustrates. The lower portion of this boiler is at right angles to the upper portion, and is contained in a pit of firebrick; the products of combustion on leaving the flue, surround the lower part of the boiler,

Fig. 389.

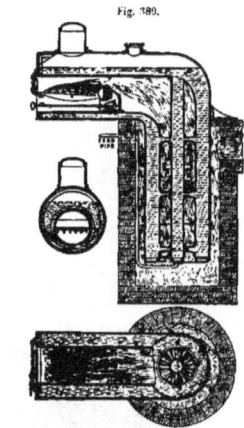

Dunn's Horizontal and Vertical Boiler, combined; the vertical portion being fitted with a vertical water tube, having branch tubes connecting the main water space. Patented in the year 1867.

now shown open to admit air behind the bridge, but it can be closed from the front of the boiler by the handle and rod shown in the drawing. There is in the vertical main flue tube a central pipe, with radiating branches, to obstruct the flame as it descends.

And another example, for the same purpose, is shown by Fig. 390; which consists of angular pipes in the flue tube.

Mr. Dunn also patented the old water tube Cornish boiler, as illustrated by Fig. 391, but was invented in the year 1811.

Next came a Mr. Pollit, with an arrangement, shown by Fig. 392; that is a crossbreed of the Lancashire Return-tube and Breeches-tube boilers. The shell is cylindrical, and contains a fire box that is open through the shell to the brick flue below, and beyond the box is a return flue tube, fitted with cross water tubes of ordinary arrangement.

The action of the flame is, of course, due to the setting in this case, which is such, that the flame, after leaving the grate, passes to the brick flue under the shell; the flame then enters the return flue tube, and passes through it to a brick side flue; then returns

Fig. 390.

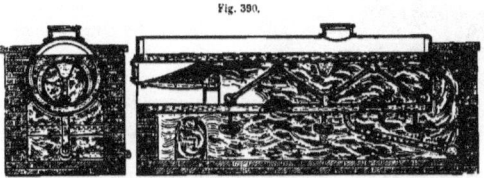

Dunn's Horizontal Cylindrical Boiler, fitted with angular water pipes, and a cup in the flue tube. Patented in the year 1867.

Fig. 391.

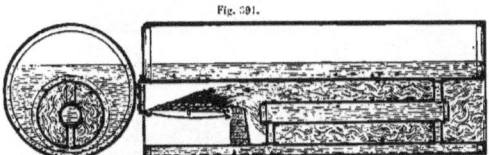

Dunn's Cornish Boiler. Patented in the year 1867, but invented in the year 1811, as shown by Fig. 285, on page 105 of this work.

Fig. 392.

Pollit's Horizontal Cylindrical Boiler, containing a fire box, with a downtake opening; and beyond, in the shell, a return flue tube, fitted with cross tubes. Patented in the year 1867.

forward to the flue under the fire grate, next ascends to the other side flue, and passes back through to the chimney.

A Mr. Holt came forward next, with his

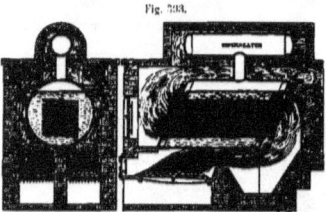

Fig. 393.

Holt's Horizontal Cylindrical Boiler, fitted with transverse sheet flues above the fire box, for the return action of the flame. Patented in the year 1867.

idea of flame tubes for generating steam, as shown by Fig. 393; which is, that an ordinary cylindrical boiler is fitted with flat flame tubes above the fire box, to cause a return action for the flame, with a superheater above the shell.

There are certain virtues in the use of the sheet tube, which consist of a spreading out of heat and fluid, and Mr. Holt evidently knew that when he introduced them, but perhaps forgot that he sacrificed strength at the same time; because the sheet is the weakest shape possible.

Mr. Kendrick not being satisfied with his arrangements of boilers, as shown by Figs. 314 and 315, on page 117 of this work, thought of other methods for obstructing the flame, as illustrated by Figs. 394 and 395. The base line of each water chamber forms about a right angle with the base line of the next water chamber, and all the base lines are at

Fig. 394.

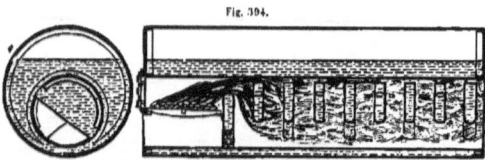

Kendrick's Horizontal Cylindrical Boiler, fitted with a flue tube, that is fitted with semicircular water chambers. Patented in the year 1867.

Fig. 395.

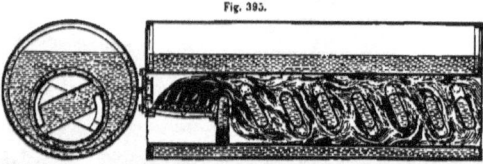

Kendrick's Horizontal Cylindrical Boiler, fitted with a flue tube, that is fitted with alternate angular parallel water chambers, centrally situated. Patented in the year 1867.

an angle of about 45° with a horizontal line carried across the flue. It will also be seen that each parallel water chamber fills about half the width of the flue, and by the alternating arrangement of the water chambers, a serpentine course is left for the products of combustion, and a clear opening is left along the bottom of the flue to enable a man to go through to clean or repair.

The curved edges of the semicircular water chambers are shown as being a short distance from the surface of the flue, leaving openings to allow the products of combustion to act upon as much surface as possible, and to cause the gases to become mixed and the smoke to be consumed. From each corner of the water chambers is a short tube, to connect these chambers with the main water space; these tubes pass through holes in the flue tube, and are secured by expanding their ends. The chambers can be cleaned through the tubes, and any of the chambers may be easily removed and replaced when repair is necessary.

A French engineer, named M. Guyet, next came forward with a combined vertical and horizontal arrangement of tubes, as shown in sectional elevation by Fig. 396. There is a large combustion chamber above the vertical tubes, in which chamber is a tubular feed water heater, which can be used as a superheater alone, also, when required.

Fig. 396.

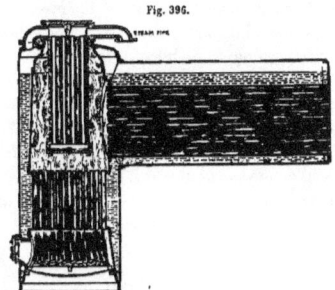

Guyet's Horizontal and Vertical Tubular Boiler, combined, with a feed water heater in the combustion chamber. Patented in the year 1867.

Fig. 397.

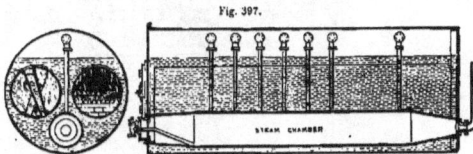

Hopkinson's Horizontal Cylindrical Boiler, fitted with two flue tubes containing angular water tubes; and below the flue tubes is fitted a steam chamber in the water space. Patented in the year 1867.

Fig. 398.

Chamberlain's Horizontal Cylindrical Boiler, fitted with flue tubes, containing perforated brick cylinders, in connection with an outside brick furnace and air chamber. Patented in the year 1868.

Next came Mr. Hopkinson, who claimed the idea of putting the steam chamber in the water in the boiler, as illustrated by Fig. 397, but was preceded by Townsend, as shown on page 140 of this work.

A Mr. Chamberlain ushered in the list of horizontal boiler inventions in the year 1868, with the arrangement shown on the preceding page by Fig. 398; which is a Lancashire boiler, having in each flue tube a brick perforated cylinder, in connection with an outside brick furnace and air chamber — this plentiful use of brickwork being, in Mr. Chamberlain's opinion, the best means for generating steam, but which in practice would be found to be very costly, and constantly out of repair, and equally requiring clearing.

Mr. Whittle put mud in his boiler, he did not state; however, the arrangement of the internal patented improvements is as follows:— The shell of the boiler is fitted inside with curved plates, uniform to the water level, and near the base, where they are "turned up," and constitute troughs for the reception of the mud and other deposits.

There is a flue passing through the boiler, surrounded by a casing, forming a circulating space, also open at the top and bottom by the turned edges of the plates.

When heat is applied to the outside of the boiler, the water in the shell space is first

Fig. 399.

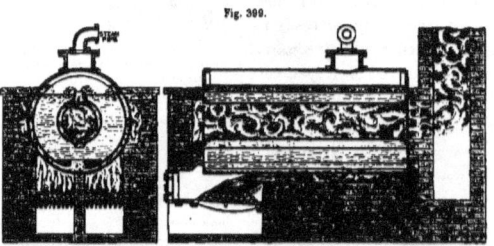

Whittle's Horizontal Cylindrical Boiler, fitted with a central flue tube; the shell and the tube being fitted with curved plates, to form circulating spaces for the water to boil in, and thus more rapidly raise the steam. Patented in the year 1868.

We now refer to the fact again, that the most economical method for raising steam is to have the least practical amount of water on or against the heating surface: and similarly thought Mr. Lawes, in the year 1852, as shown by the illustration, Fig. 299, on page 109 of this work; and equally so did a Mr. Whittle, in this year of inventions, 1868, as illustrated by Fig. 399, adding also that by the circulating water spaces the "mud" would be constantly collected also—but why

heated, and commences to ascend, and, as the heat increases, a rapid circulation of water and steam takes place up the heated sides of the boiler.

When the water and steam reaches the "lip" of the casing—which extends a little above the water level—the steam is separated from the water, and occupies the upper part of the boiler; but the water boiling over the edge of the lining into the boiler, it descends through the lower opening into the circulating space again, and thus maintains a continuous circulation so long as heat is applied.

The mud carried over settles in the troughs at the bottom of the lining, as shown in the drawing, where it is retained, thus preventing the formation of deposit on the plates of the boiler, and by this arrangement the boiler is said to be kept clean and free from incrustation.

and, not being contented thus far, he introduced also a perforated water tank, "hung" from a cylindrical shell with egg-ends, as shown by Fig. 401; in fact, he proposed that the heat should travel as much through water as possible.

Next, a Mr. Arnold informed the world that he had discovered something worthy of protection by the English patent law, which

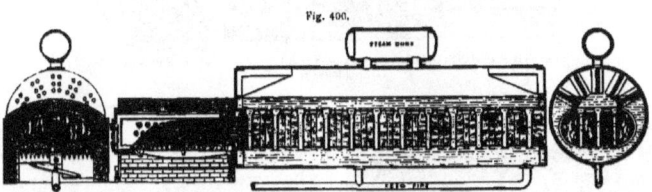

Fig. 400.

Hepworth's Horizontal Cylindrical Boiler, containing an oval flue tube fitted with vertical water tubes; and projecting from the front of the boiler is a water space box, perforated with flame tubes, situated over a fire grate, contained in a brick combustion chamber. Patented in the year 1866.

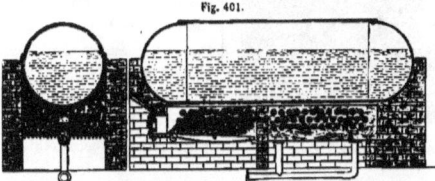

Fig. 401.

Hepworth's Egg-ends Horizontal Cylindrical Boiler, fitted with a "hung" water tank, perforated with flame tubes. Patented in the year 1866.

After Mr. Chamberlain had produced his outside brick furnace, as shown by Fig. 398, on page 149 of this work, a Mr. Hepworth thought he would introduce one also, as illustrated by Fig. 400, with a projecting water box—over the fire grate—perforated with flame tubes in connection with a horizontal cylindrical shell containing an oval flue tube, fitted with vertical water tubes;

something is illustrated by Fig. 402, on the next page, and shows a Lancashire boiler, having the tubes fitted internally with pear-shaped water tubes, connected to the main flue by washers and lock nuts; but how those are disconnected after being exposed to the flame, the inventor explains not — but we do, by stating that it involves their destruction.

After Arnold, a Mr. Kinsey qualified himself for legal protection by introducing the arrangement of horizontal boiler, shown in sectional elevation and plan, by Fig. 403, that consists of a series of corrugated shells forming flame and water spaces, connected at the bottom and top by tubes, as illustrated.

A Mr. Smart wound up the list of horizontal boiler inventions for the year 1868,

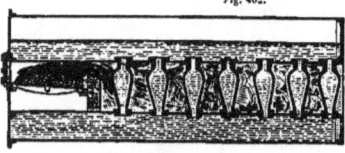

Fig. 402.

Arnold's Horizontal Cylindrical Boiler, containing two flue tubes fitted with pear-shaped water tubes. Patented in the year 1868.

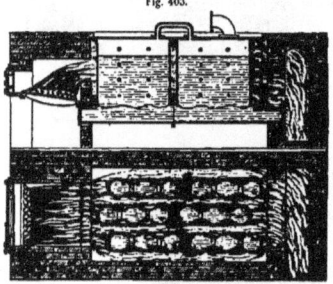

Fig. 403.

Kinsey's Horizontal Corrugated Water and Flame-spaced Boiler. Patented in the year 1868.

by introducing his arrangement, shown by Fig. 404, in which a wagon-shaped shell contains a fire box of the same form, fitted with three tubular water chambers, suspended by branch tubes, between which are small syphon water tubes; and above the fire box are return flame tubes. What advantage there is in all this complication is not apparent, except perhaps to the inventor, who does not express it; and he is equally silent about his tubular arrangement for Cornish boilers, as shown by Fig. 405; but we will here give him credit for that, because there is an evidence of

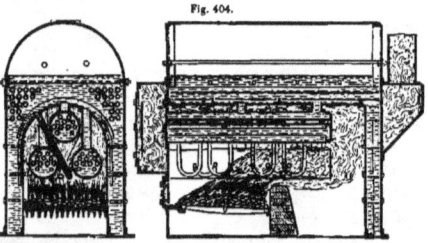

Fig. 404.

Smart's Horizontal Wagon-shaped Shell, containing a similar shaped fire box, in which are three cylindrical water chambers, fitted with flame tubes, and a series of syphon water tubes. At the front end of the shell is a combustion box, connected by a series of horizontal flame tubes, above the fire box, to the chimney. Patented in the year 1868.

forethought, of water and heat circulation in the position and shape of the tubes.

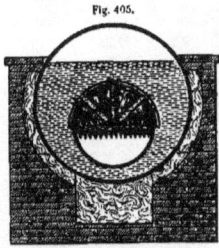

Fig. 405.

Smart's Horizontal Cylindrical Boiler, having the flue tube fitted with circular and radiating water tubes. Patented in the year 1868.

With the commencement of the year 1869, commenced a Mr. Foster to invent horizontal boilers; and how he succeeded is illustrated by Fig. 406, and sufficiently explained underneath it.

spaces, perforated with flame tubes, suspended from horizontal pipes enclosed in brickwork —and a very good boiler it is.

Next appeared two arrangements of the

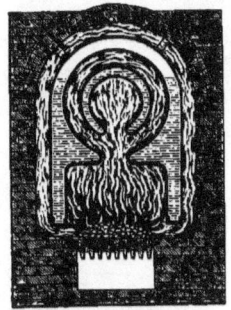

Fig. 407.

Ormson's Horizontal Cast Wagon Flue Boiler, shown in transverse sectional elevation. Patented in the year 1869.

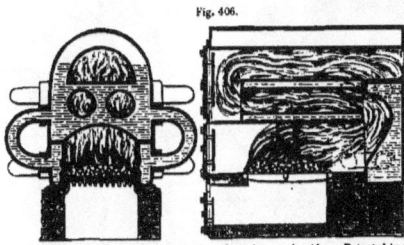

Fig. 406.

Foster's Horizontal Cast Wagon Flue Boiler, with syphon pipes at the sides. Patented in the year 1869.

Then came a Mr. Ormson, with three arrangements, as shown by Figs. 407 to 409, in transverse sectional elevation only.

The American improved boilers were represented next by a Mr. Miller, whose ideas on the subject are illustrated by Fig. 410. This arrangement consists of a series of water "Elephant" boiler, by a Mr. Crosland, as illustrated by Figs. 411 and 412, which are really but a very close modification of those arrangements shown on page 107 of this work, and again illustrate how our Patent law is made use of, as also does the next example, shown by Fig. 413; being also an arrangement of the "Elephant" class of boiler, fitted with flame tubes in the lower and upper shells, as introduced by a Frenchman named M. Gemmell, who was followed by

Fig. 408.

Ormson's Horizontal Cast Wagon Flue Boiler, shown in transverse sectional elevation. Patented in the year 1869.

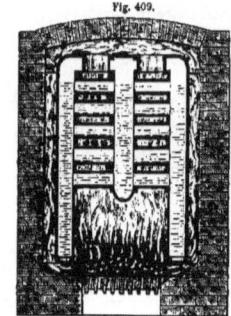

Fig. 409.

Ormson's Horizontal Cast Wagon Flue Boiler, shown in transverse sectional elevation. Patented in the year 1869.

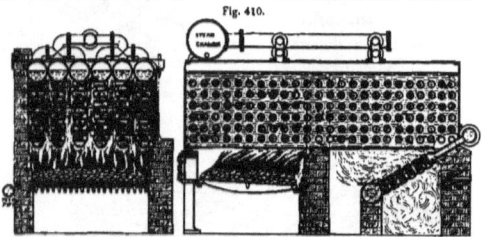

Fig. 410.

Miller's Horizontal Boiler, composed of hanging flat water spaces, perforated with flame tubes, suspended from horizontal pipes. Patented in the year 1869.

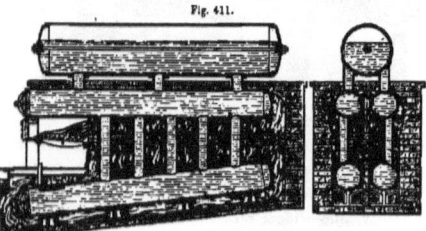

Fig. 411.

Crosland's Horizontal "Elephant" Boiler. Patented in the year 1869.

a Mr. Horton with his arrangement of a square flue fitted with water tubes in a cylindrical shell, as illustrated by Fig. 414.

Another "Elephant" boiler next appeared, by a Mr. Hawksley, which might be termed three combined boilers, because they are in claimed merit; and accordingly a Mr. Cockey might have thought that metal bridges, alternately secured, were equally patent, as illustrated by Fig. 417, which he termed "heat condensing plates." Such success, doubtless, prompted a Mr. Hargreaves to

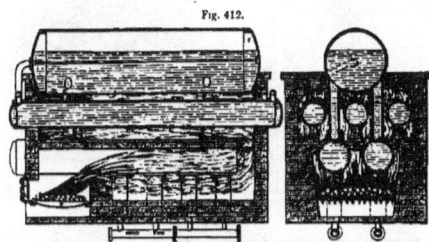

Crosland's Horizontal "Elephant" Boiler. Patented in the year 1869.

Gemmell's Horizontal "Elephant" Boiler, fitted with flame tubes. Patented in the year 1869.

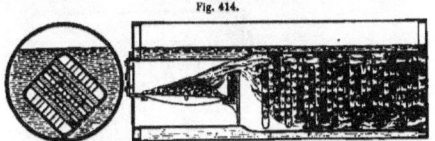

Horton's Horizontal Cylindrical Boiler, containing a square flue tube, fitted with cross water tubes. Patented in the year 1869.

separate groups, connected on the top by pipes, not shown in the illustration, Fig. 415. An equal arrangement as a patent was next patented by a Mr. Fraser, as illustrated by Fig. 416, in which tubes of a peculiar form patent the bending of a pipe in a flue tube of a Cornish boiler, as shown by Fig. 418.

As a wind up of the year, 1869, there issued forth from the brain of Mr. Hamilton an idea to obstruct the flame in the flue tubes of Lancashire boilers, by securing water spaces or pockets in the flue tubes, as illustrated in four views by Fig. 418.

But Mr. Hamilton forgot—if he ever knew of this work; which illustrates also, with Fig. 419, that our Patent law is very accommodating to "all whom it may concern."

Early in the year 1870, a Mr. Hopkinson

Fig. 415.

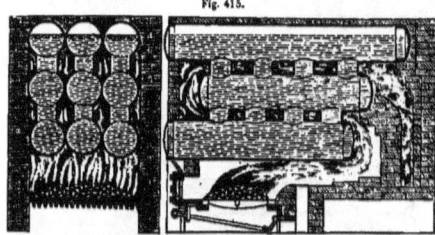

Hawksley's Horizontal Boiler. Patented in the year 1869.

Fig. 416.

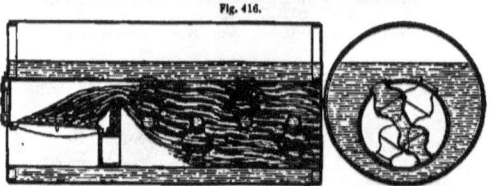

Fraser's Horizontal Cylindrical Boiler, containing a flue tube fitted with corrugated cross water tubes. Patented in the year 1869.

Fig. 417.

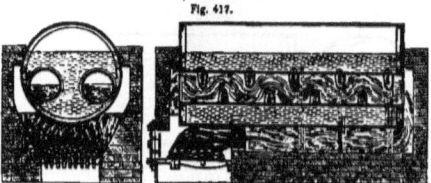

Cockey's Horizontal Cylindrical Boiler, fitted with two flue tubes containing metal bridges. Patented in the year 1869.

—that Mr. Kendrick, not more than two years previously, also invented and patented water spaces secured at an angle of 45° in flue tubes, as illustrated by Fig. 394, on page 148 brought forward his boiler improvements, illustrated by Fig. 420, which is an arrangement of a cylindrical boiler, fitted with flame tubes, in connection by pipes with a common shell, of a lesser diameter, termed a "vessel," not exposed to the fire.

The fire grate, it will be noticed, is shown at an acute angle across the shell; but how

the generation, or rather the formation, of steam.

About this period was patented also an arrangement of cylindrical boilers, by Sir

Fig. 418.

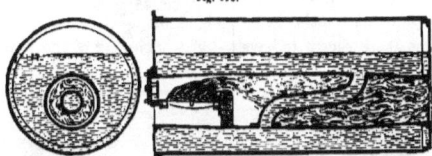

Hargreave's Horizontal Cylindrical Boiler, in which the flue tube has a bent pipe in it. Patented in the year 1869.

Fig. 419.

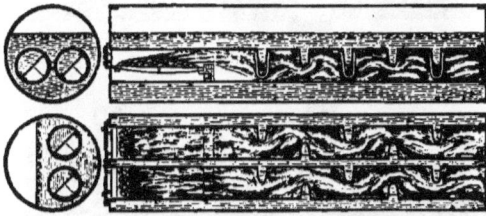

Hamilton's Horizontal Cylindrical Boiler, in which the flue tubes are fitted with water pockets. Patented in the year 1869.

Hopkins's Horizontal Cylindrical Boiler, fitted with small flame tubes; the shell being in connection with a common horizontal cylindrical shell, termed a "vessel," not exposed to the fire. Patented in the year 1870.

the fuel remains there, as illustrated, we have nothing to do with, our present purpose being to make known, occasionally, impossible, as also possible arrangements of boilers, for

W. Fairbairn, Bart., as illustrated on the next page by Fig. 421, where two fire tubes are surrounded by water, enclosed in two larger tubes, in connection with a shell directly above; the three shells being set in brickwork, as generally done.

We here pause in our chronological path, and step aside to point out again how our Patent law in this special instance is worse than worthless. To begin with, see Fig. 298, on page 109 of this work, and the very same means for raising steam by an annular water space, as Fairbairn's, is there shown to have a flue arrangement, in connection with a cylindrical boiler, which is illustrated by Fig. 422, and sufficiently explained by the description under it.

Mr. Crosland then put in appearance again, with a little variation in what he did in the previous year, which variation is shown by Fig. 423.

The beginning of the year 1871 was signi-

Fig. 421.

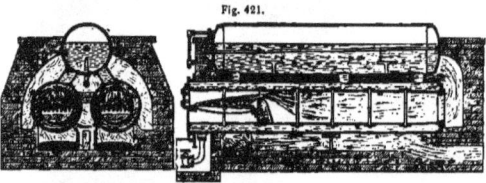

Sir. W. Fairbairn's Horizontal Cylindrical Boilers, in which the flame tubes are jointed with tram-rail section rings, and surrounded by water, and the water spaces are in connection with one shell or tube, centrally situated above. Patented in the year 1870.

Fig. 422.

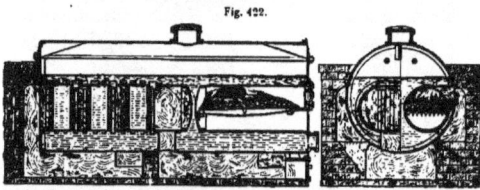

Arnold's Horizontal Cylindrical Boiler, containing two short cylindrical fire boxes, opening to the outer sides of the shell, into brick flues; the flame then descends, and passes up through the bottom of the shell, into an elliptical flue — fitted with vertical water tubes — in a line with the fire boxes; the flame next descends at the back end of the shell underneath it, and then passes to the side flues leading to the chimney. Patented in the year 1870.

been patented eighteen years previously. Next, see Fig. 355, on page 133 of this work; and a veritable similar means for the expansion and contraction of the flame tube is there shown, as patented ten years before the date of Fairbairn's patent, and notwithstanding that the same law protects the inventors (*sic*).

A Mr. Arnold was induced next to patent fied by the invention of a Mr. Mack's boiler, which is shown by Fig. 424; and being an American production, claims notice here. The arrangement of the boiler is, that a cylindrical shell is connected to two arched water boxes, situated at each end under the shell, and connected by horizontal tubes, thus forming an arched or semicircular tubular flue. The flame from the fire grate, after circulating amongst the tubes, passes forward through the flue tube in the boiler, and from thence splits, and passes back at the sides, and finally up the chimney.

—which, by the way, had been proposed several years before.

A Mr. Atkins next appeared with a horizontal cylindrical boiler, shown on the next page by Fig. 426, the arrangement of which

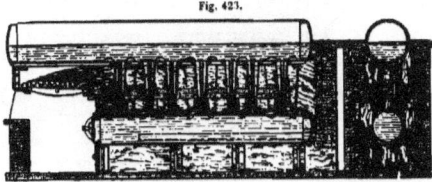

Fig. 423.

Crosland's Horizontal "Elephant" Boiler. Patented in the year 1870.

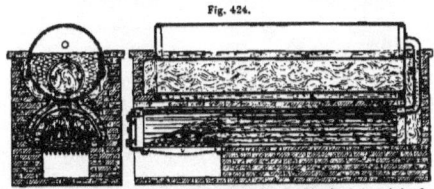

Fig. 424.

Mack's Horizontal Cylindrical Boiler, in connection with a semicircular water tubular flue under the shell, connected by water pipes. Patented in the year 1871.

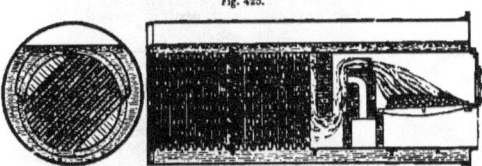

Fig. 425.

Boulton's Horizontal Cylindrical Boiler, in which the portion of the flue tube that is fitted with cross tubes is composed of flanged cylinders, connected by bolts and nuts. Patented in the year 1871.

On page 155 of this work, is shown, by Fig. 414, an arrangement of cross tubes, secured in the flue tube, and by Fig. 425, a similar arrangement is illustrated, with the addition of flange rings, to form the flue tube is, that there are two flue tubes, fitted with horizontal and vertical tubes, at the top and sides, for nearly the length of the boiler, the remainder being a single flue tube with a globular water pot in it.

The Belgian engineers next entered in the list of inventors, and the arrangement of boiler illustrated by Fig. 427, made its appearance; which is a wagon-shaped shell, fitted with

The North British engineers were represented next by a Mr. Davidson; his boiler being shown by Fig. 428, which consists of a cylindrical shell, containing two flue tubes,

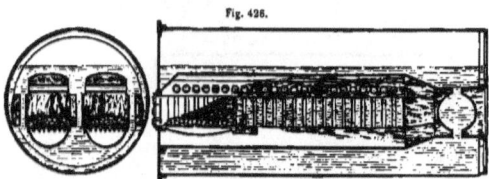

Fig. 426.

Atkins's Horizontal Cylindrical Boiler, containing two flue tubes, fitted at the top and sides with horizontal and vertical tubes, for nearly the length of the boiler; the remainder being a single flue tube, fitted with a globular water pot. Patented in the year 1871.

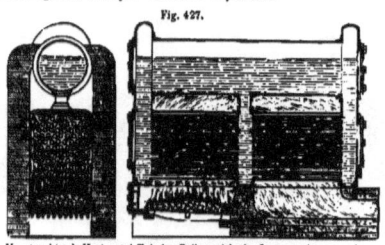

Fig. 427.

Vansteenkiste's Horizontal Tubular Boiler, with the flame passing up and down amongst a twin set of tubes. Patented in the year 1871.

Fig. 428.

Davidson's Horizontal Cylindrical Boiler, containing two flue tubes, fitted with pipes, forming fire bars at one part, and pipes to generate steam in the remaining part. Patented in the year 1871.

tubes and water spaces, with the flame passing up and down amongst the tubes. There is nothing in this boiler to claim either protection of notoriety.

fitted with a double set of pipes, the front set being for fire bars, and the back set, to generate steam. As there are three views, we need not explain further, but direct attention next to

LAND STATIONARY HORIZONTAL BOILERS. 161

Figs. 429 to 431, which are the productions of M. Laharpe, a French engineer, who combined the "Elephant" and tubular classes of boilers to make up, as he thought, a novelty; which, after all, is but a sorry result, and the same amount of thought it cost might certainly have been far better expended in a more sensible manner by increasing the heating surface in the least possible space, instead of spreading it out to make a display for the sake of personal vanity.

Fig. 429.

Laharpe's Horizontal and Cylindrical Boilers, arranged over each other, with the flame passing down through the shells. Patented in the year 1871.

When Mr. Arnold had completed his arrangement of boiler illustrated by Fig. 422, on page 158, he might with some credit have considered that, as a flue arrangement, the matter was complete; but not so, thought a Mr. Edge, who introduced the arrangement shown by Fig. 432, proving that the same effect is possible with a reverse arrangement.

Mr. Crosland not having exhausted his ideas on "Elephant" boilers, introduced next the arrangement illustrated by Fig. 433.

Fig. 430.

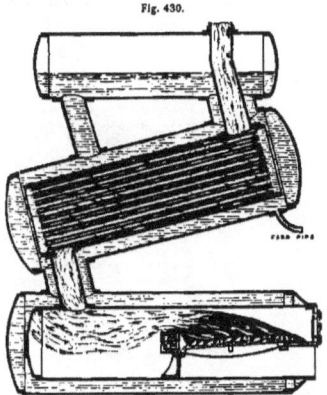

Laharpe's Horizontal Cylindrical Boilers, arranged over each other, with the flame passing up through the shells. Patented in the year 1871.

Fig. 431.

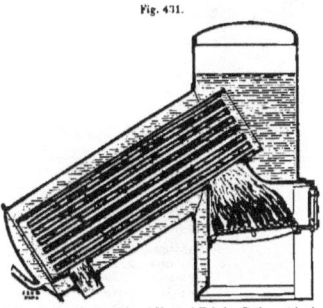

Laharpe's Semi-horizontal and Vertical Tubular Boiler, with the flame passing down through the angular shell. Patented in the year 1871.

The next novelty in horizontal land boilers that we direct attention to, is a peculiar arrangement, invented and patented by Mr. Hawksley, as illustrated by Fig. 434. This boiler is cylindrical, fitted with a flue tube that is constructed, for about a third of its length, telescopic in form, connected by flanges, and the remainder of the tube is parallel, surrounded with T-iron to strengthen it. The parallel portion contains a puddling furnace, and the telescopic part contains cross water tubes; but the pith of the patent is the obstruction of flame by the flange connections, and the puddling furnace in combination with a boiler for raising steam—both forming very good claims for a patent.

A Mr. Heywood next brought out an arrangement of several ideas blended into a

Fig. 432.

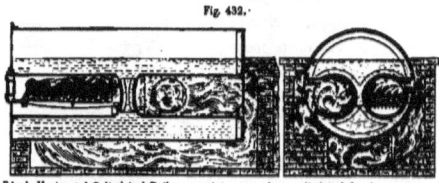

Edge's Horizontal Cylindrical Boiler, containing two short cylindrical fire boxes, open to the outer sides of the shell; the flame then descends into a flue under the shell, and then ascends at the back end into two short flue boxes in the shell in a line with the five boxes, having side flue tube openings into two side flues leading to the chimney. Patented in the year 1871.

Fig. 433.

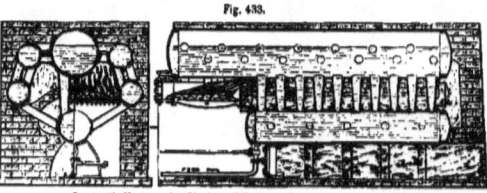

Crosland's Horizontal "Elephant" Boiler. Patented in the year 1871.

Fig. 434.

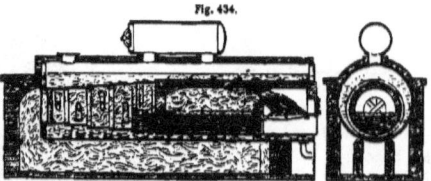

Hawksley's Horizontal Cylindrical Boiler, in which is a flue tube, telescopic in shape, at the back end, and there fitted with cross-water tubes, and at the front end fitted with a puddling furnace. Patented in the year 1871.

combination for the generation of steam, as illustrated by Fig. 435. The shell is cylindrical, with the fire grate under it at the front end; the grate is divided centrally by a water chamber, and similarly, on each side, each grate is bounded by a similar chamber, all three of which are connected by water tubes to the shell: that contains the ordinary flue tube, fitted with a side flame opening,

The action of the flame in Heywood's boiler is thus: on rising from the grate, the flame passes through side flues into the bottom flue, and from there ascends at the back end into the boiler flue tube, from which it passes out at the side flue tube opening, into the upper flues that lead to the chimney.

Now if this "side flue tube opening" in the flue tube is a part of Heywood's patent—

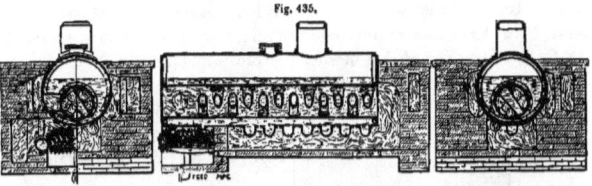

Fig. 435.

Heywood's Horizontal Cylindrical Boiler, having a flue tube fitted with cross-water tubes and a side end flue; also water chambers in a line with the fire grates, central, and on each side: and at the bottom of the shell, beyond the chambers, three rows of water cups. Patented in the year 1871.

and beyond it, in the tube, are cross-water tubes.

At the bottom of the shell are fitted three rows of water cups, for the flame to act on, which we think belongs to Barran, as shown by Fig. 318, on page 118 of this work; only there the cups are for flame action inside, instead of outside—but the "cups" are there, nevertheless.

which he claims as such—where, in the name of Justice, is the use of the inventions, as patents, as shown by Figs. 422 and 432, on pages 158 and 162 of this work; because there are illustrated that the very side tube flue openings referred to are the claims protected.

We end this chapter with the statement on our part, that the present Patent law appears to be a national snare.

CHAPTER III.

LAND STATIONARY TUBE BOILERS.

THE first tube boiler proper was invented and patented by a Mr. Moore, as illustrated by Fig. 12, being a repetition of that illustration from page 5 of this work, and proving

Fig. 12.

Moore's Vertical Tube Boiler, with horizontal pipe rings, surrounded with brickwork. Patented in the year 1824.

thereby that whatever credit has been given by writers and speakers to Dr. Alban, a German machinist, as the originator of high-pressure boilers, that credit was misplaced, and most certainly should rest on an Englishman; because it was in the year 1824 that Moore introduced his tube boiler, or sixteen years before Dr. Alban invented his. We preface our chapter thus, because "fair play" is an Englishman's precept always.

Dr. Alban's boiler is illustrated by Fig. 436,

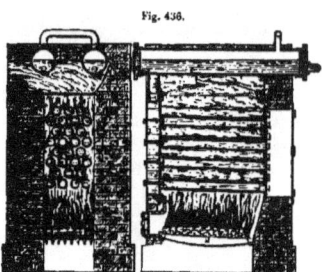

Fig. 436.

Dr. Alban's Tube Boiler, composed of horizontal water tubes, connected to a rectangular water chamber, in connection with two cylindrical chambers above, containing water and steam. Constructed in the year 1840.

and is explained in his practical work on "The High-pressure Engine," as follows:—

I lay the tubes in eight rows or tiers, one over another, and in such wise that the tubes of each row stand over the interstices between those of the row immediately below. There are eight rows of tubes in the position above described; four alternate rows consisting of one tube less than the other four, this inequality being a consequence of the position: I arrange them in such manner that the lowest row has the greater number. The space between the tubes I have made about 1½ inch. Between the outside tubes of the widest rows, and the vertical walls of the furnace, I allow ¾ inch space. The manner in which the fire current plays among the tubes is easily seen in the figure.

The *hearts*, as I term them, are flat chambers, from 6 to 8 inches deep. Their height in the clear should in all cases reach 40 or 42 inches; their width depends on the number of tubes in the several rows; the rule obtains that they should be from 8 to 12 inches wider, in the clear, than the outside width of one of the widest rows. The object of this will appear presently. The hearts are constructed of iron; their sides I make usually of cast iron, of such strength as to remove all danger: wrought iron, however, may be used. The front and back plates are of very strong wrought iron plate, the former ½ inch, the latter ¾ inch thick. They are so tied together by several rows of strong iron bolts, that no bending or bulging out is possible. They are also screwed to the side plates with a proportionate number of bolts, equally strong. The joint is made for the back plate with the ordinary iron cement, and for the front plate with lead, as the latter has to be opened for cleaning.

The generating tubes fit into an annular groove sunk in the back plate of the heart. The oval openings which form the communication between the heart and the tubes must come as close as possible to the upper and lower surfaces of the interior of the tube: this is particularly necessary with the upper opening, in order that the steam may pass freely away. The size of these openings is 1½ inch in the longer and 1 inch in the shorter diameter. The manner in which the tubes are secured to the heart is that the opposite or front end of the tube is fixed upon the back plate of the heart, and this end is surrounded with a wrought iron ring 1½ inch wide and ¼ inch thick, fast brazed on, in order to give the requisite strength and firmness to this part of the tube, and to present a wide surface of iron for the purpose of fastening the joint between the tube and the heart plate with iron cement. On the inner surface of the tube are riveted firmly two iron ears, set about ¼ inch from the end, each having a square recess of about ⅜ inch wide hollowed in its back end, in which lies the hinder part or arm of the T-shaped tie-bolt, so as to hold by these recesses upon the tube without turning round. The bolt itself is in the screw about 1 inch diameter; it passes through the heart plate between the two oval openings, and is screwed up on the front side by a strong nut. The tube is thereby drawn firmly into a groove ¼ inch deep, prepared for it in the heart plate, and the joint is made tight by iron cement. By loosing this bolt, any tube may be easily removed, and repaired or replaced by a new one when necessary.

The next inventor who is noticed here, is a Mr. Brayshays, whose boiler is illustrated by Fig. 437; the arrangement being nine horizontal tubes, connected by eighteen vertical tubes, and set in brickwork, as shown. This arrangement of tubes is the best possible, where strength is required, as they are "bound" together by their connections.

The next example of tube boiler is rather more complicated, but being a French production, is excusable on that account, as the Fig. 438 illustrates. The arrangement consists of a set of hanging tubes, containing water to the top, in connection with another

tubes, while the steam generated escapes from the boiler tubes by ascending towards the plate between the interior surfaces of the boiler tubes and the exterior of the dipping tubes. A groove made in the length of each hood gives easy passage to the steam to pass into the conducting tubes, where it is superheated before passing to the chamber.

The steam tubes, like the boiler tubes, are

Fig. 437.

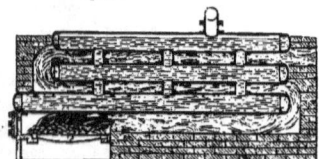

Brayshay's Tube Boiler, composed entirely of tubes—being not unlike the "Elephant" boiler on a small scale. Patented in the year 1856.

Fig. 438.

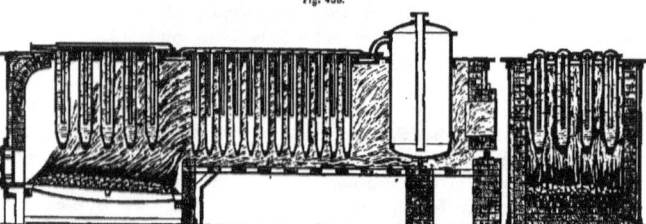

Joly's Vertical Tube Boiler, composed of hanging tubes and circulating pipes, in connection with a cylindrical superheating chamber. Patented in the year 1857.

set of hanging tubes, to contain the steam. The openings of each row of boiler tubes are covered with a metal hood or bonnet; this hood carries a tube descending into each of the boiler tubes; this dipping tube extends nearly to the bottom of the boiler tubes, and it is through this descending tube that the water in the feed pump enters into the boiler

hermetically closed at their lower part, and are suspended from and supported by a metal plate in like manner. All the upper openings of each superheating tube are covered by a metal hood pierced with two holes facing each of the said openings. A small dipping tube is fixed firmly and immovably in one of the two holes; it is through this dipping tube

LAND STATIONARY TUBE BOILERS.

that the steam enters and descends to the bottom of the superheating tube, and in reascending to the hood it is superheated by contact around the exterior of the superheating tubes; it leaves the hood by the second opening, which is entirely free. These small dipping tubes placed in the boiler and superheating tubes, through which the water and steam enter, are employed in order that the water and the steam shall enter into larger areas, and consequently flow at a slower rate as they progress, which permits the one being more easily vaporized and the other more easily superheated.

In the year 1858, a native of North Britain, known by the name of Meiklejon, patented a combined boiler, that is illustrated by Fig. 439; the arrangement consists of four water tubes, disposed in two horizontal planes, being united by a narrow cross-water space. The lower twin tubes rest upon the curved ledges of the brickwork, and from each of the pipes the hollow water web extends upwards in an angular and converging direction, the junction of the lower surface of the web being somewhat below the centre of the fire chamber. From this central part of the boiler the web again diverges, until its junction with the

Fig. 439.

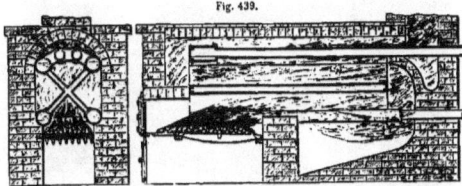

Meiklejen's Horizontal Tube Boiler, in connection with sheet water spaces. Patented in the year 1858.

The hood of the superheating tubes covers a whole length of the plate comprising a series of tubes; a channel made in the thickness of the hood extends throughout the whole row of one series of superheating tubes; at the opening of each superheating tube in the channel itself, and between the two openings, there is a partition, which causes the steam to enter through the small dipping tubes into the corresponding superheating tubes, and leave by the other opening, and pass by means of another part of the same groove into the superheating tube contiguous thereto, and so on until it reaches the main chamber, whence it is employed for the engine.

upper twin tubes, which are parallel with the lower tubes.

The upper tubes are united at the front end of the boiler by the union pipe, on the back part of which are cast or otherwise connected the two sockets, which receive the ends of the pipes, that are carried backwards below the crown of the arch that forms the upper part of the chamber, so that they receive the heat of the gaseous and non-combustible matters before they escape into the chimney.

The lower tubes at the back part or end of the boiler extend out beyond the web, forming two sockets, which receive the ends of the feed water pipes.

The next example we notice is a tube boiler proper, as illustrated by Fig. 440, invented by Mr. Perkins, the well-known high-pressure steam advocate.

Steam boilers constructed according to his invention, have the water contained entirely in welded wrought or drawn tubes, and these are principally arranged in a nearly horizontal position. The lower tubes serve as fire-bars, and for this purpose they are arranged side by side, with a suitable distance between them for the passage of air to the fuel; this distance may be obtained by drawing down the centre portions of the tubes smaller than the ends. The walls of the furnace on each side of the fire-bars are formed by placing a number of horizontal tubes one above another, and in close contact the one with the other; and the top of the furnace is formed of tubes, arranged side by side in a manner similar to the fire-bars, but close together, in place of with spaces between them. Over the top of the furnace a flue is formed, which is also enclosed on the top and two sides by tubes arranged in a similar manner to those which form the top and two sides of the furnace; and the smoke and heated products of combustion from the furnace are caused to traverse this flue before they escape into the chimney. In order to provide for the escape of the steam generated in these tubes, they are connected together by other tubes of smaller size, which, by preference, are screwed at their ends, and enter tapped holes in the larger tubes. These connecting tubes are also, by preference, arranged in a diagonal position; thus the fire-bar tubes are connected by diagonal connecting tubes, with the tubes forming the walls of the furnace, and these again by other diagonal tubes, with the tubes forming the top of the furnace, in a similar manner these latter are connected with the tubes of the side walls of the flue, and, lastly, the steam is led to the tubes forming the top of the flue, from which it is taken away for use by small pipes connected with each of these tubes, and leading into the main steam pipe. The feed water may be introduced into the tubes forming the fire-bars by means of small tubes, one in connection with each of them, and which at their other ends communicate with the main feed pipe from the engine or donkey pumps.

Fig. 440.

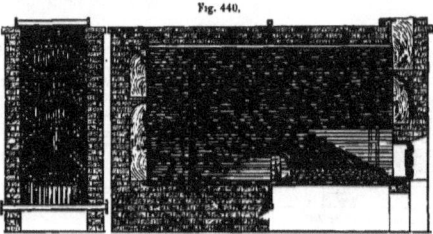

Perkins's Horizontal and Vertical Combined Tube Boiler, composed entirely of tubes set in brickwork. The action of the flame is, first, forward—second, back—third, forward—fourth, back—and up the chimney. Patented in the year 1859.

LAND STATIONARY TUBE BOILERS.

Next Irish boilers were represented by the arrangement illustrated by Fig. 441, which shows that the main portion of the boiler is a horse-shoe shaped figure, the lower part of which rests upon the sole of the arched part of the brickwork. The front part of the furnace is enclosed by a door in the ordinary way, and above this an opening is formed in the brickwork, which gives access to the front end of the boiler for the purpose of cleaning the tubes. This aperture is also closed by a door, as also is a corresponding opening formed in the brickwork at the opposite end of the boiler, which permits access to the tubes of the boiler and to the flue or space surrounding it.

The boiler is formed of an external and internal shell of boiler plates, which are curved to the horse-shoe figure, and arranged parallel to each other, so as to leave a comparatively thin water space between the inner and outer shells. The ends of the boiler are flat, and form vertical water spaces corresponding to the arched portion.

Extending longitudinally to each of the inner end plates is a series of tubes, and arranged within each of these tubes is an internal tube, which is attached to the outer end plates or external shell of the boiler. With this arrangement it follows that the water within the boiler flows along the thin water space formed between the tubes, and thus completely surrounds the inner tubes.

Water is supplied to the boiler from the side inlet pipe.

The introduction of the double tubes in this

Fig. 441.

Traye's Horizontal Saddle Shell Double Tube Boiler; the action of the flame being to surround the shell and the tubes. Patented in the year 1860.

boiler is the only part worthy of notice or comment.

A Mr. Matheson then came on the scene, with a tube boiler composed entirely of pipes in the form of coils, as shown by Fig. 442. The working of this apparatus is thus described by the inventor:—Steam as it is generated is allowed to escape from the open upper ends of the coiled pipes into the upper part of the chamber which forms the water reservoir above, and which therefore forms a steam chest; the pressure of steam in the boiler and in this chamber is therefore equalized, while the supply of water to the bottom of the coils is from the small tank at the end of the fire grate. This water reservoir must, of course, be kept supplied with water by means of a force pump, and as the upper part of the coils will by the rapid evaporation of the water be filled with steam, only this steam will be superheated and made dry by the action of the fire on this part of the coils.

This, of course, is a step in the right direction, with pure feed water. And so, undoubtedly, thought a Mr. Green, who—in the fire grate in the first four rings, instead of in front, as in Matheson's boiler.

Mr. Green also introduced a vertical tube

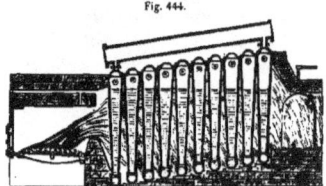

Fig. 444.

Green's Vertical Tube Boiler, entirely composed of tubes; set in brickwork. Patented in the year 1861.

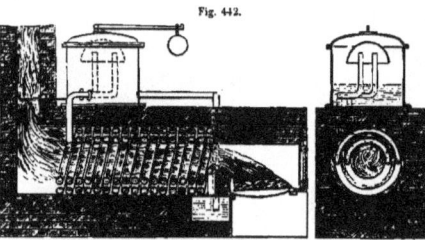

Fig. 442.

Matheson's Horizontal Coil Tube Boiler, set in brickwork. Patented in the year 1861.

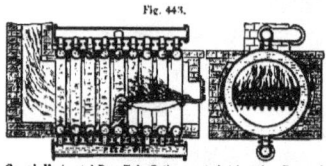

Fig. 443.

Green's Horizontal Ring Tube Boiler, set in brickwork. Patented in the year 1861.

same year as Mr. Matheson—brought forward a ring tube boiler, as illustrated by Fig. 443. In this case a series of ring tubes are attached, by flange connections, to two pipes, over and under them, and set in brickwork, with the boiler, as shown by Fig. 444; the arrangement being a series of vertical water pipes, partly filled, in connection with a larger pipe, angularly situated above them, for the steam chamber.

In the ensuing year, 1862, the American opinion on tube boilers was represented by a Mr. Harrison, as illustrated by Fig. 445. This boiler is formed of short duplicate portions, bolted together by long stay bolts and nuts: thus forming pipes, closed at each end, which are arranged to form the sides or walls of the furnace and lower part of the generator in columns, rising at an angle from a vertical line, and falling back from the front of the boiler, whilst the units of four spheres, forming the upper or main steam generating part of the boiler, are disposed in tiers, of any given length, at or about right angles to the sides or wall columns. These tiers are inclined, such inclination being downwards from the front of the generator, and they are also disposed one above the other in two or more regular tiers.

Between, and resting upon the top of each tier, a partition of fire-clay or other material vertical lines to the uppermost part of the upper half of those tiers, also, without interfering with the current of water in any great degree, following the necks and enlargements of the tier at the top, until the steam is discharged at the upper end of these tiers.

The several rows of tiers are connected rigidly together only at the end nearest the furnace, thus all harm from irregular expansion is prevented; the disconnection being made by breaking joints with one of the units; a unit of four spheres from the upper range of one tier to the lower range of the

Fig. 445.

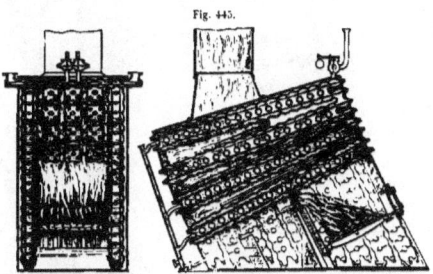

Harrison's Angular Tube Boiler, composed of duplicate portions, secured together by long stay bolts and nuts, set in brickwork. Patented in the year 1862.

is introduced, a free passage-way being left at each end of the partition alternately for the passage of the flame and gases from the furnace, their course being thus rendered zig-zag through the several tiers of units until the gases reach the opening to the funnel at the top.

The units are also so disposed and connected that the current of the water contained therein will flow down the lower slope of each tier, and upwards on the upper slope, whilst the bubbles of steam at all the lower points of the tiers filled with water, ascend in nearly tier above, and a half unit being used at the bottom and top for filling the vacant spaces.

When the boiler is in action, water is forced so high in the two lower tiers that they are quite filled with water, and a portion of the water, as it rises by ebullition, is carried into the tier just above, which tier is the "steam tier," and is allowed to pass down towards the lower end, depositing a portion of this overflow in the small pools formed in the bottom spheres of this tier. This steam tier is only connected with the water spaces of the boiler at its highest end, and not by the connecting pipes at the back of the boiler, as is the case with the other tiers, consequently no water enters therein except what boils over from the tier below. The steam generator in the lower tiers passes also down the lower half of the steam tier not filled with water, until it reaches a joint or junction neck, through which it passes to the upper half of the said tier, and from there is discharged at the upper end into the next tier above, through which the steam again descends a short distance, and again turns to the point of the outlet to the engine cylinders.

Should too great a quantity of water be allowed to pass from the lower tiers, a portion of it would be deposited at the lower end of the steam tier, which would continue to receive such surplus until it reached the opening through which the main body of steam must pass. When this occurs, and so as to prevent inconvenience from this cause, a series of ball valves, shown in the second tier down, are placed in a portion of the vertical necks of the units in this tier, which are arranged to lift when water impedes the regular opening, beginning to lift at the bottom valve first, the one nearest the rising water being always open. These balls cover the same area of opening, and are made lightest towards the lower part of the tier, so that the lower balls must lift first.

This is rather a complicated arrangement; but it, for all that, shows an evidence of forethought, and has, to the present time, been pretty largely used in the United States with some success.

Mr. Elder next appeared with a water coil tube boiler, as illustrated by Fig. 446; the arrangement being that a hollow coil is set in a vertical position in brickwork, and at both ends is connected with a pillar, containing water. The upper branch contains a screw

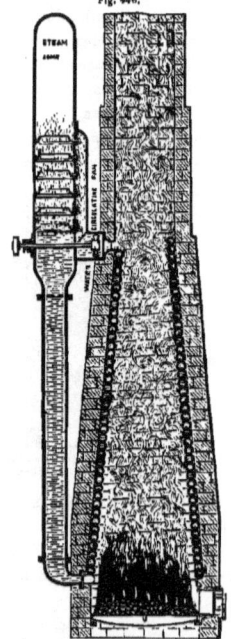

Fig. 446.

Elder's Spiral Coil Tube Boiler, in connection with a water and steam pillar, fitted with a propeller to cause a circulation of the water through the coil. Patented in the year 1862.

propeller or fan, that by its motion causes the water to circulate down in the coil, and above that branch is a steam chamber, fitted with priming rings and valves.

The worst feature in this arrangement is, that the steam chest is not in the chimney, so that the steam could be dried; because it will probably be very "wet," owing to the action of the propeller.

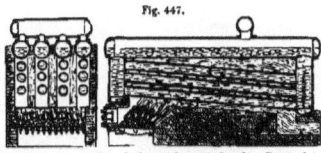

Fig. 447.

Inglis's Horizontal Tube Boiler, with vertical ends. Patented in the year 1863.

as illustrated by Figs. 447 and 448, which are to a great extent reproductions; before patented by other inventors.

Mr. Twibill, however, represented N.B. in the year 1865, and with something novel too in the way of tube boilers, as illustrated by Fig. 449. The arrangement being: that there are two cylinders vertically situated on two opposite supports of brickwork at different levels; and that connected on each side of the cylinders, longitudinally, are a series of tubes, connected across at the positive angle by tubes of the same diameter; thus

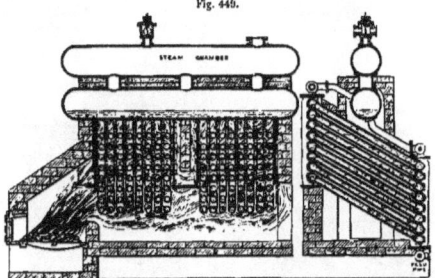

Fig. 449.

Twibill's Angular Tube Boiler, formed with flanged pipes bolted together, and thus forming a rectangular group for the flame to act in between. Patented in the year 1865.

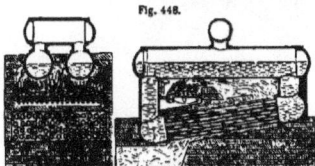

Fig. 448.

Inglis's Horizontal Tube Boiler; with the fire-grate over the multitubular portion; with a return action for the flame. Patented in the year 1863.

Mr. Inglis, in the year 1863, introduced two arrangements of horizontal tube boilers,

forming a rectangular group of tubes amongst which the flame passes. The two cylinders are connected at their tops to a longitudinal chamber also in connection, by branches, to a second upper chamber situated in the same direction.

Another of Twibill's arrangements is illustrated on the next page by Fig. 450, but in this case a perfect right-angle connection for the tubes is maintained throughout.

Next came a French engineer, named Belleville, with varied positions for horizontal syphon tubes, enclosed in a brick furnace, for the generation of steam, arranged as shown by Fig. 451.

taneously into the generating tubes until its level is the same in the tube as in the globe, and consequently in the gauge glass on it. The lowest tubes, that is to say, those nearest

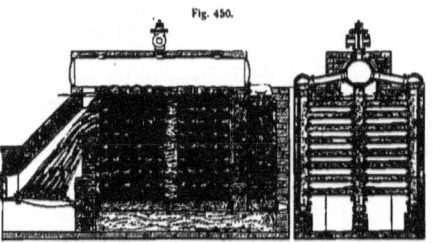

Fig. 450.

Twibill's Vertical and Horizontal Tube Boiler, formed with vertical and horizontal pipes, connected together in all directions. Patented in the year 1865.

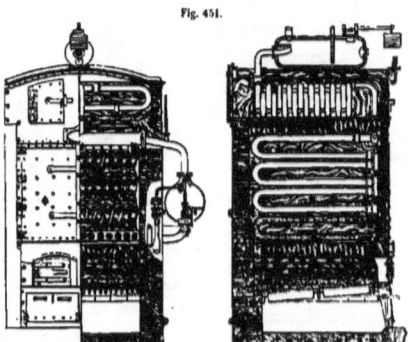

Fig. 451.

Belleville's Horizontal Syphon Tube Boiler. Patented in the year 1865.

The working of this boiler is as follows:— The water furnished by the feed pump enters the upper part of the globe by the graduated feed cock and by the valve; the globe being in communication with the upper and lower syphons. From the globe the feed water passes into the lower syphon, and rises simul- the fire, are traversed by the current of water at the least elevated temperature; they are thus not liable to injury from the fire. It is in the second and third rows of tubes that the boiling is the most active, and that the steam bubbles as they become disengaged draw with them a relatively large quantity of water;

those bubbles, to which the steam acts as a vehicle, are rapidly vaporised by contact with the tubes of the upper rows, then the whole of the steam thus formed is perhaps dried before even reaching the superheating tubes above, into which it issues by passing through the tubes in connection.

Early in the year 1867 came a Mr. Howard with his idea of a tube boiler, as shown by Fig. 452.

water in the boiler to the horizontal pipes; those pipes connect with the lower ends of the vertical tubes, and constitute with a given length of those tubes the water space of the boiler.

A Mr. Lochhead next introduced a twin tube boiler as illustrated by Fig. 453, and sufficiently explained underneath it.

The French engineers were represented next by M. Carville, whose invention is shown

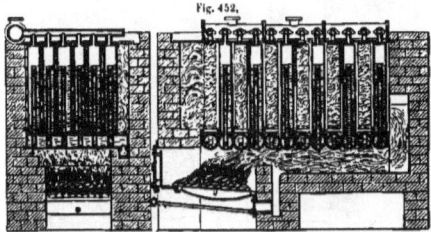

Fig. 452.

Howard's Vertical and Horizontal Tube Boiler, in which the tubes are fitted with internal circulating pipes. Patented in the year 1867.

The boiler is composed of groups of vertical tubes, which groups are formed by being connected together at top and bottom by transverse tubes or pipes, and those transverse tubes or pipes are themselves brought into connection by means of longitudinal pipes, the upper longitudinal pipe forming a steam chamber, and the lower pipe constituting a water supply pipe; and the boiler thus formed is set in a brick furnace, the roof being iron plates which cut off the upper transfer tubes and longitudinal pipe from the action of the heated gases. In order to quicken the circulation of the water in the vertical tubes there are internal tubes of peculiar construction, which form channels for conducting down the upper portion of the

on the next page by Fig. 454. The boiler is composed of a series of tubes formed by two

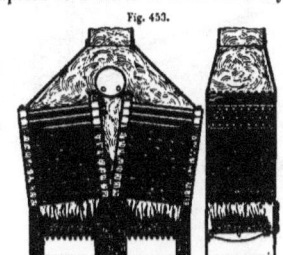

Fig. 453.

Lochhead's Angular Tube Boiler, arranged as a twin boiler with two fire grates. Patented in the year 1867.

plates of metal bolted together, and arranged vertically in connection by pipes with a horizontal cylindrical shell.

Towards the end of the year 1867, a Mr. Root, of New York, America, laid claim for English protection for the use of his tube boiler as shown by Fig. 455, which is a series of tubes angularly placed on brickwork.

connections of the tubes with each other in each vertical series are established. To establish the connection between the tubes each plate is provided with apertures, the one above the other, and around each of such

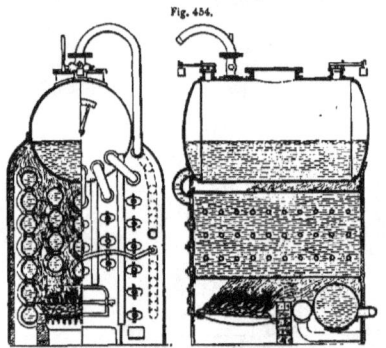

Fig. 454.

Carville's Horizontal Tube Boiler in connection with a cylindrical shell, and a water-bridge. Patented in the year 1867.

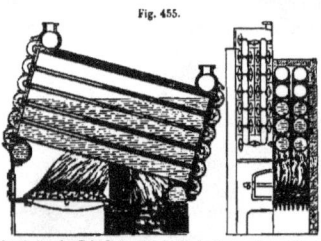

Fig. 455.

Root's Angular Tube Boiler, fitted with duplicate end connections, but the tubes situated in a vertical line over each other. Patented in the year 1869.

The mechanical arrangement consists of the tubes being screwed into independent plates of square or parallelogramic form, which when combined and in their places constitute the ends of the boiler, and through which the

apertures there is an annular socket within which may be inserted an India-rubber ring, and into these sockets the ends of return bend pipes are fitted, and thus connect the upper aperture at each end of one tube with the lower aperture at the same end of the next tube above. They are secured or held to their places by clamping bars, lapping over on the faces of lugs or projections, and fastened by nuts and stud bolts connected with the plates, the clamping bars being preferably so constructed and arranged that either one bar serves to bear on or hold two of the return bend pipes or more as required.

Cross water and steam pipes are arranged above and below the tubes at opposite ends. Those cross pipes are connected with the upper and lower rows of the tubes by bend pipes, thus connecting the nearest of the apertures of the tubes with apertures in a similar socketed manner to that used in establishing the connection of the return tube pipes, and similarly held or secured by clamping bars and stud bolts connected with the cross pipes.

The horizontal tubes are drilled and tapped, and the short vertical tubes are screwed to fit therein; the lowest row of tubes forming the group are the roof of the fire box, and the bottom row of the grate. The flame passes first back, second forward, and third back to the chimney.

Mr. Inglis followed Perkins in the same

Fig. 456.

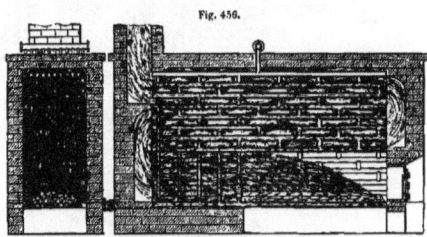

Perkins's Tube Boiler, composed entirely of small tubes connected horizontally and vertically in a group. Patented in the year 1868.

Fig. 457.

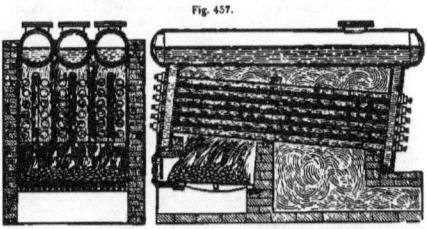

Inglis's Horizontal Tube Boiler, composed of horizontal tubes connected to narrow water chambers in connection with three egg-end cylinders forming water and steam chambers. Patented in the year 1868.

The annals of tube boilers of the year 1868 next claim notice, and the first on the list of inventors is Mr. Perkins again; with a much more grouped arrangement of horizontal and vertical tubes than before, as illustrated by Fig. 456.

The mechanical construction is as follows:

year with an arrangement of tube boiler as shown by Fig. 457, which is sufficiently explained by the description underneath the illustration.

Next a Mr. Mackie considered that tubes in the main flue beyond the fire grate and tubes beyond that in a brick chamber, and both sets of tubes in connection with an egg and boiler set in brickwork as shown by Fig. 458, was worthy of a patent for heating feed water, which really meant generating steam as well. accomplished it is shown by Fig. 459. It will be seen Mackie's pipes are outside the shell, but Loader's are inside, and there the difference ends; but we must give Mackie the

Fig. 458.

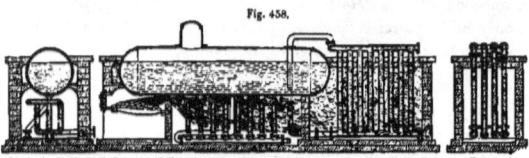

Mackie's Horizontal Egg-end Boiler, in connection with tubes for the generation of steam. Patented in the year 1868.

Fig. 459.

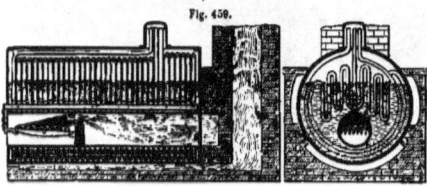

Loader's Horizontal arrangement of vertical syphon tubes fitted in the shell of a Cornish Boiler. Patented in the year 1869.

Fig. 460.

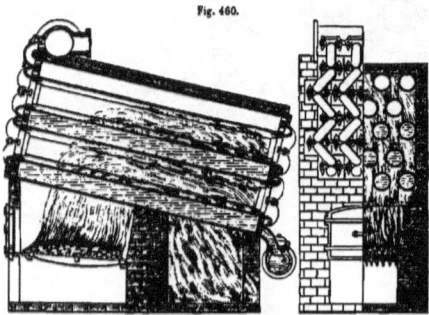

Root's Angular Tube Boiler, fitted with duplicate end connections; but the tubes situated between each other. Patented in the year 1869.

At the commencement of the year 1869, came Mr. Loader with a parody on Mackie's arrangement, and the manner in which Loader accomplished it credit of exposing his water pipes to the fire, which Loader did not.

Mr. Root next again made his appearance,

or rather his modification did, as illustrated by Fig. 460. In this case he proposed the angular tubes should be situated between each other vertically rather than in a line vertically, as shown on page 176, by Fig. 455. America was again represented, but this time by a Mr. Babbitt, as shown by Fig. 461.

Fig. 461.

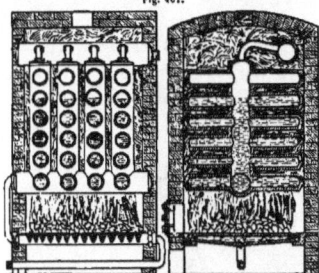

Babbitt's Vertical and Horizontal Cast Tube Boiler, with hollow cast fire grate filled with water. Patented in the year 1869.

angle branch tubes closed at their outer ends, and are relative to each other, to form tiers of tubes with free circulation for the flame between them both in vertical and horizontal directions, as also around them.

The fire grate is also cast in one piece, and is made hollow with an intermediate main bar, and hollow branch bars projecting from the main bar on either side of it: this construction allows of a free supply of water throughout the whole body of the grate, with provision for independent expansion and contraction in the direction of their length of the several bars of which it is composed, by reason of the outer ends of the bars being left free or disconnected from each other.

Water is pumped through the hollow pipe into one end of the main bar to supply the grate, and after passing through it the water is conveyed by another bent pipe from the

Fig. 462.

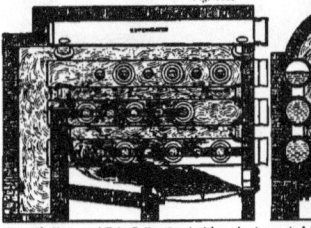

Wigzell's Horizontal Tube Boiler, fitted with projecting conical tubes closed at one end, between the two side tiers. Patented in the year 1869.

The boiler is made of cast iron, in a single casting, by making it in "flasks" arranged to form opposite halves.

The body consists of an intermediate lower horizontal tube, connected to which are four parallel vertical main tubes formed with right

opposite end of the main bar to the horizontal main tube of the boiler, thus feeding the latter with water heated by its passage through the grate.

A Mr. Wigzell's scheme next appeared as illustrated by Fig. 462, the arrangement being that a series of pipes are connected together to form a parallelogramic cage open at each end, and each horizontal pipe is fitted with conical tubes, closed at their small ends—projecting over the fire grate; this arrangement makes a good boiler for generating steam undoubtedly, but how the insides of the conical tubes are cleaned out without disconnecting them we do not know, because the inventor appeared to have ignored that important feature altogether.

boiler is illustrated by Fig. 463. In this case a group of tubes are angularly situated and connected at each end to cross tubes, that are in connection with three steam chambers, secured in the roof of the brick structure. The tubes are connected together by bolts and nuts, thus forming a ready means for disconnection.

A much more angular arrangement next was patented by a Mr. Miller, which is shown by Fig. 464. The inventor in this example

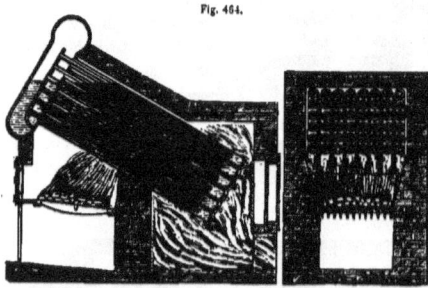

Fig. 464.

Miller's Angular Circulating Tube Boiler. Patented in the year 1869.

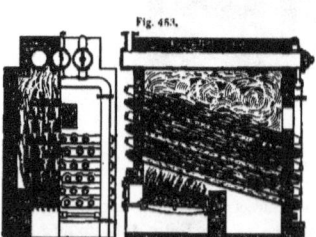

Fig. 465.

Luders' Angular Tube Boiler, composed of tubes connected at each end with cross tubes, the connection being made with bolts and nuts. Patented in the year 1869.

A Mr. Luders' arrangement of tube boiler followed the previous example, and Luders'

evidently carried out the idea that circulating tubes was the main point to accomplish, which was done by securing tubes between two plates and importing in those tubes others of a smaller diameter that are secured in a plate a little distance from the larger tube plate, the entire set being set in brickwork as shown.

Mr. Rowan's idea of what a tube boiler is, is illustrated by Fig. 465, as our next example, which consists of the arrangement of a set of vertical tubes connected at each end to four chambers, and there the matter terminates as a patent, as also, very similar, does another arrangement of Rowan's patent terminate, as shown by Fig. 466.

A Mr. Ashbury came forward next, and introduced his notions of tube boilers by three examples, as illustrated by Figs. 467 to 469.

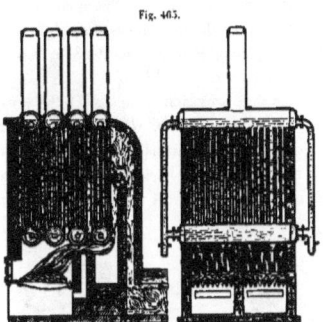

Fig. 465.

Rowan's Vertical Tubular Boiler, composed of vertical small tubes connected at each end to four horizontal chambers. Patented in the year 1865.

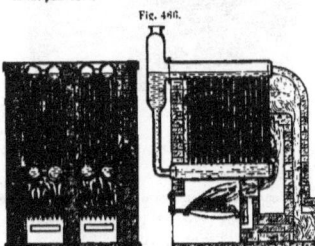

Fig. 466.

Rowan's Vertical Tube Boiler, entirely surrounded with brickwork, with the steam chamber at the front of the structure. Patented in the year 1869.

It occurs to us that Ashbury's and Root's boilers are very nearly related, if not twins in arrangement.

Following Ashbury came Mr. Howard with two arrangements of tube boilers; the first

example being illustrated by Fig. 470. The arrangement is, that the front and back ends of the boiler are composed of vertical tubes

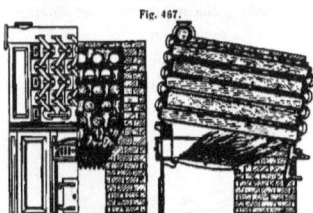

Fig. 467.

Ashbury's Tube Boiler, consisting of tubes connected at each end by small bend pipes. Patented in the year 1869.

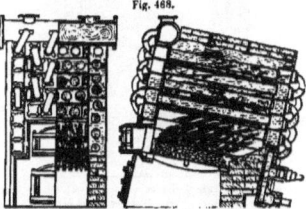

Fig. 468.

Ashbury's Tube Boiler, consisting of tubes connected at each end by large bend pipes. Patented in the year 1869.

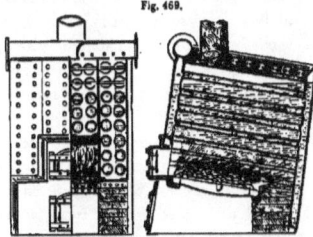

Fig. 469.

Ashbury's Tube Boiler, composed of tubes connected at each end to flat water chambers. Patented in the year 1869.

connected together by inclined water and steam tubes. The water tubes are severally fitted with an internal water circulating tube, the upper ends of which are cut off at a sharp angle. The lower circulating tubes extend nearly across the vertical tube, but the others that lie nearer the water level are shorter or so as not to extend into the tube. By this arrangement of the internal tubes they will not be required in the vertical tubes for facilitating the circulation.

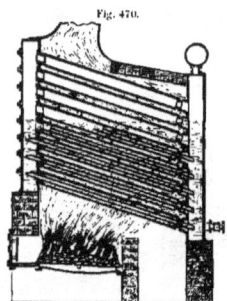

Fig. 470.

Howard's Angular Tube Boiler, in which the water cross tubes contain circulating tubes. Patented in the year 1869.

Openings are made in the front side of the vertical tubes to provide ready access to the cross tubes for cleaning them and for placing and removing the internal tubes as required; these openings are closed by screw plugs as shown.

Howard's second example is illustrated by Fig. 471. In this case the ends of the boiler are composed of sockets, which are fitted together, and thus form vertical tubes, which are closed at top and bottom by cap pieces, and the several sections of the boiler forming the back tubes are connected by the steam pipe on the top, and the front end tubes are connected by the coupling pipe at the bottom,

through which feed water is supplied to the boiler.

The short socket pieces are held together by means of tie-rods, which are composed of jointed links and pass up through the middle of the compound tubes. The ends of the jointed rods of the front tubes project through the cap pieces, and are screwed to receive tightening nuts.

A similar provision is made for the lower ends of the tie-rods of the back tubes, but their upper ends or heads rest in lugs cast with the top pieces of the back tubes. This allows of the insertion of short tubes in the top caps

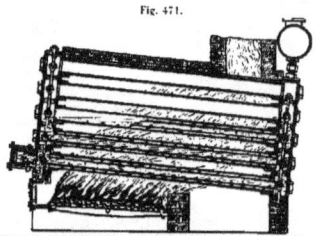

Fig. 471.

Howard's Horizontal Tube Boiler; the ends being composed of sockets connected by link tie-rods. Patented in the year 1869.

of the back tubes for connecting each section with the steam pipe.

The socket pieces are also cast with sockets to receive the ends of the cross tubes, and opposite these sockets are screwed openings closed by screws for the purpose of giving access to the joints of the tie-rods.

It will therefore be understood that by disconnecting the joints of the tie rods of any section, any selected tube may be readily withdrawn, a portion of the vertical tubes being carried away with the removed tube.

LAND STATIONARY TUBE BOILERS.

About the middle of the year 1870, Mr. Root put in appearance again with a third modification of his tube boiler, as shown by Fig. 472. In this example the cross tubes have circulating tubes fitted in them, and below that group of tubes are cross water tubes over the fire grate.

The upper cross tubes are connected at their ends by cast iron heads and bends through which the water and steam circulate.

Mr. Root not being satisfied then about those tube connections; introduced directly after, a fourth arrangement as illustrated by Fig. 473. In this instance the tubes are screwed into packing lugs formed with the ends of the boiler, and a small cover—with a rib inside

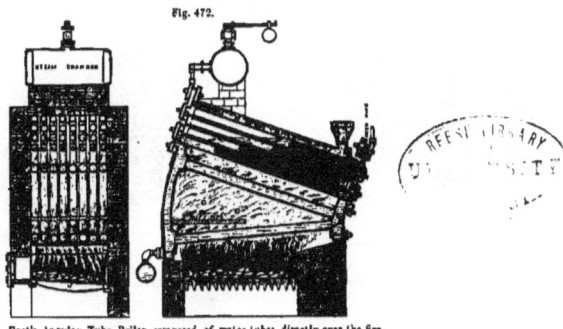

Root's Angular Tube Boiler, composed of water tubes directly over the fire grate, and above that a cluster of cross water and steam tubes, fitted with circulating tubes. Patented in the year 1870.

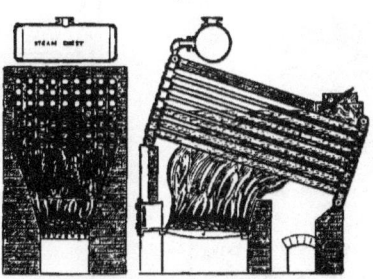

Root's Angular Tube Boiler. The main feature of which is that each end is formed in one casting with packing lugs to receive the cross tubes. Patented in the year 1870.

to assist the circulation of the water—are secured opposite each tube.

At the beginning of the year 1871, a Mr. Norton patented no less than five examples of land tube boilers; the main features being the various kinds of circulating passages at the ends of the tubes, as illustrated by Figs. 474 to 478. There is a certain amount of ingenuity displayed in those arrangements which indicate that the inventor gave great consideration to the circulation of the heat and water in the tubes.

An American engineer, named Allen, came forward next from New York with a really good boiler, composed entirely of tubes "put in the fire," as illustrated by Fig. 479. The arrangement being that a row of tubes are placed horizontally side by side; and the back ends of those tubes rest on rollers attached to the back brick wall to allow for

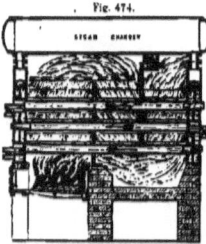

Fig. 474.

Norton's Horizontal Tube Boiler; fitted with circulating passages opposite the tubes for the water to flow as indicated by the arrows. Patented in the year 1871.

Fig. 475.

Norton's Horizontal Tube Boiler; fitted with circulating passages formed in the end portions connecting the tubes. Patented in the year 1871.

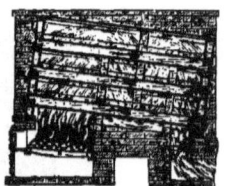

Fig. 476.

Norton's Angular Tube Boiler; fitted with circulating passages formed in the end portions connecting the tubes. Patented in the year 1871.

Fig. 477.

Norton's Angular Tube Boiler; fitted with circulating passages and pipes formed within the end portions connecting the tubes. Patented in the year 1871.

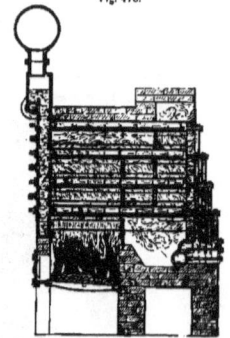

Fig. 478.

Norton's Horizontal Tube Boiler; fitted with a plain casing at the front end and circulating passages at the back end. Patented in the year 1871.

the expansion and contraction while the forward ends project partly into an opening made in the front brick wall, closed by a suitable plate to give easy access to the tubes for the purpose of cleaning them. Above these tubes and near their forward ends a cylindrical chamber is arranged crosswise, to the bottom of which the tubes are attached by means of nipples having right and left-handed threads. Small plates are riveted to the inside of the chamber for the purpose of strengthening the same where the nipples enter. Into the lower parts of the tubes pipes are securely fastened inclining backwards at about 20° inclination. These pipes are welded circular at the bottom, and each is provided with a plug screwed into the end for the purpose of cleaning the inside.

Above and near the fire these pipes are made about four inches in diameter by four feet in length, or about in that proportion of area of tubes to the surface exposed, and increase in length as the same are further away from the direct effect of the fire. By this arrangement of giving the tubes the above-mentioned inclination, and by making the same in the specified proportions a good circulation will be obtained in these pipes without the necessity of placing any internal tubes or loose plates into the same for the purpose of creating the desired circulation.

The pipes forming the last row are connected together at their lower ends by means of right and left-handed nipples screwed into suitable **T**'s fastened on the ends of said pipes, whereby a connection of the different sections is formed, the ends of said connection being connected through by the outside pipes with the bottom of the steam chamber whereby a

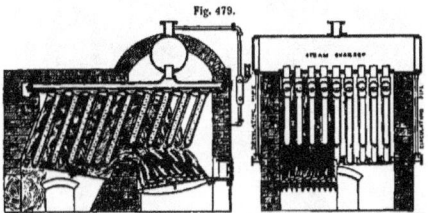

Fig. 479.

Allen's Hanging Tube Boiler, arranged over the fire grate and in the combustion chamber. Patented in the year 1871.

Fig. 480.

Howard's Horizontal Tube Boiler, in which the cross tubes are connected to the end castings by stay bolts and nuts. Patented in the year 1871.

perfect circulation between the different sections forming the steam generator is obtained.

Mr. Howard appeared next with another arrangement of tube boiler, as illustrated by Fig. 480, in which the cross tubes are secured by bolts and nuts, the bolts also acting to stay the casings that form the ends of the boiler.

Next came to light the invention of a Mr. Mirchin, in which was claimed that by arranging cross tubes at alternate angles over the fire grate and allowing *only* the straight portions of those tubes to be acted on by the flame steam would be generated; besides which, if dirty water were used, Mirchin proposed to have receptacles, in connection with the tubes, beyond the brickwork supporting them, to receive the deposit, as shown in two arrangements by Figs. 481 and 482.

The only advantage possible in those two arrangements is the circulation of the heat by the alternate angles of the pipes, but even that advantage is counteracted by the radiation due to the exposure of the ends of the pipes to the atmosphere, whereas the whole of them should be surrounded by the flame.

A Mr. Watt followed Mirchin with the arrangement of tube boiler shown by Fig. 483, the main feature of which is that the water spaces at the ends of the tubes are

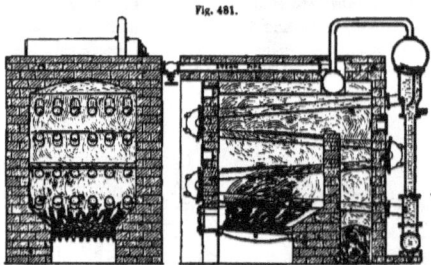

Fig. 481.

Mirchin's Return Angular or Zigzag Tube Boiler; the straight portion only of the tubes being exposed to the flame. Patented in the year 1871.

Fig. 482.

Mirchin's Zigzag Tube Boiler, in which the cross tubes are connected at each end to vertical water tubes fitted with steam chambers on their tops. Patented in the year 1871.

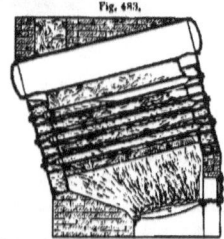

Fig. 483.

Watt's Horizontal Tube Boiler, in which the cross tubes are connected to the end casings by stay bolts and nuts. Patented in the year 1871.

stayed by the bolts that secure the cross tubes —which Mr. Howard proposed also, it will be seen, on page 185 of this work; which coincidence again proves the utility of our Patent law.

If novelty of arrangement were the main effect to be attained with tube boilers, then a Mr. Dodge might claim honours for his arrangement shown by Fig. 484, in which a cylindrical shell contains a fire box, above which are a series of short vertical tubes that

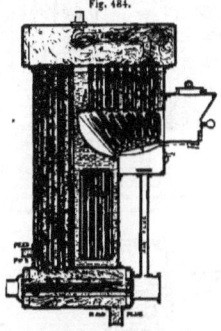

Fig. 484.

Dodge's Vertical Tube Boiler; in which the tubes are as stays throughout. Patented in the year 1871.

lead into a combustion chamber whose sides and top instead of being surrounded by water are exposed to the atmosphere, evidently to induce radiation rather than evaporation. The flame next descends through a set of long vertical tubes into a smoke box fitted with air tubes. Another peculiarity is, that at the side of the long tubes, and contained in the water, is a drum or chamber fitted with water tubes, and the exhaust steam from the engine passes amongst those tubes for the purpose—

the inventor states—of condensing it; this is the first surface condenser put in a boiler we ever heard of, and we hope it is the last also.

A Frenchman, M. Jouet Pastrés by name, next introduced his ideas of a tube boiler, in England, as illustrated by Fig. 485; the

Fig. 485.

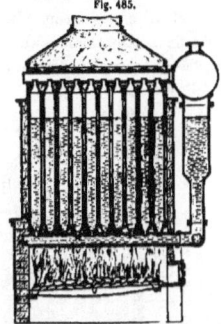

Jouet Pastry's Vertical Tube Boiler, composed of vertical tubes connected to horizontal tubes, with a water and steam chamber beyond the brickwork. Patented in the year 1871.

Fig. 486.

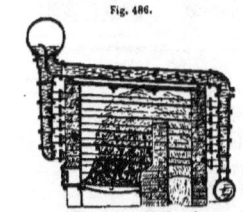

Mirchin's Horizontal Tube Boiler, with the end water casings outside the brickwork. Patented in the year 1871.

arrangement being a series of tubes in connection with an outside vertical casing or chamber beyond the fire. Bad, however, as this is, it is not so bad as Mirchin's arrangement, shown by Fig 486, because he has two casings beyond the brickwork enclosing the fire; and both those examples fully illustrate the doubt, as to whether the inventors considered that the loss of heat by radiation from the end casings must be a powerful waste of evaporation.

A Mr. Parsell next thought that by the introduction of an air or steam blast in the fire box, the flame therefrom might be, with advantage, forced through five sets of vertical tubes, as illustrated by Fig. 487, and to

Fig. 487.

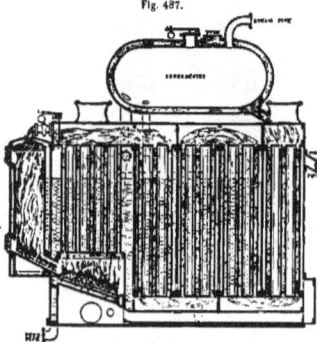

Parsell's Vertical Tube Boiler, in which the flame, accelerated by an air or steam blast, makes five up and down circuits before passing around the superheater to the chimney. Patented in the year 1871.

accomplish that two fire boxes and grates were essential.

A Mr. Westerman's ideas on tube boilers are next illustrated by Fig. 488, which is an arrangement of a group of tubes disposed at right angles in tiers—each tube being accessible by the removal of a cover secured opposite each end.

Mr. Allen, in the year 1872, again introduced his notions of tube boilers, as shown by

Fig. 489. The arrangement is that there are four vertical chambers to which are connected horizontal syphon pipes, the upper part being

Fig. 488.

Westerman's Cross Tube Boiler, forming a group of tiers of tubes at right angles throughout. Patented in the year 1871.

Fig. 489.

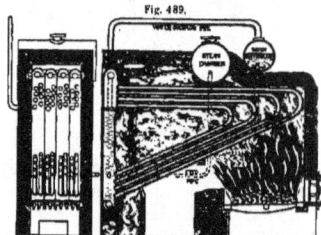

Allen's Tube Boiler, formed with syphon tubes horizontally arranged, and connected to back flat vertical chambers that are also connected to overhead-cross, steam and water chambers. Patented in the year 1872.

on the level line, and the lower part at an angle; above the pipes is a steam chamber, and near it is a water tank, and both are connected by pipes to the lower part of the vertical chambers.

The principle of this arrangement is that in modern sectional steam generators, which contain only a comparatively small quantity of water, great care is necessary to maintain a constant supply; therefore, in Allen's boiler, according to his statement, it is accomplished as follows: As soon as the water falls below the opening of the pipe that is at the water level to be always maintained in the boiler, steam will enter into this pipe pressing upon the water in the water tank, and forcing thereby the water in the same through the feed pipe into the lower part of the vertical chambers until the water rises again above the opening of the pipe, closing thereby the same, and consequently prevents the further entrance of the steam into the pipe when the pressure upon the water in the tank will diminish, and the flow of water out of the tank into the boiler will stop, until the water falls again below the water line, when the same operation will be repeated.

The feed pump if at work will replenish the water in this tank during the time the water level in the boiler is above the level line, or while the mouth of the pipe is below the water line. When the feed pump becomes deranged by accidents, or is not in operation, this tank will be capable of supplying the boiler with water for some length of time.

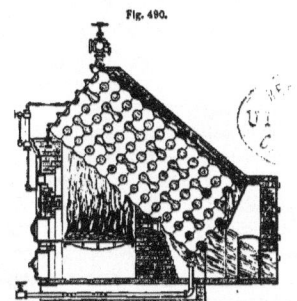

Fig. 490.

Harrison's Improved Six-Ball-Unit Cast-Iron Tube-Boiler, as constructed by him in America, in the year 1872.

Mr. Harrison also in the same year, 1872, introduced an improvement in his boiler, as illustrated by Fig. 490, which is composed of a series of duplicate portions or units, and secured together by stay bolts and nuts.

CHAPTER IV.

INJECTION BOILERS.

THE best commencement for this chapter is to explain the meaning of the term "Injection Boiler," which is a boiler containing but little more water than is requisite for the steam required at each stroke of the engine it works. The next supply of water is therefore injected into the boiler, and by its minimum quantity in proportion to the heating surface, is at once generated into steam, and thus the process of generation is repeated. With this operation, however, one great drawback has been that, as the steam is formed in jerks or flashes, similarly the engine is worked, and therefore a more steady supply is required to cause a constantly even motion.

The first practical injection boiler was invented by Mr. Jacob Perkins, an American engineer, who introduced it in England, when residing in Fleet Street, City of London, in the year 1822—as illustrated by Fig. 491. He preferred then to make the boiler of a cylindrical shape, of copper three inches thick: the principle of the invention being that, first, the water that filled the boiler was heated to a pressure of 7,500lbs. on the square inch; and, second, any more water forced in the boiler displaced an equivalent amount, which at once flashed into steam—the safety valve and weight regulating the required pressure at all times.

This was really a case where heated water generated steam, inasmuch that the water first in the boiler was made nearly red hot, and

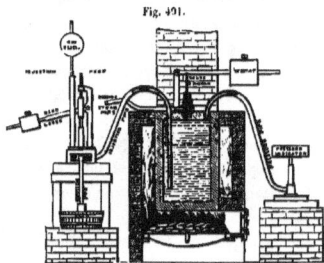

Fig. 491.

Perkins's Injection Boiler; in which, first, the boiler is filled with water; secondly, the water heated to a pressure of 7,500lbs. on the square inch; and thirdly, any more water injected, the top portion at once flashes into steam. Patented in the year 1822.

thus any more water put in with it rapidly caused an increase of pressure; the safety valve then lifted, and the amount of heated water displaced flashes into steam. We have no hesitation in stating that Perkins, considering the time he lived in, was a General in boiler warfare, for warfare it was then against the usual low pressures used in practice.

Two years elapsed, and forth came the invention of a Mr. Paul in two methods, as shown by Figs. 492 and 493. The first is a spiral coil tube boiler composed of two coils

INJECTION BOILERS.

containing the fuel between them; and the fuel is dropped through the chimney, around which is the feed injection water tank—the force pump being at the side of the shell

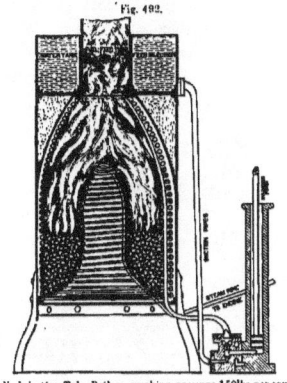

Fig. 492.

Paul's Injection Tube Boiler—working pressure 150lbs per square inch—composed of spiral coils in connection with an over head feed-water tank, and a force pump at the side below. Patented in the year 1824.

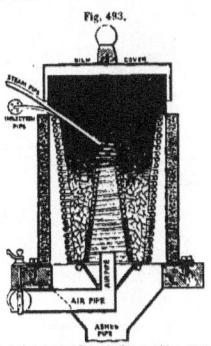

Fig. 493.

Paul's Cone Coil Injection Tube Boiler, working pressure 150lbs. per square inch, enclosed in a kiln, fed at the top, with fuel at the top, and air at the bottom. Patented in the year 1824.

enclosing the larger coil. The second method is a cone coil boiler, set in a kiln fed with fuel at the top and air at the bottom.

The year following came, and also did the invention of a Mr. Gilman, as illustrated by Fig. 494, the arrangement of which is, that there are four cylinders fitted with water

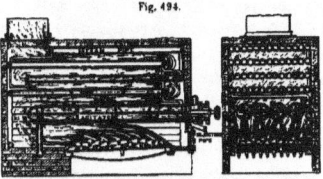

Fig. 494.

Gilman's Horizontal Injection Tube Boiler, fitted with cylinders containing water agitators over the fire grate. Patented in the year 1825.

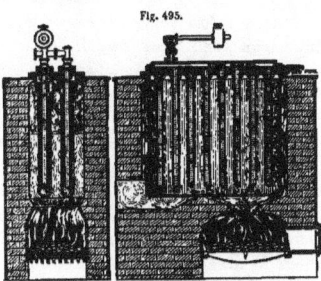

Fig. 495.

Raddatz's Molten-Boiler, containing vertical water tubes fed by injection. Patented in the year 1825.

agitators or fans, driven by bevel gearing in connection with a series of horizontal pipes above the cylinders, and another series of pipes on each side of the fire grate.

In the year 1825, the invention of a Mr. Raddatz appeared, as illustrated by Fig. 495, which possesses the peculiarity of being a molten-boiler with vertical water tubes in a cast-iron tank, containing a quantity of fusible metallic mixture of tin and lead in the proportions of one part of tin and two parts of lead. The steam-generating tubes are closed at their lower extremity and screwed at their upper ends into a brass plate, which is furnished with necks on its under surface for that purpose, while on its upper surface a groove is cut exactly over the open ends of the generating tubes. Over this plate is screwed a cover, having a corresponding groove cut in its under surface, so that when this plate is screwed down upon the under plate a hollow space is formed over the generating tubes, which constitutes a steam chamber.

The feed pipe from an ordinary forcing pump, is in a line with the grooves in the plates, with small holes bored in its under side for the emission of water, one over each of the generating tubes; then, if water be injected into the generating tubes through the small holes in the feed pipe, by means of a forcing pump, the water so injected will instantly be converted into steam, which will collect in the steam chamber, and thence escape into the steam pipe as occasion may require for the engine.

Mr. Perkins, in the year 1827, introduced another injection boiler, as shown by Fig. 496, in sectional elevation, and by Fig. 497, in complete elevation; the principles and mechanical arrangement being thus described by the inventor:—I have found by experience that to generate steam by heating water in tubes, all the parts of which are exposed to the fire, steam will become more or less surcharged with caloric, and will often show a much higher temperature for the power it is exerting than would be the case if it had had its proper proportion of water, thereby wasting much of the caloric as it passes off with the steam without a proportional power being derived, at the same time injuring parts of the engine by overheating them; to effect this important object, of ensuring to the steam

Fig. 496.

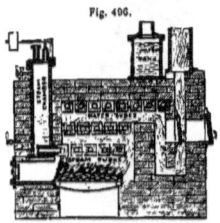

Perkins's Injection Tube Boiler, shown in sectional elevation, composed of square tubes with round holes through them, in connection with a steam chamber, force pump, and pressure valve. Patented in the year 1827.

Fig. 497.

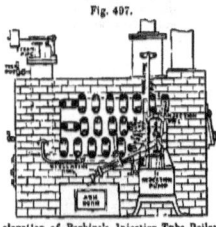

Complete elevation of Perkins's Injection Tube Boiler shown by Fig. 496.

its due proportion of water, I have caused a strong steam chamber to be fixed in some non-conducting material, having no part of it in contact with the fire. In this vessel I keep constantly more or less water. I also insert in the bottom of the steam chamber a steam tube, which is connected immediately with the last

INJECTION BOILERS.

generating tube; this steam tube enters through one side, and at the bottom of the steam chamber it runs horizontally till it comes in contact, or nearly so, with the other side; the under side of that part of the tube which is inside the steam chamber is perforated with a sufficient number of holes to allow the surcharged steam to rush downwards, and impinge on the bottom of the steam chamber; after which it rises by its own lightness through the water, taking in its ascension its due proportion of water necessary to form perfect steam, or steam of its proper density, with respect to its temperature. To ensure a sufficient supply of water to the steam chamber a small tube is connected with the pipe of the feed pump, which is attached to the bottom of the steam chamber or the steam pipe leading to it. To prevent the supply being too great, a regulating stopcock is inserted in this tube; thus, by surcharging the steam and afterwards allowing it to pass through the water which is forced into the steam chamber for that purpose, great economy is effected.

We prefer Mr. Perkins' first arrangement for its simplicity, but consider the boiler now under notice much the better for the increase of heating surface and distribution of the flame; however, in point of fact, both boilers were in advance of their time.

The next example of injection boiler that we direct attention to is of friendly interest to ourselves, because it was in connection with the first steam engine we worked, at the age of fourteen years. The engine was the "Table" kind, with a horizontal tappet motion to work the slide valve; and the boiler was then and there considered the wonder of the age; the boiler being an egg-end cylindrical casting, supported at each end, with the fire underneath it, as represented by Fig. 498.

The inventor was the son of a gentleman named Armstrong, living at Gunnislake, Cornwall; and the engine and boiler were made at Tavistock, Devon, with the most sanguine hopes from all concerned; to such an extent, however, were the hopes tortured

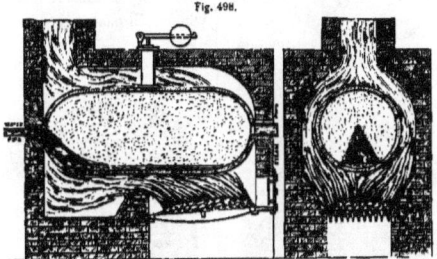

Fig. 498.

Armstrong's Egg-end, Cast Iron Injection Boiler, supported at each end on nozzles; with a fire grate at the steam end, and the flame surrounding the body of the boiler. The water entered at the left hand end and the steam escaped at the right hand end. Constructed and erected in Cornwall, in the year 1848.

INJECTION BOILERS.

that the capitalists withdrew, because the young inventor carried the matter with such a high hand; and the result was that we broke up the boiler for sale as old iron, and used the engine as required.

Nothing further worthy of record in these pages was done in injection boilers until the year 1852, when a Mr. Belleville, a French Republican, sent to England the fact that tubes, horizontal and coiled, were available for the generation of steam, as arranged according to Fig. 499; the application of the

Fig. 499.

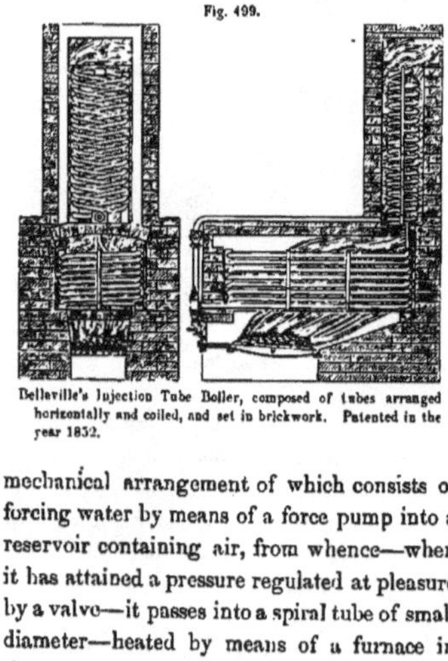

Belleville's Injection Tube Boiler, composed of tubes arranged horizontally and coiled, and set in brickwork. Patented in the year 1852.

mechanical arrangement of which consists of forcing water by means of a force pump into a reservoir containing air, from whence—when it has attained a pressure regulated at pleasure by a valve—it passes into a spiral tube of small diameter—heated by means of a furnace in the centre of the spiral at the lower part—in such a manner that one portion of the water injected becomes converted into steam by circulating in one part of the spiral, then meets a further portion of the water as it is injected at another part of the spiral, and the two portions together pass up through the upper part of the heated spiral tube, from which the steam generated is taken off at the top end.

Next came a Mr. Hyde's invention to light and knowledge, as illustrated by Fig. 500. The bottom of the boiler is dish-shaped, with a deep cover secured to it. The lower surface is covered with ribs of deep angle iron secured to the curved bottom across the fire; and the curved plate and the deep angle iron will hold a considerable quantity of heat, which will pass with certain regularity to the

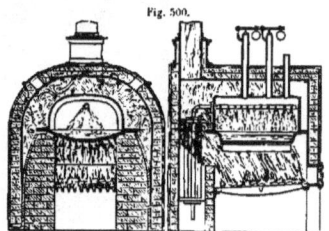

Fig. 500.

Hyde's Dish-shaped Injection Boiler, fitted with a feed water heater at the back end in the fire box. Patented in the year 1852.

inner casing. The generator is provided with a force pump of ordinary construction, capable of forcing or injecting any required quantity of water, according to the volume of steam required. The water is taken from the condenser and heated by passing through pipes; and it is then injected into the generator, and distributed over the internal surface by means of the feed pipe with numerous small holes in it. The steam so produced may be admitted into the cylinder in the ordinary way.

We need scarcely remark that this boiler

would be expensive in the consumption of fuel, besides being weak in form.

In the year 1855, Mr. Perkins, junior, brought out a very sensible arrangement of tube boiler, as illustrated by Fig. 501, for the generation of steam; and for that purpose, the circulation of water is employed to heat and evaporate successive quantities of water which are forced into one of a series of tubes —surrounding the water circulating pipe— by means of a plunger worked by steam generated in the apparatus.

This system of circulating water which is used in this invention, is that which is now well known as Perkins' system, wherein the the next, and becomes more highly heated and completely expanded into steam as it arrives in the last of the series of expanding tubes which surround the water-circulating tubes. From the last of those heating and generating tubes the steam is taken for the purposes of use, and part is employed for giving motion to a plunger or plungers, which force regulates the quantity of water required into the first of the series of heating tubes.

A French engineer named Hediard, in the year 1857, introduced in England a very complicated arrangement of tube boiler, of which it would be a charity to consider it possible of construction. The arrangement is

Fig. 501.

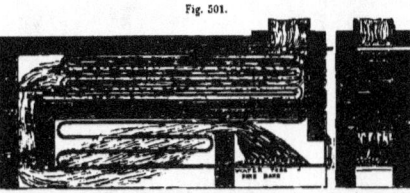

Perkins' Horizontal Tube Injection Boiler, set in brickwork. Patented in the year 1855.

circulating water is hermetically closed within tubes, leaving space only for expansion.

A quantity of the tubing containing the circulating water is subjected to the heat of a fire, by which the water contained is caused to circulate in the system of tubing from and to that portion thereof which is heated by the fire. On or around those portions of the tubing which are away from the fire are applied a certain number of tubes as steam generators, communicating in succession with each other, so that the water pumped into the first, which is on the return tube nearest the fire, becomes partly heated, passes to the next, and a series of vertical tubes in connection with a set of vertical coils with a steam chamber beyond at the chimney end, as shown by Fig. 502.

Very soon after this event, and in the same year, an American engineer, named Scott, invented and introduced a revolving tube boiler, as illustrated by Fig. 503 ; the arrangement being a tube of iron coiled in the form of a helix, each end being straight in a line with the centre of the coil. Those ends are suitably finished to constitute journals fitted in bearings mounted on the ends of the surrounding masonry, containing a furnace with its fire door, fire bars, and bridge.

INJECTION BOILERS.

The products of combustion ascend among the convolutions of the coiled tube and pass down through the flue to the chimney.

The coil is enclosed at the upper part by a semi-cylindrical double casing, which serves for heating the water preparatory to forcing it into the coil. This casing is supplied with water by a feed pump, and the heated water is drawn therefrom and forced into the coil as required.

The water enters in small jets through the meshes of one or more diaphragms of wire gauze or perforated discs in the tube leading to the end of the coil, and thence into the coil, where it is converted into steam. The pumping apparatus, which is to take the water from the casing and force it into the generator, is connected with the front end of the coil by a suitable stuffing box to make a water and steam-tight joint, and yet permit the coil to rotate.

The back end of the coil is in like manner connected by a stuffing box or steam-tight joint to the lower end of a steam drum or chamber provided with a safety valve and other usual appendages of a boiler. This end of the coil is also fitted with a toothed wheel, which is driven by a pinion on a shaft receiving motion from any suitable means which will rotate the coil at the rate of about one revolution per minute.

As the coil rotates, every part of the circumference is exposed in succession to the action of the heat in the furnace and flue, so that the heat—states the inventor—will be equalized instead of exposing some portions

Fig. 502.

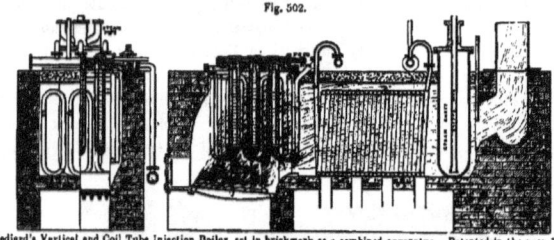

Hediard's Vertical and Coil Tube Injection Boiler, set in brickwork as a combined apparatus. Patented in the year 1857.

Fig. 503.

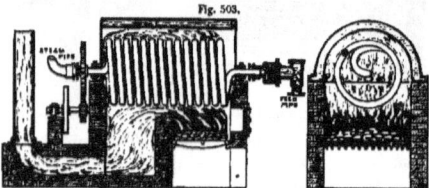

Scott's Revolving Coil Tube Injection Boiler; the water enters at the right hand end, and steam escapes at the left hand end to the engine. Patented in the year 1857.

INJECTION BOILERS. 197

to too high a degree of heat and others to a much lower temperature.

This boiler has worked well and with some advantage; but the difficulty of cleaning the inside finally led to its overthrow.

The next year, 1858, rolled on, and also enrolled the invention of a boiler by a Mr. Benson.

The boiler is composed of an inner and outer shell; the space between them serves as a water and steam room, while the chamber which the water space surrounds is occupied at the upper part by the arrangement of serpentine pipes, through which the water to be generated into steam circulates, and the lower part of this chamber is the fire box. The outer and inner shells are stay-bolted in the usual manner, and furnished with fire-door openings similar to other boiler shells of a more general structure.

The feed water is injected by a pump into the serpentine pipes, and the supply is taken from the boiler water casing enclosing those pipes, through a tank in connection directly below the injection pipe.

There is an obvious absence of means for ready cleansing or repair in this arrangement, and it is also complicated, because a tank is used beyond the boiler casing.

In the year 1859, a Mr. Carr from America sent to England his ideas about an injection boiler, believing, no doubt, that they were the best of the period; but how far he was mistaken is obvious from the illustration, Fig. 505, illustrated on the next page. The arrangement is a shell enclosing a fire chamber divided horizontally into two parts; the lower part forms the fire-box where the grate is situated, and the upper part is a steam chest, directly above which is a water chamber, and above this chamber is an air chamber.

A coil pipe is placed within the fire box, and the upper end of this coil communicates with the water chamber, and its lower end is

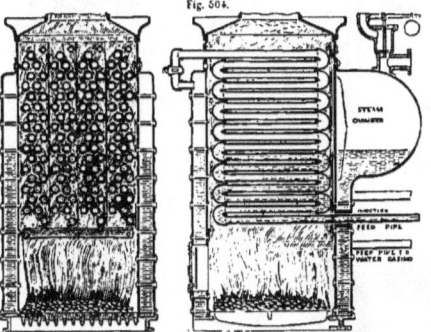

Fig. 504.

Benson's Horizontal Tube Injection Boiler, in connection with steam and water spaces.
Patented in the year 1858.

attached to a force pump. A vertical tube passes down through the air chamber into the water chamber, and has a loaded valve at its upper end as a pressure valve.

To the bottom of the steam chest there are bolted a series of inverted cone-shaped tubes, which extend down over the fire. Inside the steam chest there are placed a series of perfo-

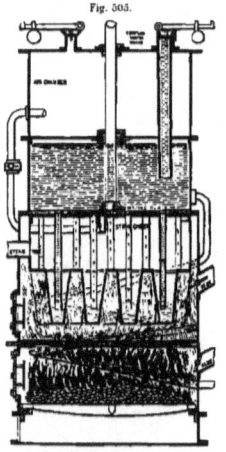

Fig. 505.

Carr's Coil Tube Injection Boiler, in connection with vertical water tubes perforated to permit the water to fall on conical tubes that are surrounded by flame, and thus steam is generated. Patented in the year 1859.

rated vertical tubes, which extend down within the cone-shaped tubes. The upper ends of the perforated tubes pass through the top plate of the steam chest, and are made flush with it, the tubes being open at their upper ends, and perforated at their lower parts.

Over the top of the steam chest is placed a circular disc, attached to the lower end of a vertical shaft, which passes up through the water and air chambers, and is connected in any suitable way to the engine, so as to cause the disc to be partially rotated or receive a vibratory motion. The disc is perforated with holes, the diameter of which is equal to that of the perforated vertical tubes, and concentric with them.

The operation is as follows:—Water is injected through the coil pipe into the water chamber, and is heated during its passage through the fire chamber. The disc, having a vibrating or partially rotating movement, is so arranged that its holes will register with the orifices of the perforated vertical tubes at the termination of each stroke, and water will be forced down the tubes, due to the pressure of the water in the water chamber, which is regulated by adjusting the weight on the lever of the valve.

The water issues in jets through the perforations in the lower parts of the tubes, is forced in spray form against the hot cone-shaped tubes, is immediately converted into steam, which fills the steam chest, and is then conveyed by the steam pipe to the engine.

In the year 1860, a Dr. Grimaldi, an Italian, introduced in England a revolving boiler very much the same as Armstrong's stationary boiler, shown by Fig. 498, on page 193 of this work. The Doctor's boiler is illustrated by Fig. 506, and is also placed in the furnace, and supported by two axles; the right hand axle is mounted on the bearing, and is provided with a pulley driven by a strap to give the generator a rotatory motion of a suitable velocity.

The other axle consists of a tubular shaft

INJECTION BOILERS.

opening into the generator, and connected with the steam tube by a stuffing box, so as to make a steam-tight joint, and yet not impede the boiler from revolving. This tube is plunger is set in motion. During the up-stroke, the top passage being open and the side passage shut, water flows into the barrel through the outer foot valve opening; and

Fig. 506.

Dr. Grimaldi's Revolving Cylindrical Injection Boiler, in connection with a superheating pot. Patented in the year 1860.

supported on the bearing, and it conveys the steam to the superheating steam chest.

Two small pipes also pass through the tube within this tubular shaft into the generator. One of those pipes is straight till it reaches the centre of the generator, where it curves a little downwards; that is the feed pipe, and is provided with a horizontal disc under its opening, in order to let the water spread in all directions. The other pipe approaches the bottom of the boiler more directly and shows the level of the water, being fitted outside with a glass gauge, two stop cocks, one above the other, and a blow-through cock.

A Frenchman named Sautter followed, the Doctor, with a coil tube boiler, as shown by Fig. 507, with an injection pump and a steam and water chamber at the side of the coil casing. The action of this apparatus is as follows:—The chamber being partly filled with water as indicated, and the coil tube being full, the fire is lighted, and at the same time, or preferably immediately before, the

during the down stroke it is sent, through the inner foot valve opening, into the lower part of the coil, displacing an equal quantity from its upper extremity through the side passage

Fig. 507.

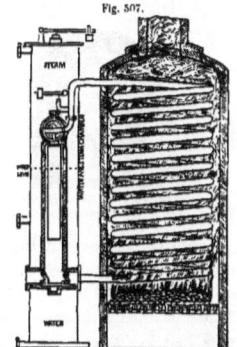

Sautter's Injection Coil Tube Boiler, fitted with a self-acting steam plunger injection pump at the side. Patented in the year 1860.

into the top of the coil. On the succeeding ascent of the plunger, the side passage is closed, and the top passage open, the water will be discharged into the steam space of the chamber, while a fresh supply is taken into the pump from below, to be in its turn sent into the heater on the down stroke of the plunger.

At the latter end of the year 1861, Dr. Grimaldi evidently considered that his first boiler

The front trunnion of the boiler projecting inside it, is fitted with six pairs of radial steam pipes; the main steam pipe tightly fitting into the trunnion, which revolves round it. One of the radial pipes communicates with the blow-off pipe, while all the remaining pipes communicate with that pipe, so that they return any water that may be thrown

Fig. 508.

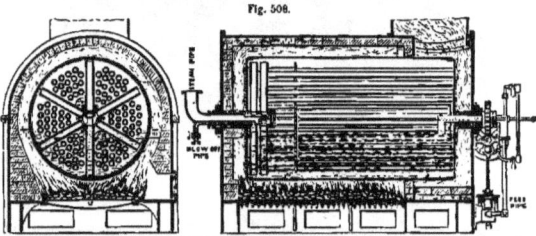

Dr. Grimaldi's Revolving Injection Cylindrical Boiler, fitted with horizontal flame and radial steam tubes. Patented in the year 1861.

Fig. 509.

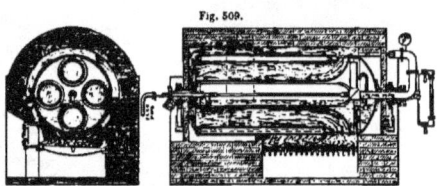

Dr. Grimaldi's Stationary Injection Boiler, comprising a cylindrical shell fitted with flame tubes, opening through the sides at the front end. Patented in the year 1861.

was not perfect: he consequently then introduced a revolving cylindrical shell tube injection boiler, as illustrated by Fig. 508, showing that the boiler is fitted with flame tubes two inches and a half in diameter; the front end of the boiler is provided with two doors for access to the radial steam pipes; and the other end is similarly provided with mudholes and doors not shown.

into them on their way up to replace the top pipes.

The back trunnion of the boiler is fitted outside the casing with a toothed wheel, by means of which the donkey engine below imparts to the boiler the rotatory motion.

The feed water pipe enters the boiler through this trunnion, which is fitted also with the water and steam pipes of the water gauge.

INJECTION BOILERS.

But the Doctor, not being contented even with that, introduced at the same time an improved stationary injection boiler, as shown by Fig. 509; in this case the shell is fitted with four large flame tubes opening through the sides at the front end. The injection water enters the shell at the back end, and is passed through the nozzle pointing downwards directly inside the shell—the steam pipe and fittings being at the opposite end.

fourths into the boiler; this flue at the parallel length of the boiler is fitted with seven radial sheet flues that conduct the flame from the flame space around the boiler to the central flue, and from there to the chimney, a portion of which is only shown. The result is, states the inventor, that the plates forming the radial flues heat the water and superheat the steam: a fact we rather doubt, because during the motion of the shell the steam and the water

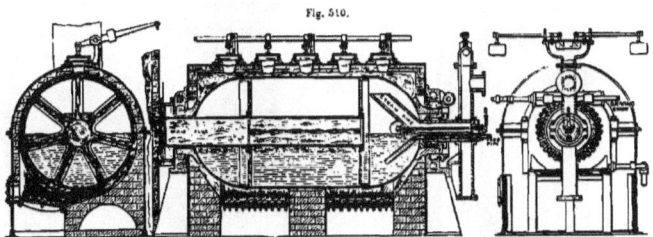

Fig. 510.

Brown's Egg-end Shell Revolving Injection Boiler, fitted with a central flue that is fitted with radial flame arms or sheet flues that conduct the flame from the fire grate to the central flue. Patented in the year 1845.

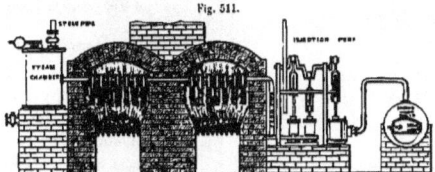

Fig. 511.

McCurdy's Flat Surface Water and Steam Casings Injection Boilers, each set of casings being suspended over the fire grate. Patented in the year 1838.

From the period 1861 until the year 1865 nothing worthy of notice here was done in injection boilers, and the first to break the silence again was a Mr. Brown, who sent from St. Petersburg the arrangement illustrated by Fig. 510, which consists of an egg-end shell having a central flue extending about three-

will be pretty well mixed together, and thus saturation instead of superheating will be mostly carried out.

The idea of generating steam by the exposure of flat surfaces to the flame was first introduced by Mr. McCurdy as far back as the year 1838, as illustrated by Fig. 511, the arrangement consists of a series of flat casings connected by pipes suspended over the fire grate with the flame surrounding them.

American injection boilers were represented also in the year 1865, by the arrangement shown by Fig. 512, which consisted of a false bottom fire grate situated over a coil of piping filled with water from the boiler, and also in connection with the steam chamber.

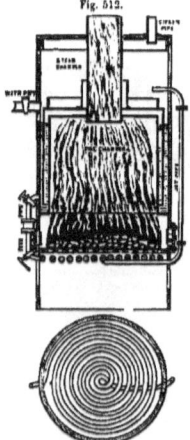

Fig. 512.

Thayer's Injection Boiler, fitted with a coil water tube fire grate. Patented in the year 1865.

The shell of the boiler is cylindrical, containing a fire box and a central chimney flue. The fire box is also formed with a false roof plate, which is attached to the upper end of a casing loosely set in an annular space around the furnace, and in this space water is conveyed through a pipe, and as fast as steam is formed it passes into the steam chamber above.

This false roof, when the feed water is properly supplied, rests at all times upon the water in the chamber between the same and the furnace roof proper, and is consequently raised or lowered according to the agitation thereof caused by boiling and the generation of steam.

Attached to the upper surface of the false roof is a circular raised lip forming a cup around the annular space enclosing the chimney, in which cup the sediments contained in the water are deposited, they being carried over from the boiler through the annular

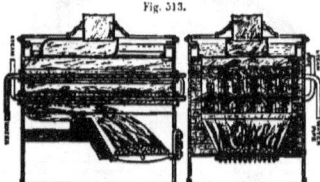

Fig. 513.

Romminger's Injection Boiler, composed of horizontal cast iron pipes lined with wrought iron tubing. Patented in the year 1865.

space by the steam in passing from the boiler, but more fully by the priming of the boiling water through the space.

A German production next appeared, by a Mr. Romminger, illustrated by Fig. 513, which consisted of two tiers of heavy cast iron pieces laid on each other, and by their shape horizontal flues were formed between them. The boiler proper was a wrought iron tube passing through each casting, and connected at each end, the water entering through the first tier, and the steam escaping from the second tier below—which is really a copy of Perkins' arrangement, shown by Fig. 496, on page 192 of this work.

INJECTION BOILERS.

A Mr. Sturgeon next introduced two spherical injection boilers, as shown by Figs. 514 and 515. In the first example the flame surrounds the body, and in the second it passes through it, and on those facts the patent rests.

It will be remembered that, on page 194 of this work, we illustrate an injection boiler by Fig. 500, which generates steam by the fall of water on the bottom of the boiler, and we now direct attention to a similar system, introduced in the year 1867, by a Mr. Gould, from America, as illustrated by Fig. 516, and explained underneath it.

A revolving injection boiler next appeared in the same year—also from America—fully illustrated by Fig. 517, and equally well explained by the description below it.

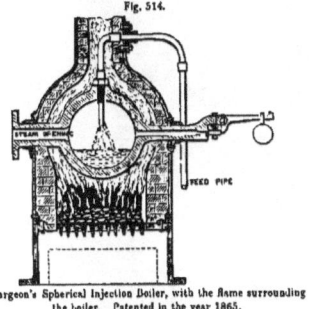

Fig. 514.

Sturgeon's Spherical Injection Boiler, with the flame surrounding the boiler. Patented in the year 1865.

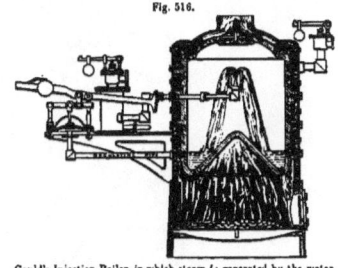

Fig. 516.

Gould's Injection Boiler, in which steam is generated by the water falling on the upper side of the bottom of the boiler, and the pressure of the steam regulates the supply of water by the lever and valve gear outside. Patented in the year 1857.

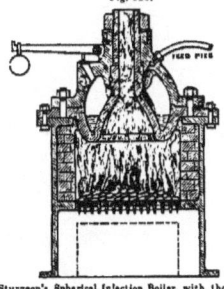

Fig. 515.

Sturgeon's Spherical Injection Boiler, with the flame passing through the boiler. Patented in the year 1865.

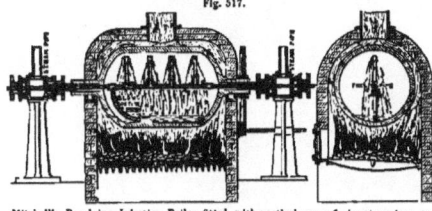

Fig. 517.

Mitchell's Revolving Injection Boiler, fitted with vertical spray feed water pipes, to cause the steam to mingle with the water as it is generated. Patented in the year 1867.

But a very much better revolving boiler is illustrated by Fig. 518, on the next page, being the invention of a Mr. Duncan, introduced by him as far back as the year 1853, and is the best example of revolving injection boilers recorded in this chapter, because the boiler proper is tubes over the fire, and those tubes are connected to hollow discs at each end;

INJECTION BOILERS.

and thereby a continual circulation of heat and water is readily effected. The water enters the front end trunnion, and the steam passes out through the opposite trunnion, the steam chamber being a cylinder centrally connecting the tube plates.

We have, however, no faith in the revolving or stationary injection boilers, for the reason that the gush of feed water is followed by a gush of steam, which, as we stated at the commencement of this chapter, produces a "jerking motion for the engine."

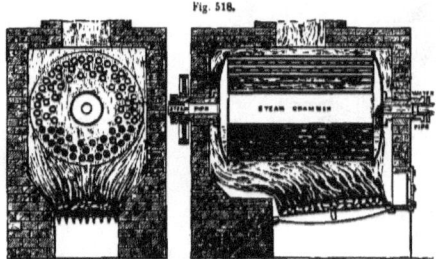

Fig. 518.

Duncan's Revolving Tube Injection Boiler, composed of two hollow discs connected by tubes, and surrounded by the flames from the fire grate situated below. Patented in the year 1853.

CHAPTER V.

MARINE BOILERS.

THE first marine boiler was, according to history, similar in shape to that shown by Fig. 9, on page 4 of this work, and from that form it grew into the square shell flue boiler. For

In the year 1827 a Mr. Steenstrup introduced a water tube boiler, as illustrated by Fig. 519, which is very ingenious and practical, and more so considering when he invented it.

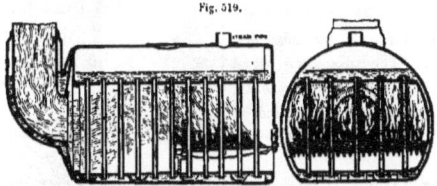

Fig. 519.

Steenstrup's Marine Boiler, being a cylindrical shell with a flat bottom, containing a fire box fitted with five rows of water tubes, and the fire grates between them. Patented in the year 1827.

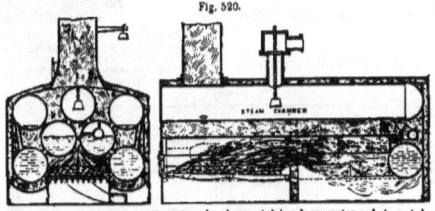

Fig. 520.

Church's Marine Boiler, being a square chamber containing large water and steam tubes. Patented in the year 1829.

the present purpose it is not requisite to illustrate examples in use before the occurrence of marked improvements, but to commence with the best boilers from that period.

The arrangement is a cylindrical shell placed on the side which is flattened, and the fire box and combustion chamber are one common chamber fitted with five rows of

MARINE BOILERS.

vertical water tubes that act as stays also, the fire grate being between them.

The next marine boiler we notice is illustrated by Fig. 520, being invented in the year 1829 by a Mr. Church, who considered that a cluster of horizontal cylinders, containing water and steam, enclosed in a chamber containing also the fire grate, would generate steam with advantage on "board ship;" which, however, we think doubtful.

A very sensible arrangement of marine boiler for high pressure steam was introduced in the year 1830 by a Mr. Summers, as shown by Fig. 521, composed of annular tubes, the water being in the annular space, and the flame passing around the larger tube and through the smaller. The tubes are grouped and connected top and bottom by horizontal tubes, as seen in the plan, and surrounded by a square chamber containing the fire grates under the tubes.

In the same year, 1830, Mr. Church came to the front again with a worse arrangement of marine boiler than before, as illustrated by Fig. 522, and sufficiently explained underneath it.

In the year 1833 Mr. Joseph Maudslay invented and patented a marine boiler, as shown by Fig. 523, the principle of which

Fig. 521.

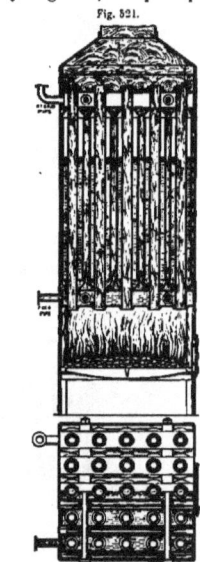

Summers' Marine Annular Vertical Tube Boiler, being a high square shell containing a group of annular tubes, with the water in the annular spaces, and the flame passing around and through the tubes. Patented in the year 1830.

Fig. 522.

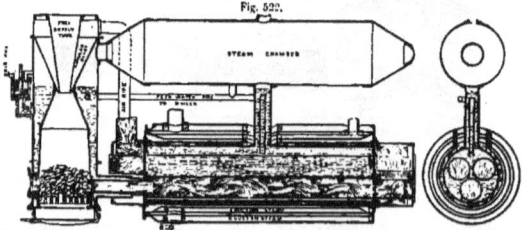

Church's Marine Boiler, being a series of cylindrical shells to form air casings around a cylinder containing water and flame tubes in connection with a separate boiler containing the fire box and chimney. Patented in the year 1830.

is that the flame heat should act on the least depth of water possible over the plate, or around it, in this case, which is effected by a series of annular spaces for the flame to pass through, thus dividing the water into cylinders enclosing the flame.

A Mr. Rickard, in the year 1835, had symptoms of the same nature for generating steam, and he thought that an arrangement of backward and forward flues, with water tubes at each end, would be an improvement, as illustrated by Fig. 524.

In the year 1838 a Mr. Price employed some of his time to prove to the world that if flame could be made to pass forward, backward, and forward again to the chimney through tubes connected from the fire box, that arrangement should be patented; which he did accordingly, as illustrated by Fig. 525, and which we think might have been left alone for the value it contains, because the flame tubes are much too large to "split" up the water sufficiently; besides the draught being blocked at each end of its circuit.

Inventors of marine boilers then halted for a time; but, in due course, forth budded a new flue marine boiler, as illustrated by Fig. 526, being the result of a Mr. White-

Fig. 523.

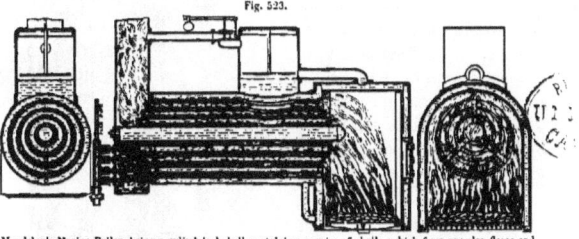

Maudslay's Marine Boiler, being a cylindrical shell containing a series of shells, which form annular flame and water spaces in connection with a deep fire box and water space. Patented in the year 1833.

Fig. 524.

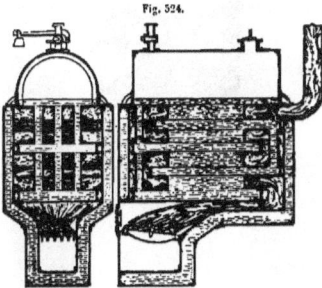

Rickard's Marine Boiler, being an arrangement of backward and forward flues, in consecutive connection over each other, surrounded by a water casing below the water level. Patented in the year 1835.

Fig. 525.

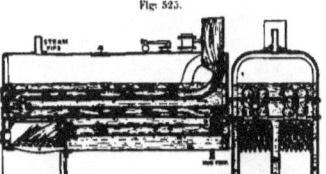

Price's Marine Boiler, being a rectangular shell containing flame tubes within the water space, to cause the flame to pass forward, backward, and forward again. Patented in the year 1838.

vertical water tubes that act as stays also, the fire grate being between them.

The next marine boiler we notice is illustrated by Fig. 520, being invented in the year 1829 by a Mr. Church, who considered that a cluster of horizontal cylinders, containing water and steam, enclosed in a chamber containing also the fire grate, would generate steam with advantage on "board ship;" which, however, we think doubtful.

A very sensible arrangement of marine boiler for high pressure steam was introduced in the year 1830 by a Mr. Summers, as shown by Fig. 521, composed of annular tubes, the water being in the annular space, and the flame passing around the larger tube and through the smaller. The tubes are grouped and connected top and bottom by horizontal tubes, as seen in the plan, and surrounded by a square chamber containing the fire grates under the tubes.

In the same year, 1830, Mr. Church came to the front again with a worse arrangement of marine boiler than before, as illustrated by Fig. 522, and sufficiently explained underneath it.

In the year 1833 Mr. Joseph Maudslay invented and patented a marine boiler, as shown by Fig. 523, the principle of which

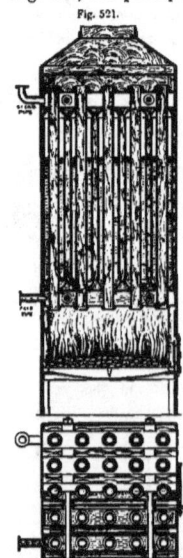

Fig. 521.

Summers' Marine Annular Vertical Tube Boiler, being a high square shell containing a group of annular tubes, with the water in the annular spaces, and the flame passing around and through the tubes. Patented in the year 1830.

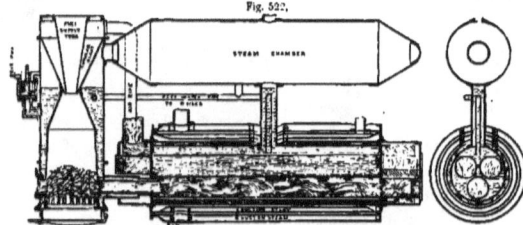

Fig. 522.

Church's Marine Boiler, being a series of cylindrical shells to form air casings around a cylinder containing water and flame tubes in connection with a separate boiler containing the fire box and chimney. Patented in the year 1830.

MARINE BOILERS.

is that the flame heat should act on the least depth of water possible over the plate, or around it, in this case, which is effected by a series of annular spaces for the flame to pass through, thus dividing the water into cylinders enclosing the flame.

A Mr. Rickard, in the year 1835, had

In the year 1838 a Mr. Price employed some of his time to prove to the world that if flame could be made to pass forward, backward, and forward again to the chimney

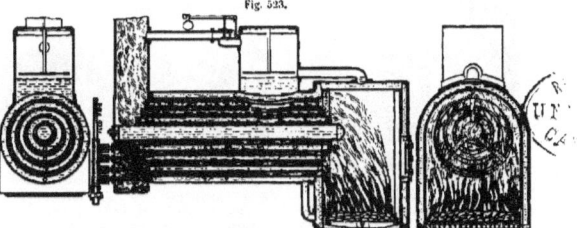

Fig. 523.

Maudslay's Marine Boiler, being a cylindrical shell containing a series of shells, which form annular flame and water spaces in connection with a deep fire box and water space. Patented in the year 1833.

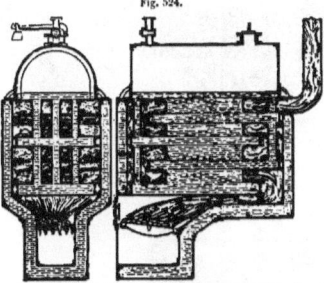

Fig. 524.

Rickard's Marine Boiler, being an arrangement of backward and forward flues, in consecutive connection over each other, surrounded by a water casing below the water level. Patented in the year 1835.

symptoms of the same nature for generating steam, and he thought that an arrangement of backward and forward flues, with water tubes at each end, would be an improvement, as illustrated by Fig. 524.

through tubes connected from the fire box, that arrangement should be patented; which he did accordingly, as illustrated by Fig. 525, and which we think might have been left alone for the value it contains, because the

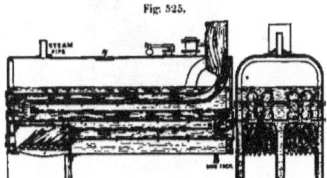

Fig. 525.

Price's Marine Boiler, being a rectangular shell containing flame tubes within the water space, to cause the flame to pass forward, backward, and forward again. Patented in the year 1838.

flame tubes are much too large to "split" up the water sufficiently; besides the draught being blocked at each end of its circuit.

Inventors of marine boilers then halted for a time; but, in due course, forth budded a new flue marine boiler, as illustrated by Fig. 526, being the result of a Mr. White-

comprised the patent, as illustrated by Figs. 535 and 536.

A Mr. Knowles also thought of subdividing

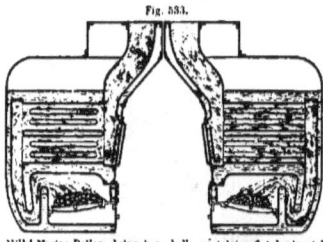

Mills' Marine Boilers, being box shells containing flat horizontal sheet flues over the fire boxes, and double water bridges at the end of the grates. Patented in the year 1851.

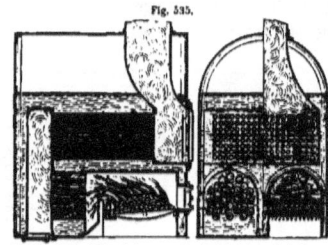

Selby's Marine Boiler, being a semi-cylindrical topped shell containing a water box fitted with flame tubes behind the fire grate, and the usual return tubes above. Patented in the year 1852.

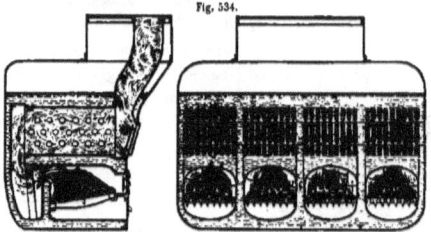

Mills' Marine Boiler, being a box shell containing vertical flat sheet flues over the fire boxes, and vertical water tubes behind the water bridge. Patented in the year 1851.

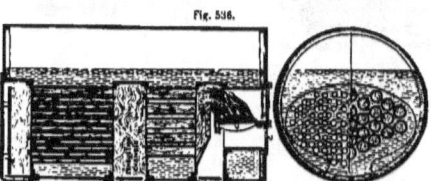

Selby's Marine Boiler, being a cylindrical shell containing an oval fire box and smoke box fitted with horizontal tubes in a line with each other, with an intermediate combustion chamber. Patented in the year 1852.

the tubes longitudinally, as illustrated by Fig. 537; but he introduced two fire boxes centrally, with a set of tubes between them—for what reason, is a mystery, because the flame has no tendency to enter those tubes—and tubes beyond each fire box in connection with

MARINE BOILERS.

outside smoke boxes that lead into a long flue under the shell; in fact, the entire affair as a boiler is so far removed from correct principles that we decline further explanation of it. vours, as Fig. 538 represents, were not of a very ambitious, but rather of a humble, nature, that relates only to fixing a water chamber fitted with flame tubes in the combustion

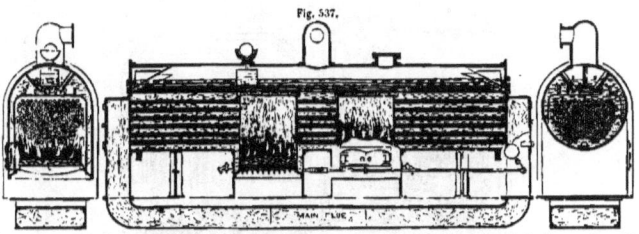

Fig. 537.

Knowles' Marine Boiler, being a cylindrical shell containing a twin arrangement of fire box, tubes, and smoke box in connection with a long flue under the boiler. Patented in the year 1852.

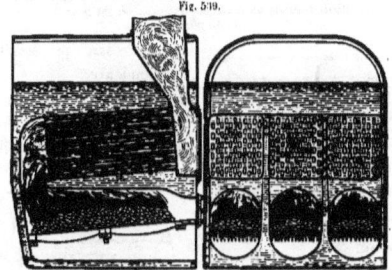

Fig. 539.

Glasson's Marine Boiler, in which horizontal oval tubes are the invention. Patented in the year 1852.

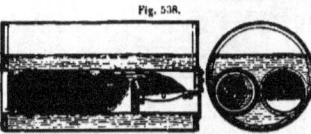

Fig. 538.

Whytehead's Marine Boiler, being a cylindrical shell containing a cylindrical water box fitted with flame tubes, fixed by short water tubes in the combustion chamber. Patented in the year 1852.

Appearing next on the scene of boiler actions was a Mr. Whytehead, whose endea-

chamber—which, by the way, Mr. Dickson did in the year 1842, as shown by Fig. 528, on page 208 of this work, and also shows the value of our patent law again.

This was followed by the introduction of oval tubes, horizontally and vertically secured, as illustrated by Figs. 539 and 540, as the result of a Mr. Glasson's thoughts.

Mr. Adamson next brought forth his boiler for public use, if the users paid him royalties, for which he permitted the arrangement shown by Fig. 541, and fully explained under it, to repay them, if possible.

That being a patent, and a very simple one too, perhaps induced Mr. Kendrick to worry himself into inventing six marine

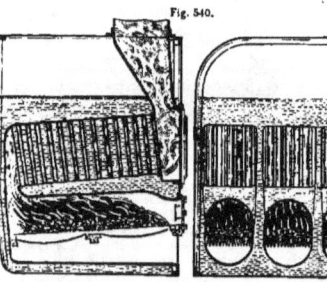

Fig. 540.

Glasson's Marine Boiler, in which vertical oval tubes are the invention. Patented in the year 1852.

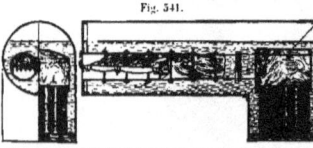

Fig. 541.

Adamson's Marine Boiler, being a cylindrical shell containing two flue tubes in connection with a combustion chamber fitted with vertical flame tubes. Patented in the year 1852.

boilers, each possessing a certain merit independently.

Fig. 542 illustrates the first example, showing an amalgamation of water spaces and tubes. Fig. 543 illustrates the second

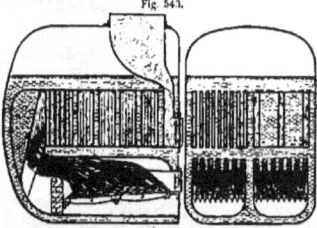

Fig. 543.

Kendrick's Marine Boiler, being a flat, curved ends shell, containing fire boxes with corrugated crowns, and the combustion chamber similarly formed at the back, in connection with sheet flues transversely and longitudinally situated over the fire boxes. Patented in the year 1853.

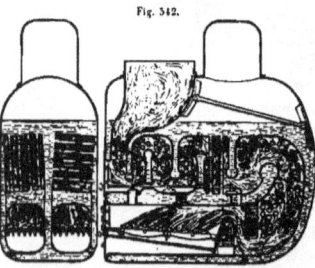

Fig. 542.

Kendrick's Marine Boiler, being a flat, curved ends shell, containing flat water spaces, horizontal and vertical water tubes in the combustion chamber, extending from the bridge to the up-take. Patented in the year 1853.

example, in which transverse and longitudinal sheet flues are the main acquirement claimed. Fig. 544, being the third arrangement, shows by the plan that only one combustion chamber is connected direct to the up-take. Fig. 545, as the fourth claim, possesses the advantage of cross sheet water spaces over the fire boxes. Fig. 546 embodies much

Fig. 544.

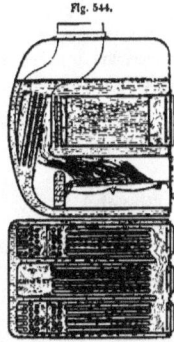

Kendrick's Marine Boiler, being a shell containing tubes in the combustion chamber, and longitudinal sheet flues over the fire boxes, the central chamber only connected to the up-take. Patented in the year 1853.

Fig. 545.

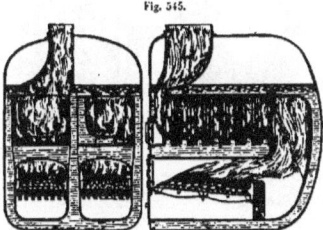

Kendrick's Marine Boiler, being a flat, curved ends shell, containing cross sheet water spaces over the fire boxes. Patented in the year 1853.

of the preceding boilers, with the addition of a return action for the flame over the fire boxes; while Fig. 547 is a vertical boiler with a zigzag combustion chamber fitted with

Fig. 546.

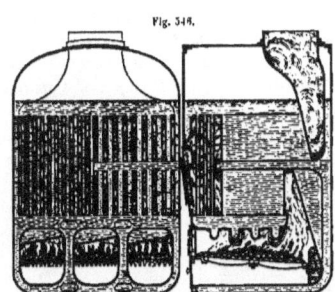

Kendrick's Marine Boiler, being a box shell containing cross water spaces in the crowns of the fire boxes, corrugations in the back of the combustion chamber, vertical sheet flues over the fire boxes, vertical water tubes in the smoke box, and a return action for the flame through the double set of sheet flues. Patented in the year 1853.

Fig. 547.

Kendrick's Vertical Marine Boiler, being a cylindrical shell containing a fire box corrugated at the back, and a zigzag combustion chamber above it fitted with vertical water tubes. Patented in the year 1853.

water tubes; and, altogether, Mr. Kendrick showed skill in boiler engineering.

A French engineer, Erard by name, next came forward with his invention of a marine boiler, as shown by Fig 548, the principle of which is that the water tubes should be as near the fire as possible, which he accomplished by putting them in the fire boxes.

A Mr. Bristow's idea next appeared, as

Fig. 548.

Erard's Marine Boiler, being a box shell containing a water tube fire grate, and water tubes and water boxes over the fire grates, so that all the tubes are in the fire boxes. Patented in the year 1853.

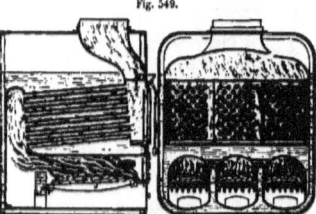

Fig. 549.

Bristow's Marine Boiler, in which the patent is the connecting of the fire boxes near the crowns by flame tubes. Patented in the year 1853.

shown by Fig. 549, being the connection of the fire boxes near the crowns by flame tubes.

The ensuing year, 1854, is remarkable from the fact that a Mr. Hyde patented the idea of putting two boilers with one set of fire grates in one shell, as illustrated by Fig. 550; but why he did so we cannot explain, and neither does he.

Mr. Henley, the well-known telegraphic engineer, invented rhombus-shaped tubes in section early in the year 1855, and arranged them for marine boilers, as illustrated by Fig. 551; and he was followed by a Mr. Lee, who ignored the fact, that to generate steam economically the flame should be as close to the water as possible, but patented an arrangement of boilers as a combination, by which

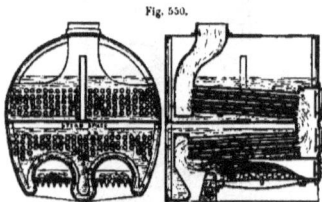

Fig. 550.

Hyde's Marine Boiler, being a curved shell containing two boilers with one set of fire grates. Patented in the year 1854.

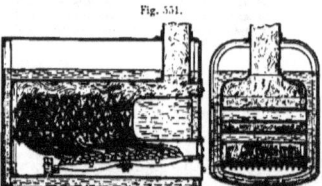

Fig. 551.

Henley's Marine Boiler, fitted with short cross rhombus-shaped water tubes. Patented in the year 1855.

the steam was first raised in an ordinary boiler, reheated in a coil tube superheater, and then used to raise steam in what he termed a "working boiler," which steam, thus generated again, worked the engine. The arrangement is shown by Fig. 552. We presume Mr. Lee thought by "burning" the steam, it would, by immediate contact, make the water boil quicker than flame acting on plates of metal containing the water; but in thinking that, Mr. Lee forgot the difference of the temperatures of the materials, as well as the loss by radiation.

In the ensuing year, 1856, Mr. Dunn issued his two marine boilers, as shown, in side sectional elevation only, by Figs. 554 and 555, the main features being that the grates

Fig. 552.

Lee's Marine Boilers, being an amalgamation of an ordinary cylindrical tubular boiler, a common egg-end working boiler, and a coil tube superheater, the water in the working boiler being heated by steam from the superheater, space in the hull in this case being of no consideration. Patented in the year 1855.

The prolonged action of flame by extending it before reaching the smoke box through a twin set of tubes, was introduced by a Mr. Pim, as shown by Fig. 531, on page 209 of this work. But that did not deter a Mr. Stevens from

Fig. 553.

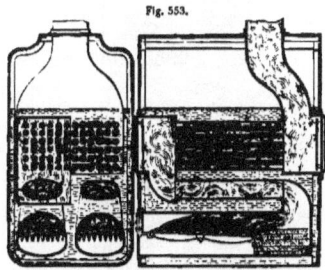

Stevens' Marine Boiler, being a square shell containing an oval flame tube over each fire box to extend the combustion chamber and to connect it with small tubes to the smoke box at the back end. Patented in the year 1855.

patenting a similar arrangement, as illustrated by Fig. 553, in which one lower flame tube to each fire box was introduced in the place of a duplicate of those above it—illustrating, also, how our patent law guardians take notice of their duties.

Fig. 554.

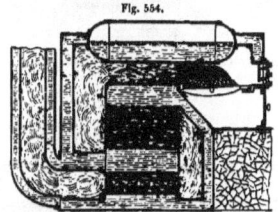

Dunn's Marine Boiler, being an arrangement of water spaces and tubes to cause a return action of the flame under the level of the fire grate, which is near the roof of the shell. Patented in the year 1856.

Fig. 555.

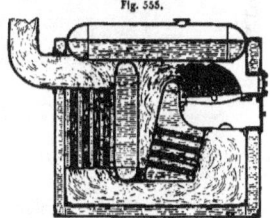

Dunn's Marine Boiler, being an arrangement of water spaces and tubes to cause a right angle action of the flame under the level of the fire grate, which is near the roof of the shell. Patented in the year 1856.

are near the roof of the shell, instead of at the base, as more general.

Mr. Galloway next sent forth his marine boiler to the world for public utility, as illustrated by Fig. 556. The patent part relates to the combustion chamber being subdivided and fitted with conical flame and water tubes.

A Mr. Morrison next introduced a marine boiler fitted with horizontal conical flame tubes

Fig. 556.

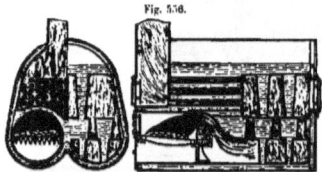

Galloway's Marine Boiler, being a curved sides shell containing vertical cone water and flame tubes in the combustion chamber in connection with horizontal flame tubes over the fire boxes. Patented in the year 1856.

Fig. 557.

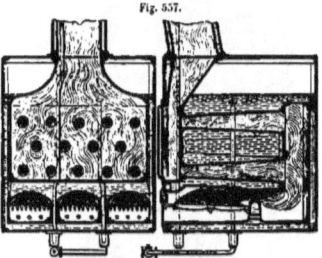

Morrison's Marine Boiler, being a square shell containing horizontal conical flues over the fire grate. Patented in the year 1857.

or flues over the fire grate, as illustrated by Fig. 557.

It occurs to us, that if Galloway's conical tubes are a patent, how can Morrison claim them also—or, rather, why does the law permit him to introduce them without reference to Galloway?

Morrison also patented conical fire boxes for vertical cylindrical marine boilers arranged in a group, as shown by Fig. 558, which is a further fact that he had every confidence in his right to the conical tube used as a flue.

A Mr. Parsons next appeared, in the year 1858, with his invention in marine boilers, as shown by Fig. 559, which consists in taking the steam from the ordinary steam

Fig. 558.

Morrison's Marine Boilers, being cylindrical vertical shells containing in each one conical fire box. Patented in the year 1857.

chest of the boiler, and causing it to pass through and across the furnace, so that it may be further heated thereby before it enters the ordinary steam pipe leading to the engine. For this purpose there is connected to the ordinary steam chest a pipe leading to another pipe which enters the back of the furnace through the bridge, a regulating valve being

MARINE BOILERS. 217

inserted at the junction of these two pipes. At the inner end of the latter or inlet pipe is connected another pipe extending along one side of the furnace from the bridge towards the front of the furnace, and there is another similar pipe on the opposite side of the furnace. These pipes are suitably supported by brackets, and have formed on their upper side sockets for the insertion of a series of arched pipes or branches extending across the fire bars. Stops are inserted in the two side-bearing pipes at the proper intervals, to cause the steam as it enters from the inlet pipe to pass across the fire by the first hollow arch, return by the second, and so on alternately, through as many arches as may be required, into the further end of the bearing pipe, and thence through the outlet pipe out of the furnace. By this continuous pipe or succession of pipes the steam is made to pass from the ordinary steam chest, enter the furnace, cross and recross the fire, and pass out from the furnace in a superheated state to be used as may be required.

With the outlet pipe is connected a pipe having a regulating valve in it leading to an additional steam chamber for receiving the highly heated steam as it leaves the furnace, the ordinary steam pipe for supplying the engine being connected with this chamber.

Mr. Rowan's marine boiler is the next that appeared, as shown by Fig. 560, the arrangement being a series of long water boxes,

Fig. 559.

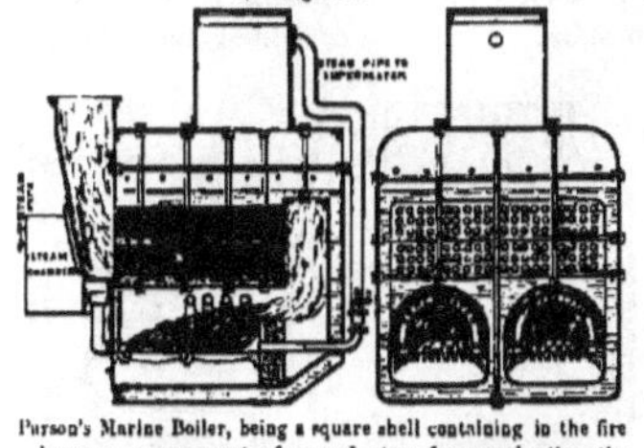

Parson's Marine Boiler, being a square shell containing in the fire boxes an arrangement of curved pipes for superheating the steam. Patented in the year 1858.

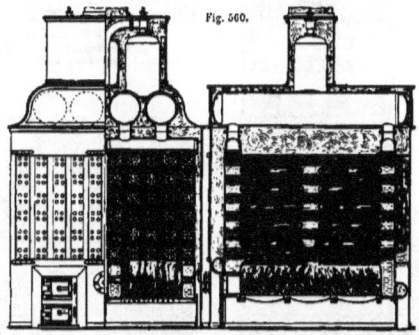

Fig. 560.

Rowan's Marine Boiler, being a square shell containing water boxes fitted with flame tubes, arranged so that the flame surrounds the boxes and passes through the tubes. Patented in the year 1858.

2 F

square in section, fitted with flame tubes—four in each box, and between each pair of boxes are transverse flame sheet flue openings, to enable the flame to make a more tortuous circuit than otherwise. The flame also passes through the tubes, so that the boxes are heated inside and out. This, doubtless, is a most expensive boiler to make and keep in repair, and we fail to see the advantage of the boxes in the place of tubes.

A Mr. Robson then made known the fact that his notion of a marine boiler was a square shell containing a serpentine flue fitted with vertical water tubes over the fire box, as illustrated by Fig. 561; but how the flue

Fig. 561.

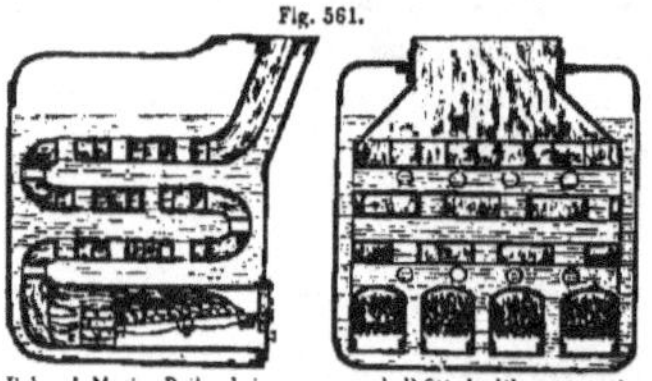

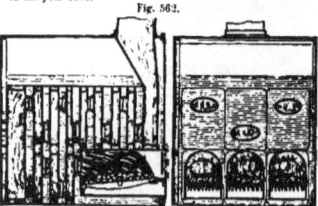

Robson's Marine Boiler, being a square shell fitted with a serpentine flue connecting the combustion chamber and up-take. Patented in the year 1858.

Fig. 562.

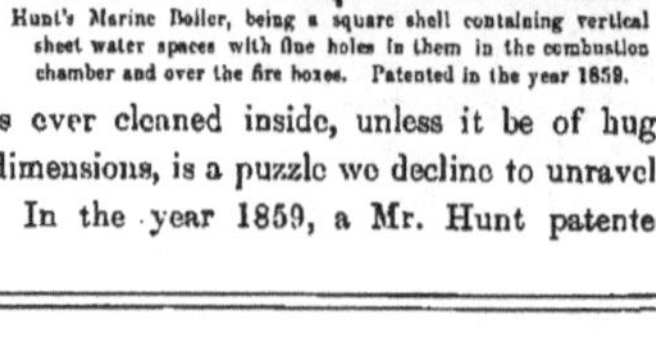

Hunt's Marine Boiler, being a square shell containing vertical sheet water spaces with flue holes in them in the combustion chamber and over the fire boxes. Patented in the year 1859.

is ever cleaned inside, unless it be of huge dimensions, is a puzzle we decline to unravel.

In the year 1859, a Mr. Hunt patented a marine boiler, as shown by Fig. 562, in which the combustion chamber and the flame chamber above the fire boxes were formed with vertical water and flame spaces; and flue openings through the water spaces at alternate positions in each direction, permitted the flame to pass tortuously, and thus gave time for evaporation. There is a certain amount of reason

Fig. 563.

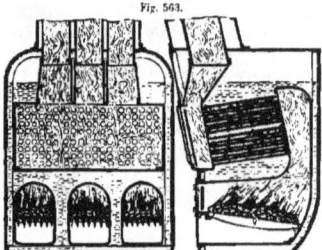

Randolph's Marine Boiler, in which the usual rake for the tubes are reversed or inclined down at the smoke box end, and a separate chimney to each fire box. Patented in the year 1859.

Fig. 564.

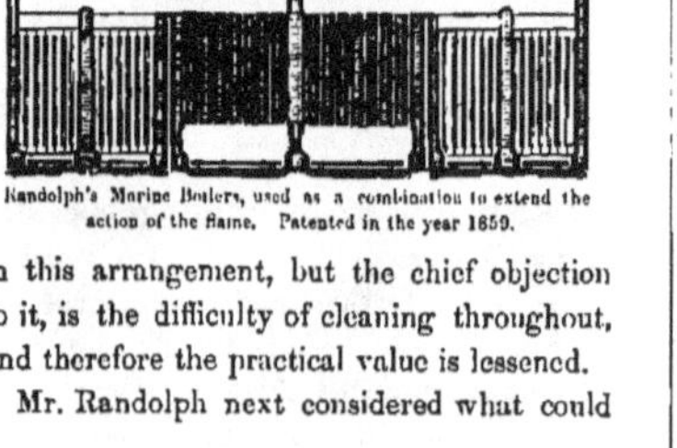

Randolph's Marine Boilers, used as a combination to extend the action of the flame. Patented in the year 1859.

in this arrangement, but the chief objection to it, is the difficulty of cleaning throughout, and therefore the practical value is lessened.

Mr. Randolph next considered what could

be improved on in marine boilers, and the result was the arrangement shown by Fig. 563, to begin with, which was extended immediately to the combined boilers, as illustrated by Fig. 564.

Following Randolph, came a Mr. Miller, who introduced three examples of marine boilers, as illustrated by Figs. 565 to 567, in which the main feature as a patent in two examples is, that the tubes are arranged at

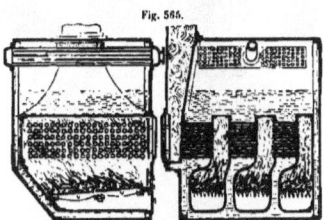

Fig. 565.

Miller's Marine Boiler, being a square shell containing three fire boxes, connected with each other directly above the fire grates by transverse flame tubes, and at the roof of the steam space is a tubular superheater. Patented in the year 1859.

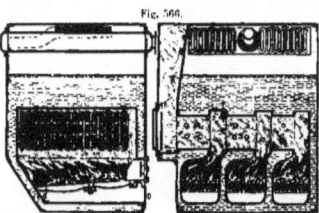

Fig. 566.

Miller's Marine Boiler, being a square shell containing three fire boxes, connected with each other directly above the fire grates by transverse sheet flues, and at the roof of the steam space is a sheet-tube superheater. Patented in the year 1859.

right angles with the fire grates, and thus connect the spaces above them. The third example has an intermediate combustion chamber between the tubes that are in a line with the fire great above it, being the same in principle as Selby's, shown by Fig. 536, on page 210 of this work.

A Mr. Tapp next appeared with the arrangement of marine boilers shown by Fig. 568, and explained underneath it.

Next came the year 1860, and so did Mr. Macnab's ideas on marine boilers, as illustrated by Fig. 569, on the next page, and sufficiently explained at the same place.

Fig. 567.

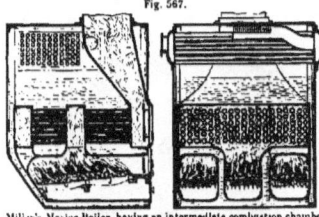

Miller's Marine Boiler, having an intermediate combustion chamber between the tubes above the fire boxes. Patented in the year 1859.

Fig. 568.

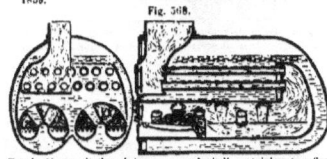

Tapp's Marine Boiler, being a curved shell containing two fire boxes, each having a curved crown, but the bottom of each raised to form a longitudinal water space connected by angular water tubes to the crown. Patented in the year 1859.

This example was soon followed by another, as shown by Fig. 570. In this case, Mr. Macnab exercised his ingenuity more by arranging his new boiler as follows:—The shell is cylindrical, with a dome top; the fire box is cylindrical also, with a central air space surrounded by a water space, which is connected by vertical tubes to the water space over the crown of the fire box. The bottom part of the central water space is also connected by outside tubes to the upper water space. The flame, after doing duty in the fire box, escapes by the central flue up-take to the chimney.

Mr. Galloway's vertical marine boiler followed Macnab's; but Galloway preferred water tubes connecting the upper water space to a shallow water chamber—which was con-

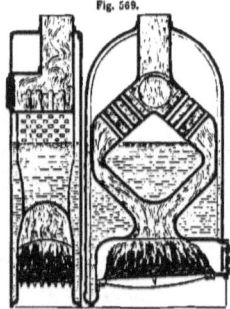

Fig. 569.

Macnab's Marine Boiler, being a vertical rectangular shell containing flues and water tubes above the fire grates. Patented in the year 1860.

patented his invention, as illustrated by Fig. 572, which consists of the fire boxes being fitted on their crowns with vertical tubes in

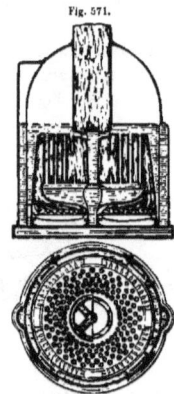

Fig. 571.

Galloway's Vertical Marine Boiler, being a cylindrical shell, with a dome top, containing in the fire box a shallow water chamber over the fire grate, connected by water tubes to the water space above. Patented in the year 1860.

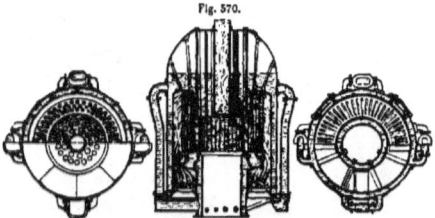

Fig. 570.

Macnab's Vertical cylindrical Marine Boiler, in which the shell contains an annular fire box fitted with vertical water tubes, and a central flue up-take; the fire box has also a central air space surrounded by a water space that is connected to the main water space above the fire box by outside pipes. Patented in the year 1860.

nected centrally by a tube also—over the fire grate, as illustrated by Fig. 571.

About the middle of the year 1861, an Irish engineer, Patrick O'Hanlon by name,

connection with horizontal tubes above, so as to divide the flame before it enters the combustion chamber.

The Hibernian was followed by a Cornish

engineer, Anthony by name, with two arrangements for extending the passage of the flame from the combustion chamber to the up-take, as illustrated by Figs. 573 and 574, and fully explained under each Fig.

Fig. 572.

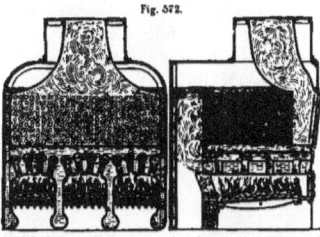

O'Hanlon's Marine Boiler, being a box shell containing horizontal flame flue tubes, connected and supported by vertical flame tubes to the roof of the fire boxes. Patented in the year 1861.

Fig. 573.

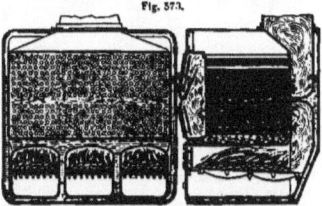

Anthony's Marine Boiler, being a box shell containing a water space in the combustion chamber between the tubes, to extend the passage of the flame to the up-take. Patented in the year 1861.

Fig. 574.

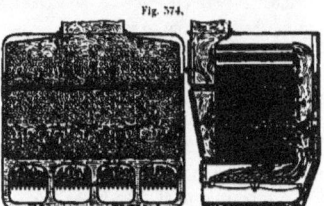

Anthony's Marine Boiler, being a box shell containing water spaces in the combustion chamber and smoke box, to extend the passage of the flame. Patented in the year 1861.

Mr. Macnab's ideas on marine boilers again appeared, as shown in sectional elevation only, by Figs. 575 and 576, the arrangement

Fig. 575.

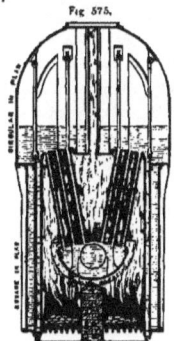

Macnab's Vertical Marine Boiler, being a square and round shell containing a square fire box fitted with a pan-shaped water pot connected by water tubes to the roof of the fire box, which is connected by flame tubes to the top of the boiler. Patented in the year 1861.

Fig. 576.

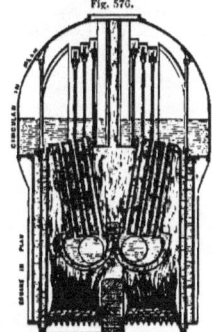

Macnab's Vertical Marine Boiler, fitted with two pan-shaped water pots over the fire grate, the remainder being as in Fig. 575. Patented in the year 1861.

being that the shell is a most peculiar shape in plan, because the lower portion is square in horizontal section, whilst the top is hemispherical or dome-shaped. A comparatively narrow water space encloses the lower square part enclosing the fire box, that has two doors in one of the sides. The fire box is crossed from front to back at a short distance above the fire bars by a horizontal pan-shaped water pot, from which tubes, with the water inside them, pass up in directions slightly inclined outwards to the nearly horizontal plate forming the roof of the fire box. The flame being made to diverge laterally by the middle water pot, pass inwards again across the tubes, and meet in a central space, from whence they pass up through a series of flue tubes that connect the fire box and dome. A man-hole is made through one of the side spaces to give access to the flame space above the water pan, to facilitate the cleaning of the tubes.

To provide for the circulation of the water, pocket-shaped casings are fixed to the sides, and communicate with the internal spaces at the top and bottom, being rivetted to the sides throughout their length. To enable the boiler to sustain the internal steam pressure, vertical stay rods are connected from the top of the dome to the middle parts of the side plates of the lower square portion, whilst the corners of the latter are continued up within the dome, being rivetted thereto at their edges, whereby a very strong connection is obtained. Stay rods also pass from the top of the dome to the crown or roof of the furnace space, and a few of these stays pass down through the innermost tubes to the lower pan space. The narrow water spaces forming the square casing are also strengthened by numerous cross stays, and the middle pan space is strengthened by vertical stays and by horizontal stays extending the whole way of it.

In the beginning of the year 1862, a Mr. Howden petitioned for the protection of his invention in marine boilers, as illustrated by Fig. 577, which was granted, because the arrangement consisted of three chambers fitted with water tubes; two chambers are vertically secured, and the third horizontally between them. There are also three cylindrical chambers secured above. The action of the flame is to rise amongst the upper tubes and then pass down amongst the side tubes to the main

Fig. 577.

Howden's Marine Boiler, being two vertical chambers situated on each side of a narrow horizontal chamber, which are fitted with water tubes connecting said chambers that are attached to cylindrical steam chambers situated above. Patented in the year 1862.

flue under the ash pit. Where the advantage is in this boiler we fail to see.

Directly after Howden, a Mr. Meriton appeared with two arrangements of vertical water and flame tubes fitted in the combustion chamber, as illustrated by Figs. 578 and 579.

In the year 1863, Mr. Howden put in appearance again as the inventor and patentee of the arrangement of tube boiler shown by Fig. 580, which is a decided improvement over his previous example; the present one being an arrangement of tubes secured horizontally in a casing forming a water chamber enclosing the tubes at the ends and sides in connection with a cylindrical water and steam chamber on the top.

England, to be patented, an arrangement of tubes in marine boilers, as illustrated by Fig. 581, by which it is shown that—according to

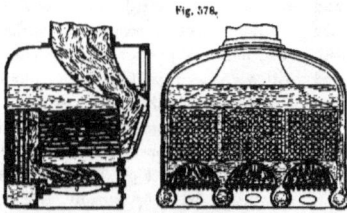

Meriton's Marine Boiler, in which the water spaces are omitted under the fire grates, and the combustion chamber fitted with vertical water tubes. Patented in the year 1862.

Meriton's Marine Boiler, in which the combustion chamber is fitted with vertical flame tubes. Patented in the year 1862.

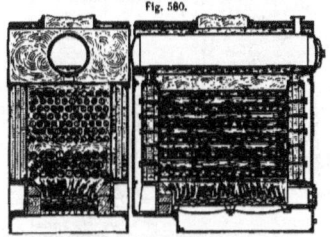

Howden's Marine Tube Boiler, composed of horizontal tubes connecting flat water chambers. Patented in the year 1863.

the inventor—the tubes are arranged in direct vertical and horizontal rows, and the vertical water spaces between the tubes gradually increase in depth from the bottom side to the top side of the tubes. The object of this is to allow the steam which is generated below at the lowest row of tubes; and upon the surface of each successive tube above, and which adds volume to the upper current of steam and water which is constantly flowing upwards at these parts when the boiler is in operation; to

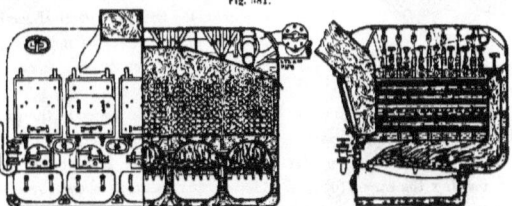

Stimers' Marine Boiler, in which the tubes are of unequal diameters, and cause unequal vertical spaces between them. Patented in the year 1864.

Towards the latter end of the year 1864, an American engineer, a Mr. Stimers, sent to

have a constantly increasing space, corresponding, if not exactly, at least in part with the increasing volume. The results of this are to permit a more quiet disengagement of the steam from the water as it comes to the surface, thereby preventing priming, and supplying a larger proportion of water to the upper tubes than would be in contact with them in the ordinary tubular boiler, and generating a proportionally larger quantity of steam.

Fig. 582.

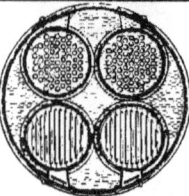

Adamson's Marine Boiler, being a cylindrical shell containing four cylindrical fire boxes with vertical tubes over them. Patented in the year 1866.

This increase in the breadth of the water spaces between the tubes at the upper part is obtained by placing the tubes in regular vertical rows, and making the upper tubes of less diameter than the lower.

In boilers that have been constructed according to this invention each tube is made one-eighth of an inch smaller in diameter than the one next below it. This diminution of the diameters of the upper tubes, without increasing their number, causes a greater quantity of the heated gases to pass through the lower tubes as compared with the upper ones, and this greater quantity also passes with a greater rapidity because the tubes are of greater area with the same length.

Advancing now to the year 1866, we notice, first, that Mr. Adamson patented the arrangement shown by Fig. 582, and explained underneath it; and, secondly, that Captain Hall, R.N., next laid claim for protection as

Fig. 583.

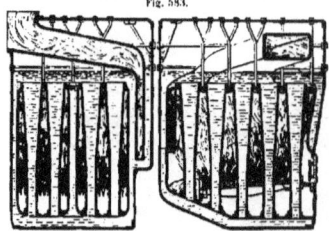

Hall's Marine Boiler, in which the fire box is fitted with vertical conical water tubes with the fire grates between them, and a cross flue up-take in the steam space. Patented in the year 1866.

a patent, the fixing of vertical conical water tubes in the fire boxes of marine boilers, and cross flue up-takes in the steam spaces, as illustrated by Fig. 583.

Now, if there is justice in the protection of the vertical tube arrangement in this case, how can the same law protect Steenstrup's tube arrangement, as shown by Fig. 519, on page 205 of this work?

Another naval captain's ideas next appeared, as illustrated by Fig. 584, relating to the fixing of steam pipes in the combustion

MARINE BOILERS.

chambers of marine boilers for the purpose of, by jets of steam, accelerating the draught.

Directly after that, Mr. Holt's improvements made their appearance, which particularly referred to the combustion chamber being extended in the form of flues over the fire boxes, which were formed with deep and narrow vertical flame sheet spaces in the

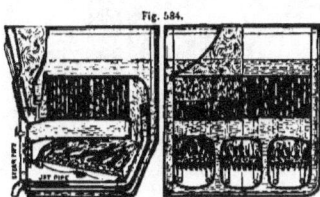

Fig. 584.

Cochrane's Marine Boiler, in which the combustion chamber is fitted with a steam pipe for the purpose of, by jets of steam, accelerating the draught. Patented in the year 1866.

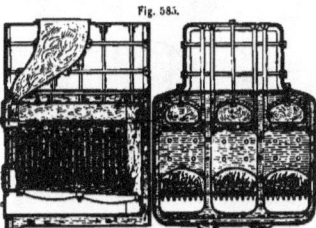

Fig. 585.

Holt's Marine Boiler, being a box shell containing fire boxes fitted with deep and narrow sheet water and flame spaces in the crowns, and combustion chamber flues above them. Patented in the year 1866.

crowns, to increase the heating surfaces. The illustration, Fig. 585, is an example of this class; and Fig. 586 illustrates another boiler, in which the combustion flues are similarly formed as the fire boxes. Now, if we turn to page 210 of this work, we shall see how far Mr. Holt was anticipated.

Two arrangements of marine boilers, as shown by Figs. 587 and 588, and explained

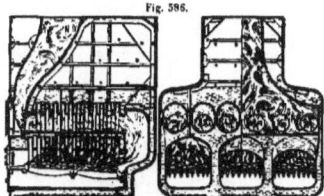

Fig. 586.

Holt's Marine Boiler, in which the fire boxes and combustion flues are corrugated. Patented in the year 1866.

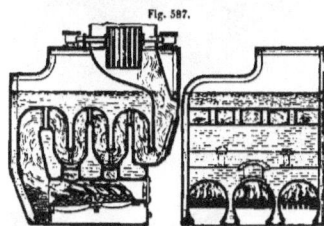

Fig. 587.

Lewis's Marine Boiler, being a curved shell containing an extended combustion chamber formed with a wide flue and narrow water spaces over the fire boxes, with vertical flame tubes connecting the central fire box; connecting vertical water tubes in the flue, and a superheater in the up-take. Patented in the year 1866.

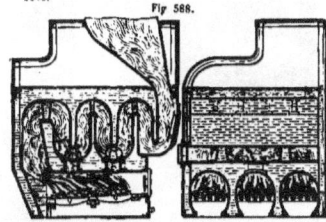

Fig. 588.

Lewis's Marine Boiler, fitted with side door openings in the main flue, and no superheater as in Fig. 587. Patented in the year 1866.

below them, were next introduced by a Mr. Lewis, as shown above.

In the year 1867, a Mr. Storey patented a very peculiar arrangement of marine boiler, as illustrated by Fig. 589, which consists of a cylindrical shell with a dome surrounded by a flame casing that is surrounded also by a

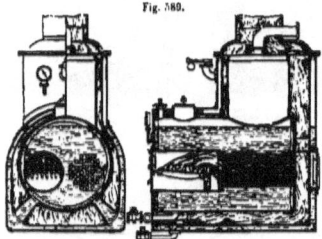

Fig. 589.

Storey's Marine Boiler, being a cylindrical shell with a dome and two fire boxes and tubes enclosed by a flame space that is surrounded by a water space casing. Patented in the year 1867.

have been limited, supposing it was ever constructed and fitted in a ship.

A Mr. Mace appeared next as a boiler

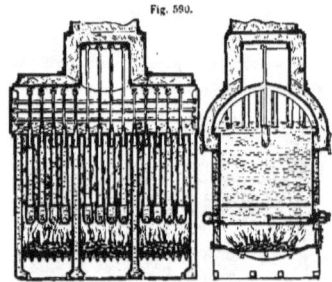

Fig. 590.

Mace's Marine Boiler, being a set of sheet water spaces connected to a semi-cylindrical steam chamber fitted with a central dome. Patented in the year 1867.

Fig. 591.

Dunn's Marine Boiler, containing a down-take combustion chamber fitted with horizontal flame tubes below the fire box, the crown of which is corrugated. Patented in the year 1867.

Fig. 592.

Dunn's Marine Boiler, containing a down-take combustion chamber fitted with a flue and sheet water spaces at the end and below the fire grate. Patented in the year 1867.

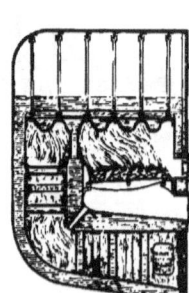

Fig. 593.

Dunn's Marine Boiler, containing a down-take combustion chamber fitted with horizontal and vertical water tubes. Patented in the year 1867.

water casing which contains feed water; this combination forms a complicated arrangement, so much so, indeed, that its use must improver, with his arrangement, as shown by Fig. 590, which consists of a set of sheet water spaces—properly stayed, but not illustrated as such—hung over the fire grate from a semi-circular steam chamber that is fitted with a central dome.

The flame, after passing between the spaces, surrounds the upper part of the boiler, and then proceeds up the chimney. The principle of this arrangement is that the water and flame spaces are equally divided, and thereby the steam is generated.

Mr. Dunn's three arrangements of downtake combustion chambers, fitted with tubes and water spaces, next appeared, as illustrated by Figs. 591 to 593.

The main feature in those arrangements is, that the stoking foot plates are much higher from the keel of the ship than in ordinary arrangements of marine boilers; the advantage of which is, that the fires will not be as soon quenched by water flooding the stoking room when the ship is "under steam."

Mr. Holt's ideas being extensive, he next brought forth three arrangements of sheet flues for marine boilers, as shown by Figs. 594 to 596, and explained beneath them.

Mr. Kendrick next put in an appearance again, and this time, to be more experimental, he introduced water spaces in the combustion chamber and also over the fire boxes, together with tubes, in marine boilers, as illustrated by Fig. 597 on the next page.

The ensuing year, 1868, commenced, and shortly after a Mr. Allibon endeavoured to show the world his abilities as a boiler inventor, according to the arrangement illustrated by Fig. 598, which is a cylindrical shell containing a water-spaced fire box surrounded by a flame space; the box is fitted directly over the fire grate, with a deep water pot that is connected by horizontal and vertical tubes to the water space enclosing the fire box; the only worthy feature in this boiler is, that it

Fig. 594

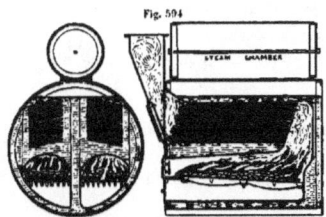

Holt's Marine Boiler, being a cylindrical shell fitted with angular sheet flues, situated horizontally over the fire boxes, with angular tube plates. Patented in the year 1867.

Fig. 595.

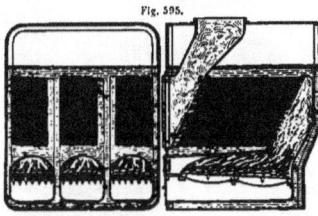

Holt's Marine Boiler, being a square shell containing sheet flues, situated horizontally over the fire boxes, with angular tube plates. Patented in the year 1867.

Fig. 596.

Holt's Marine Boiler, being a box shell containing vertical sheet flues, situated at an angle over the fire boxes, with angular flue plates. Patented in the year 1867.

can be easily constructed, but its economical working after is rather doubtful.

MARINE BOILERS.

A Mr. Whittle's marine boiler next appeared, as illustrated by Fig. 599; the shape of the shell and the arrangement of the internal parts are as general, but the patent consists of fixing plates at the sides of the fire boxes and extending them above the tubes or sheet flues, for the purpose of increasing the circulation of the water and heat to generate the steam quicker.

A French engineer, named Bèzy, next brought forward an ordinary shell fitted with horizontal water tubes over the fire box for a patent, as illustrated by Fig. 600, but water tubes had been patented in the year 1852, as shown by Fig. 540, on p. 212 of this work.

It will be apparent that the previous examples of marine boilers we have illustrated in this chapter are compound in form internally, but we introduce to notice next a real homogeneous-formed boiler, invented and patented by Mr. Hawthorn, as illustrated by Fig. 601.

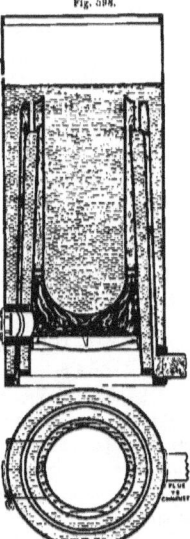

Fig. 598.

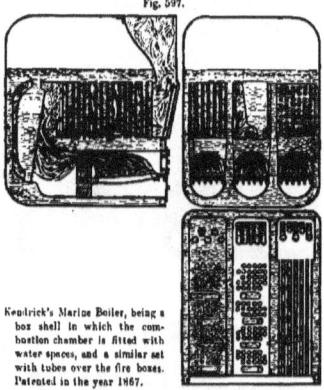

Fig. 597.

Kendrick's Marine Boiler, being a box shell in which the combustion chamber is fitted with water spaces, and a similar set with tubes over the fire boxes. Patented in the year 1867.

Allibon's Vertical Marine Boiler, being a cylindrical shell containing a cylindrical fire box having a water "pot" over the fire grate and a flame space around the fire box. Patented in the year 1868.

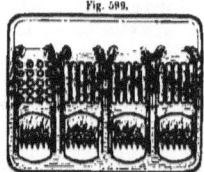

Fig. 599.

Whittle's Marine Boiler, fitted with circulating plates at the sides of the fire boxes, and extending them above the tubes. Patented in the year 1868.

The shell is cylindrical and contains two internal fire boxes, as shown, fitted with bars and fire bridges. Both those boxes communicate at the back end with a flame chamber, subdivided by the fire-brick wall situated

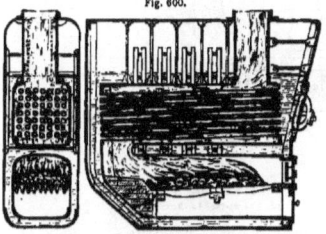

Fig. 600.

Bezy's Marine Boiler, fitted with horizontal water tubes over the fire box. Patented in the year 1868.

boiler, and pass at the back end into a back chamber communicating with the two smaller return flues, situated under the fire boxes. Those two flues cause the flame to traverse a third time the length of the boiler, and open at their front ends into the up-take there situated.

The ash pits are formed at the sides of the boiler with cylindrical air pipes, communicating with the bottom of the furnaces, and the entire set of water tubes fitted are cylindrical also.

The next example, shown by Fig. 602—invented and patented by a Mr. Mordue—is, in the main, cylindrical also; the arrange-

Fig. 601.

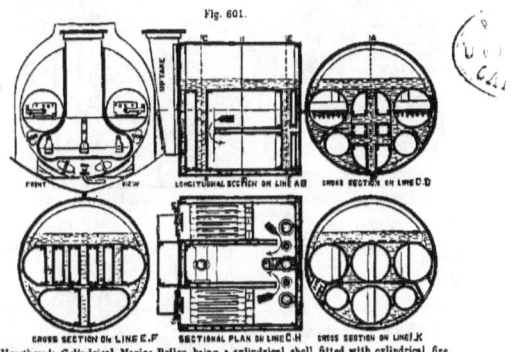

Hawthorn's Cylindrical Marine Boiler, being a cylindrical shell fitted with cylindrical fire boxes, combustion chambers, return flues, and water tubes. Patented in the year 1868.

opposite the centre line of the middle flue, which is on the same horizontal line as the centre of the boiler and of the two fire boxes, and returns towards the front of the boiler, where it communicates with the front flame chamber that opens into the bottom or second flue, that causes the flame to again traverse the greater portion of the length of the

ment being that four cylindrical shells contain in each a fire box of a smaller diameter that is connected to a combustion chamber centrally situated; and that each shell is attached by short vertical tubes to a large cylindrical main shell that contains the combustion chamber, and that two sets of small flame tubes lead to an outside smoke box at each end that is connected by a horizontal cylindrical flue—inside the main shell—to the chimney, which is situated at the centre of the main shell on the top.

In the year 1869, Mr. Howard introduced a marine tube boiler, as illustrated by Fig. 603, the arrangement of which is that horizontal tubes are connected to a flat water chamber, and, to promote a rapid circulation of heat and water, each water tube is fitted with an internal tube, and the steam tubes, situated above, are fitted with longitudinal plates connected to the chamber; this structure is mounted by cylindrical steam chambers and a chimney in the centre of their positions.

After Howard, came a Mr. Crawford with an arrangement of marine boiler, as shown by Fig. 604, which is a cylindrical shell fitted

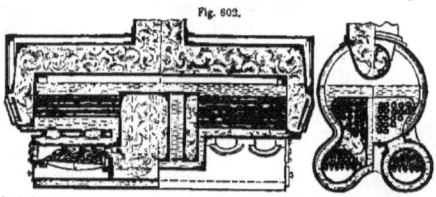

Fig. 602.

Mordue's Combined Marine Boiler, consisting of four fire boxes in connection with one combustion chamber that is situated between two sets of small flame tubes that lead to the smoke box at each end and that are connected by an internal flue up-take to the chimney. Patented in the year 1868.

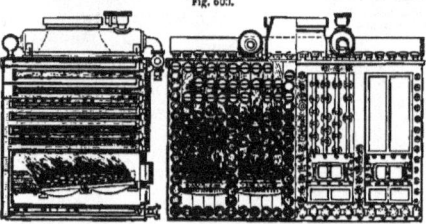

Fig. 603.

Howard's Marine Tube Boiler, composed of horizontal tubes connected to flat water chambers. Patented in the year 1869.

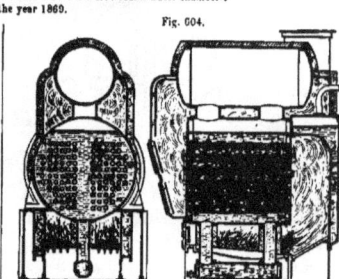

Fig. 604.

Crawford's Marine Boiler, being a cylindrical shell fitted centrally with flame tubes, and under with water "legs," between the fire grates, with a mud pocket. Patented in the year 1869.

with flame tubes, and connected to the under part of the shell are water "legs" that divide the grate, under which is a mud pocket connected by the legs to the shell.

Crawford also introduced a duplicate arrangement with one combustion chamber, as shown by Fig. 605.

A Mr. Bennett next introduced his arrangements of marine boilers, as illustrated by

Fig. 605.

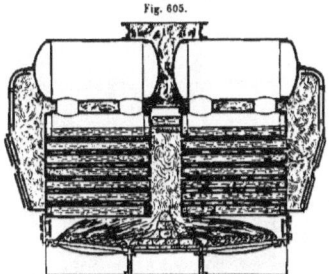

Crawford's Marine Boilers, being a twin arrangement with one combustion chamber. Patented in the year 1869.

Fig. 606.

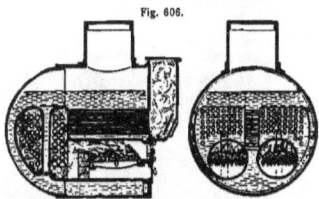

Bennett's Marine Boiler, being a curved-end cylindrical shell containing a combustion chamber fitted with vertical water tubes and a central water space fitted with cross flame tubes. Patented in the year 1869.

Figs. 606 and 607, and explained below them.

An arrangement of flame boxes and short tubes through water spaces was next brought forth by a Mr. Miller, as shown by Fig. 608, to obstruct the flame during its passage from the grate to the up-take, which also did a Mr. Fraser aim at doing, as shown by Fig. 609, and explained under it.

Fig. 607.

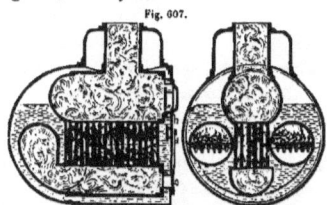

Bennett's Marine Boiler, being a curved-end cylindrical shell containing a flue combustion chamber below the fire boxes fitted with vertical flame tubes and an internal up-take. Patented in the year 1869.

Fig. 608.

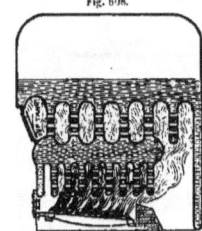

Miller's Marine Boiler, in which the crown of the fire box and the flame passage above it are fitted with water spaces, flame boxes, and flame tubes. Patented in the year 1869.

Fig. 609.

Fraser's Marine Boiler, in which the vertical water tubes are corrugated. Patented in the year 1869.

Passing to the next burst of talent, we have only one example of marine boiler worthy of recording in the year 1870, and that relates to the introduction of "cups" in the crown and sides of the fire box, as illustrated by Fig. 610.

Early in the next year, 1871, a Mr. Brough's ideas on boilers appeared, as shown by Fig. 611, which illustrates an arrangement of four boilers; each boiler is fitted with horizontal flame tubes, and placed at a sufficient distance apart from one another for fire grates to be placed between them, the several sections are also connected to one another at top and bottom by tubes, and thus they are combined together to act as one large boiler.

Fig. 610.

Lee's Marine Boiler, in which the crown and sides of the fire boxes are fitted with "cups." Patented in the year 1870.

Above each fire grate is placed an arch of fire brick, and the products of combustion from the fires in the fire grates pass through the horizontal tubes which are below the arches into a smoke box on the exterior of each of these sections, and from those smoke boxes they pass back through the horizontal tubes which are above the arches into a central smoke box fitted with a chimney.

A Mr. Ashton followed Brough, also with a vertical boiler, as illustrated by Fig. 612; the shell is cylindrical, with a dome top,

and the fire boxes are the same shape, with a vertical flame flue tube on the crown of each that leads into a combustion chamber fitted with vertical water tubes. The flame, after passing amongst those tubes issues through the side opening into the chimney. This

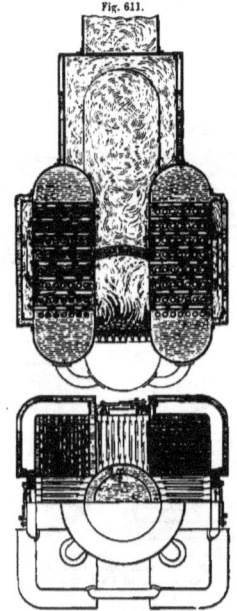

Fig. 611.

Brough's Marine Boiler, being a group of four egg-end square shells, fitted with cross flame tubes. Patented in the year 1871.

boiler is also fitted with a priming pipe, through which the steam passes before entering the main steam engine pipe.

A Mr. Crighton's boiler is next on the list

of inventors' productions, as illustrated by Fig. 613. This is a cylindrical shell containing three fire boxes, two return flues, and a set of return tubes over the fire boxes.

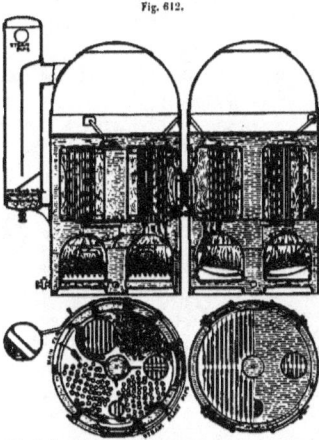

Fig. 612.

Ashton's Vertical Marine Boiler, being a cylindrical shell containing four dome-crown fire boxes, above which is a combustion chamber fitted with vertical water tubes. Patented in the year 1871.

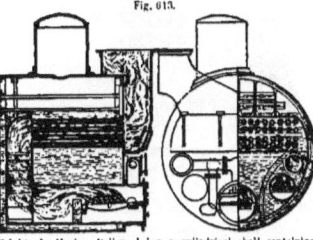

Fig. 613.

Crighton's Marine Boiler, being a cylindrical shell containing three fire boxes, two return side flues, and a set of return flame tubes above the fire boxes. Patented in the year 1871.

The two outside fire boxes are fitted with grates for the burning of coal, the flame from which passes through to a combustion chamber at the end of the boiler, which combustion chamber is separated from the main combustion chamber of the centre fire box by a thick division of fire clay, which, when thoroughly heated, will add to the combustion of the smoke. The flame then returns by two tube flues through the boiler to the central fire box, whose area is equal to the areas of the two outer boxes. The central fire box is intended to be fired with small quantities of coal and the small cinders which fall through the firebars of the two outside fire boxes, so as to keep a hot fire without much smoke. The flues of

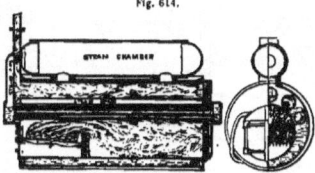

Fig. 614.

Erard's Marine Boiler, being a cylindrical shell fitted with water tubes in the fire box and return flame flue tubes around the crown. Patented in the year 1853.

the two outer fire boxes return through the boiler, and are conducted to the central fire box, and so arranged as to supply above or below the fire bars of the centre fire box the heated gases—and smoke, if unconsumed —from the outer fire boxes.

The object of this arrangement is, that the central fire·box should be supplied with the heated air and unconsumed smoke of the two side boxes only, thereby creating greater heat with less consumption of fuel, and more perfect combustion of coal and gases, which by the more general arrangement are now

234　　　　　　　　　　MARINE BOILERS.

wasted and lost by the too direct and rapid communication with the funnel.

Return flue tubes were not first introduced by Crighton, because Erard anticipated him, in the year 1853, as shown by Fig. 614. The

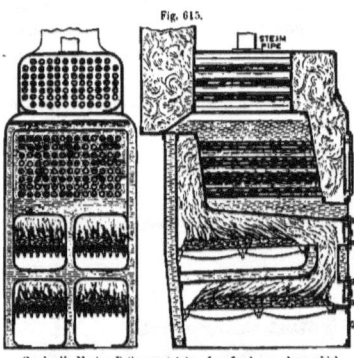

Fig. 615.

Crosland's Marine Boiler, containing four fire boxes, above which are a set of return tubes, and above these another set to be used as a superheater. Patented in the year 1856.

boilers by a French engineer, named Monteby, in the year 1855, as shown by Fig. 616; in this case the superheater was in the flue. About that period, also, water tube marine boilers were rather popular, and two arrangements were introduced by a Mr. MacFarlane, as illustrated by Figs. 617 and 618.

Returning to the events of the year

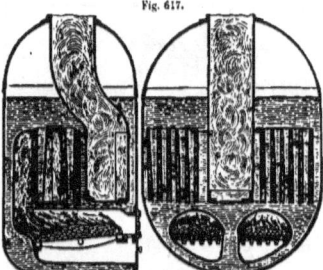

Fig. 617.

MacFarlane's Marine Boiler, containing vertical oval water tubes and an inside smoke box. Patented in the year 1854.

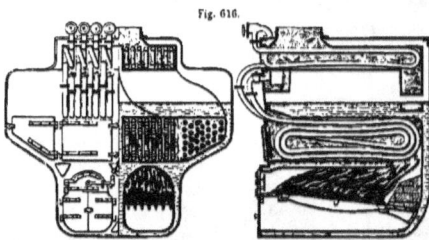

Fig. 616.

Monteby's Marine Boiler, containing tubes and a return flue in which is a coil-tube superheater. Patented in the year 1855.

arrangement of fire boxes and grates above each other in marine boilers were also patented as far back as the year 1856, by a Mr. Crosland, as illustrated by Fig. 615. And flues in combination with tubes were arranged in marine

1871, we notice next that Mr. Howard improved on his tube boiler, shown by Fig. 603, on page 230 of this work, and the improvement is illustrated by Fig. 619, consisting of cross tubes, secured by stay bolts and nuts, and fitted with an internal circulating plate at the front end.

A Mr. Bartlett also, about the same time, introduced a marine tube boiler, as shown by Fig. 620, in which the invention is, that the tubes are connected at right angles near

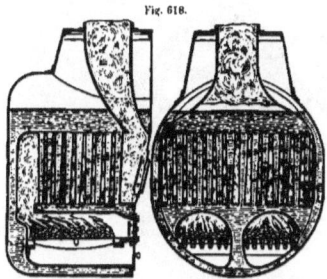

Fig. 618.

MacFarlane's Marine Boiler, containing vertical water tubes and an outside smoke box. Patented in the year 1854.

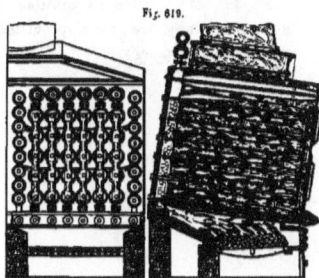

Fig. 619.

Howard's Marine Tube Boiler, in which the cross tubes are secured by stay bolts and nuts. Patented in the year 1871.

each end by short small tubes screwed into them, and fitted with screwed caps at the extremities.

A Frenchman's ideas on marine boilers next appeared, and very contrary they were too; because, in the first place, the inventor put the fire box near the base of the boiler

Fig. 620.

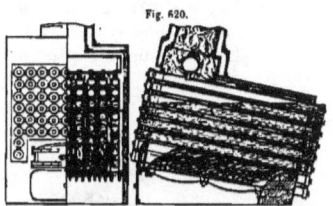

Bartlett's Marine Tube Boiler, in which the tubes are capped with screwed caps at each end, and connected together at right angles by short screw tubes. Patented in the year 1871.

Fig. 621.

Laharpe's Marine Boiler, in which the fire box is near the base of the shell, and the tubes are contained in a box through which the flame ascends. Patented in the year 1871.

Fig. 622.

Laharpe's Marine Boiler, in which the fire box is near the roof of the shell, and the tubes are contained in a box through which the flame descends. Patented in the year 1871.

shell, and next he put it near the roof, as illustrated by Figs. 621 and 622.

A Mr. Watt finished the year 1871, as far as he was concerned, with boiler inventions, as illustrated by Fig. 623, the arrangement of which is, that a series of inclined water tubes are placed over the fire grate, with sufficient space between them to allow the flames and heating gases to circulate freely. Those tubes are connected at their ends to two chambers in communication with the cylindrical steam spaces situated above. The outer plates of the chambers are formed with doors to allow the ends of any of the tubes to be easily reached, examined, and any one removed and another put in its place without disturbing the others. The outer

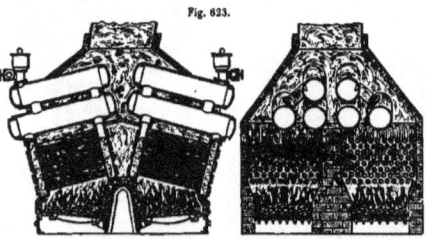

Fig. 623.

Watt's Marine Boiler, being a combination of two sets of water tubes, directly over the fire grates, with flat water chambers that are connected to six steam chambers situated above. Patented in the year 1871.

plates and the tube plates are stayed by stay bolts, which pass through the former and screw into the latter, being as sensible an arrangement as any we have previously illustrated in this chapter.

CHAPTER VI.

LIQUID FUEL BOILERS.

LIQUID fuel consists of any liquid containing carbon and hydrogen enough in it to permit combustion.

Liquid fuel boilers refer mostly to the fire boxes in which the usual coal or wood fire

Fig. 624.

Viney's Liquid Fuel Boiler, being a cylindrical shell containing a series of concentric circular conical flues, at the bottom of which is an oil chamber having a perforated bottom for the admission of air. Patented in the year 1822.

grate is assisted or displaced by an apparatus suitable for the burning of oil or any other combustible liquid.

In the year 1822, a Colonel Viney patented an arrangement of boiler as illustrated by Fig. 624. This is a cylindrical shell containing a series of concentric circular conical flues, at the bottom of which is an oil chamber having a perforated bottom for the admission of air, but how the supply of the liquid fuel is regulated to permit the admission of the air is not explained by the inventor. We presume, however, that he had not forgotten it, because he evidently, from the arrangement of his boiler, understood the principles of combustion.

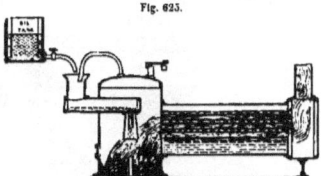

Fig. 625.

Biddle's Liquid Fuel Land Boiler, in which the fire box is fitted with two dish-shaped brick pans to contain coal, on which the oil flows through the fuel feed trough above. Patented in the year 1862.

Strange to relate, nothing worthy of record in these pages occurred again relating to liquid fuel boilers until the year 1862—being an interval of forty years—when a Mr. Biddle introduced his ideas as to what a liquid fuel boiler should be, which is illustrated by Fig. 625. The bottom of the fire box is solid, with brick dish-shaped pans to contain coals. To ignite the fuel first, the oil passes up through the central pipe between the pans, and afterwards the further supply of oil is from the feed trough situated above.

The following year, 1863, commenced, and at the same time the liquid fuel apparatus, by a Mr. Schmidt, was patented, as illustrated by Fig. 626, which is an arrangement of a

proper height, say a few inches from the top, the oil is then admitted through the pipe, and rising through the water floats on its surface: all that is then necessary is to ignite the oil, which burns with intense heat and a large amount of flame, and by properly regulating the admission of air, the smoke is almost entirely consumed. In this manner, states Mr. Schmidt, not only carbon oil, but other

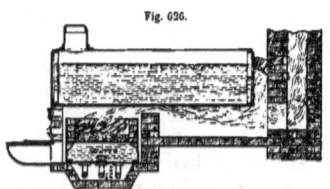

Fig. 626.

Schmidt's Liquid Fuel Land Boiler, under which the oil is ignited on the surface of water contained in an iron box with brick sides above the water line where the oil is burning. Patented in the year 1863.

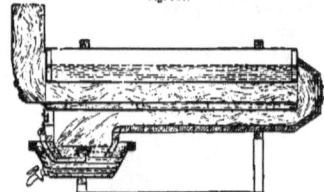

Fig. 627.

Schmidt's Liquid Fuel Marine Boiler, under which the oil and water is contained in a swivel hung box having projecting splash ribs at the insides. Patented in the year 1863.

Fig. 628.

Richardson's Liquid Fuel Apparatus, consisting of oil channels, packed with porous material, and water under the oil channel, with air and vapour pipes to assist combustion. Patented in the year 1864.

cast-iron box with brick sides above the iron; water is contained in the iron part of the box and oil in the brick part; this is fixed under the boiler at the front end for stationary land boilers; but for marine boilers, the box is hung on swivel lugs; and in other cases a compass box attachment is preferred, as illustrated by Fig. 627.

The operation of the apparatus is this:— The box being supplied with water to the

fluids rich in carbon, of less specific gravity than water, and which will not mix with it, may be burned as fuel, for which purpose a fluid composed of the pitchy residuum from the distillation of carbon and coal oils dissolved in benzole forms a suitable material.

In the year 1864, Mr. Richardson introduced his idea on liquid fuel apparatus, and the arrangement of it is illustrated by Fig. 628.

The oil is contained in four narrow channels packed at the top with a suitable porous material; below those channels are water chambers filled from the boiler with hot water and steam to heat the oil before ignition, as well as to prevent the apparatus from burning.

The working of this apparatus is as follows:—Each oil channel is packed properly with porous material as shown, the oil enters quickly, ascends, and completely soddens the porous material, which soon shows an oily surface; when paper pressed upon it becomes saturated, it can be lighted, the oil on the surface vaporizes as soon as it becomes warm, the vapour takes fire, and will continue to burn as long as there is any oil in the chambers; if sufficiently hard the porous material remains uninjured; the flame is continued at will by admitting the oil.

The vapour chamber is placed in the hottest part of the grate; it collects and forms vapour; this is conducted to a perforated tube and allowed to escape; taking fire from the flame of the grate, it gives additional heat and tends to burn the smoke; air is admitted to aid the operation.

To prevent the grate getting too hot, as well as to obtain all the heat given out by the combustion of the oil, the water from the boiler circulates through the grate and returns. There is a shield also to prevent the air and vapour tubes becoming too hot; this shield is shown in the sectional elevation.

The porous material extends at intervals to the bottom of the oil chambers; it sucks up the oil, otherwise the latter would boil and produce too much vapour, which would escape to the vapour tube.

The kind of porous material to be used, its thickness and depth, whether it should be of one, two, or more sorts, must be according to the work required and to the nature of the oil. There are several kinds of porous material suitable: thin slices or blocks of porous

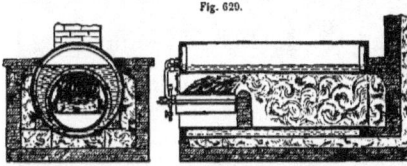

Fig. 629.

Richardson's Liquid Fuel Apparatus, fitted in the flue tube of a Cornish boiler. Patented in the year 1864.

brick, pumice stone, lime, charcoal, or moulded carbon, bath brick, natural or artificial porous stone, loam, or sand core, as used in foundries, lime, charcoal, or such like in powder, either placed upon or between solids, sponge when placed below other porous material; the thinner and more open the material, the fiercer the combustion and the greater the smoke; mineral salt with a thin top of moulded carbon or charcoal makes a good packing, but the salt by itself soon burns to cinder and will not last, according to Mr. Richardson's statement.

The illustration, Fig. 629, shows the application of this invention in the ordinary Cornish boiler.

In the ensuing year, 1865, a Mr. Sim considered that the best way to use liquid fuel was to burn it in combination with coal, and accordingly he arranged and patented a tubular retort, as shown by Fig. 630, as adapted for a land boiler.

The retort shown in transverse section is of a horse-shoe form, the interior hollow part containing the coal fire by means of which the oil is gassified. The oil is admitted through the front end of the retort by means of the pipe, a cistern containing the oil being suspended over the pipe, and the oil is allowed to fall therefrom into the recess; the column of oil in the body of the pipe serves to maintain an equillibrium of pressure of the gas in the interior of the retort. The front portion of the retort is moveable, for the purpose of access being obtained into the interior of the retort in case such may at any time be required. The upper part of the inner surface of the retort has a flange formed around it, by which the oil is prevented from running to other parts of the retort. The outer surface of the retort is perforated with several holes, into which "Bunsen" burners are fitted, and an air-tight shield is fitted all round the retort, so that no more cold air than is necessary shall be admitted, the quantity of air being at any time regulated

Fig. 630.

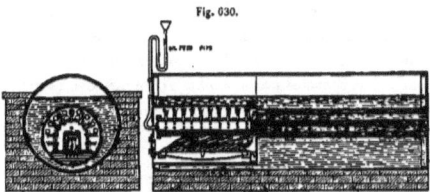

Sim's Liquid Fuel Tubular Retort Apparatus, fitted in combination with the ordinary coal fire grate in a land stationary tube boiler. Patented in the year 1865.

Fig. 631.

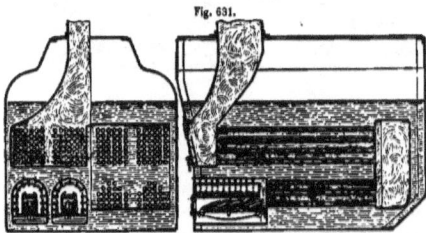

Sim's Liquid Fuel Tubular Retort Apparatus, fitted in combination with the ordinary coal fire grates in a marine boiler. Patented in the year 1865.

LIQUID FUEL BOILERS. 241

by opening the doors in the front of the furnace; and a marine boiler, similarly fitted, is shown by Fig. 631.

But the main difference in the fire box from that generally used is shown by Fig. 632, which illustrates a sectional elevation of a locomotive boiler fitted according to Sim's patent, and the mode of operating with this modification is as follows:—A small quantity that a thick layer of oil may never be present in the pan. Immediately upon the gassification of the oil, the gas itself rises through the central tube, thence into the side tubes between the first and second water spaces; these tubes being perforated, the gas issues through the openings and ignites in the flame space, and from there passes into the tubes beyond.

Fig. 632.

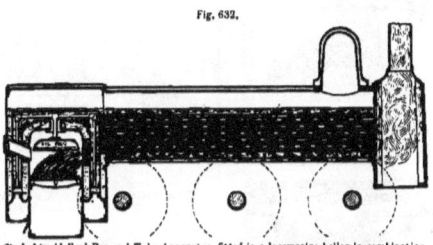

Sim's Liquid Fuel Pan and Tube Apparatus, fitted in a locomotive boiler in combination with the ordinary coal fire grate in the fire box, having a flame space around it in which are oil tubes connecting and supporting the pan over the fire grate. Patented in the year 1865.

Fig. 633.

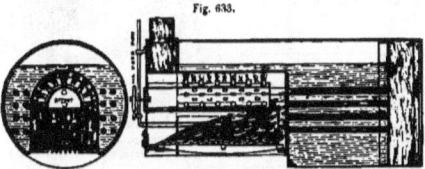

Sim's Liquid Fuel Tubular Boiler, in which the fire box is fitted with a brick retort over the ordinary fire grate, which is removed after ignition occurs, and a perforated brick slab put in its place. Patented in the year 1865.

of coal being ignited upon the fire bars, the oil pan becomes speedily heated, and when having attained a high heat, the oil to be converted into gas is admitted therein, from a reservoir, preferably in a small continuous stream or in drops, so that it may be vaporized as soon as admitted, and, further, Mr. Sim, however wise he might have deemed himself in amalgamating coal and oil for combustion, very soon after discovered his error, and introduced the arrangement illustrated by Fig. 633, which consists of a tubular brick retort that is fed with two horizontal pipes in the fire box under the flat bottom of

242   LIQUID FUEL BOILERS.

the oil chamber, and an ordinary fire grate below, which, when the fuel is first ignited, is removed, and a perforated brick slab put in its place, and the steam raised by liquid fuel only during the working of the boiler.

Next appeared the arrangement of a Mr. Wise, which consists of a "step" slab grate

Wise's Liquid Fuel "Step" Slab Grate, raised at the sides of the fire box. Patented in the year 1865.

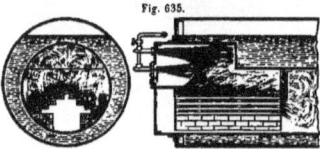

Wise's Liquid Fuel "Step" Slab Grate, raised at the centre of the fire box. Patented in the year 1865.

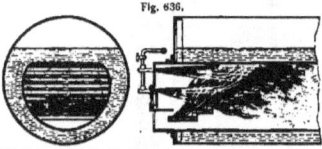

Wise's Liquid Fuel "Step" Slab Grate, raised across the fire box towards the back end. Patented in the year 1865.

for the liquid fuel to burn on, by being injected on the slabs by steam from the boiler; three examples of Mr. Wise's conception are illustrated by Figs. 634 to 636.

The injector is illustrated in sectional elevation by Fig. 637, and the arrangement of the entire fittings by Fig. 638.

Nothing else appeared, to warrant notice in this chapter, until the year 1866, when a Mr. Lees considered that liquid fuel should be dispersed on red hot brick or metal balls to promote combustion—which, in fact, was returning to Richardson's principles, if not his arrangement—as illustrated by Fig. 639.

The mode of operation is as follows:—The trough being empty, the stop-cock is opened

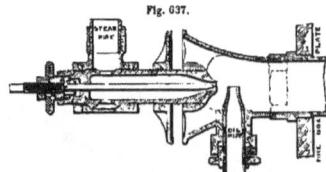

Wise's Liquid Fuel Injector, in connection with his "step" slab grates. Patented in the year 1865.

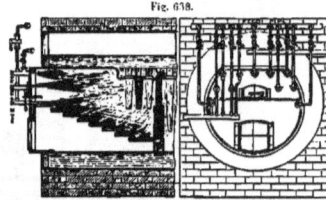

Wise's arrangement of pipes, injectors, and "step" slab grate, for burning liquid fuel. Patented in the year 1865.

and the petroleum or other hydrocarbon runs down the pipe into the trough. The smallest of the balls are then placed at the bottom of the trough through the door, after being made sufficiently hot to cause the petroleum or other hydrocarbon used to ignite; hollow and perforated balls of increasing diameters are next placed in layers in the trough until it is filled; the door being now closed, the air

LIQUID FUEL BOILERS. 243

required for combustion, and supplied through the holes, is drawn under the bottom edge of the baffle plate, and thereby brought in contact with the external and internal surfaces of the hollow balls, which, becoming red hot, effect the perfect combustion of the gases as they are evolved.

Early in the year 1867, a Mr. Barff patented the apparatus for burning liquid fuel as illustrated by Fig. 640, which is an arrangement of a cylindrical retort for using steam

Before starting the apparatus, a small quantity of oil or liquid hydrocarbon is poured upon the coke and ignited, with a view to raising the retort to the desired temperature for vaporising the liquid hydrocarbon. During this period, that is to say, before the steam has been raised in the boiler, air is injected into the retort by using a suitable pipe for that purpose, and the air jet supplies the necessary oxygen to burn the hydrocarbon in that case.

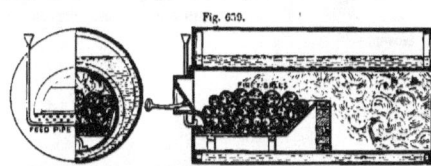

Fig. 639.

Lees' Liquid Fuel Boiler, in which the heat is assisted by brick or metal balls being piled in the fire box. Patented in the year 1866.

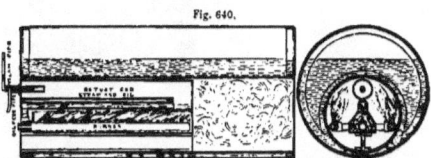

Fig. 640.

Barff's Liquid Fuel Boiler, in which the apparatus consists of a cylindrical retort, below which is a case containing an oil pipe that is surrounded by coke, and two side burners are connected to this pipe also. Patented in the year 1867.

and oil, and below it is the oil or burner pipe and charcoal case.

The burner is enclosed in the case, the two sides of which are inclined so as to contract the open mouth. In this case coke or other suitable porous material is disposed on each side of the burner, and a number of air holes along the lower part of the case are just above the surface of the coke for the purpose of combustion.

When steam has been generated the air jet is shut off, and a jet of steam is introduced through the nozzle or steam jet pipe, which induces the hydrocarbon and blows it in the form of spray into and along the retort. This mixture of steam, hydrocarbon, and vapour thus generated passes along the tube and enters the burner, whence it issues through the orifices, and is ignited by the flame from the saturated coke on each side of

2 I 2

the burner, when the apparatus thus continues in operation so long as steam and oil are supplied.

In conjunction with the central burner are two argand side burners communicating with the retort by the branch pipes, the air being induced through the central opening in these burners by the passage of the inflammable vapour through the annular space or opening of the burner. Any number of burners may obviously be used in connection with one retort, according to the capacity of such retort and size of the boiler.

When once properly started, the case with the porous material has fulfilled its function; but as its removal would in most cases be inconvenient, it may be allowed to remain. Its chief use is to generate sufficient heat on first starting the apparatus to heat the retort and generate steam.

A Mr. Dorsett's apparatus for preparing and burning liquid fuel next appeared, in the year 1868, as illustrated by Fig. 641, which shows a section of a waggon or cylindrical steam boiler, connected by a pipe to an oil boiler that is inserted in a tank containing the raw oil, that is vaporised in this boiler for the purpose of raising steam in the larger boiler; and in the event of the pressure being too great in the oil boiler, a condenser is provided at the side with a coil and water in it.

The action of the apparatus is as follows:— The creosote or mineral oil is pumped into the oil boiler through the opening in the side by means of the hand force pump. A small quantity of lighted fuel is placed on the grate in the oil boiler, and the fire is kept up therein till the vapour given off from the mineral oil has arrived at a sufficiently high pressure for the purpose required. The valve on the tube near the pump is then opened, and the vapour issues from the perforations in the tube in the fire box, and is ignited by the fire beneath. The pressure can then be maintained in the boiler by means of the ignited vapour alone without the addition of any more fuel.

The valve on the horizontal vapour pipe is next opened, and the vapour rushes through

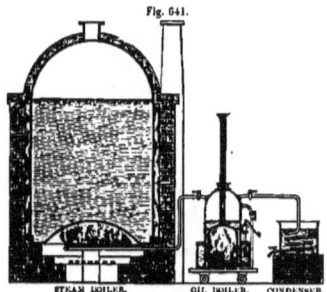

Fig. 641.

STEAM BOILER.　OIL BOILER.　CONDENSER.

Dorsett's Liquid Fuel Apparatus, being a small boiler inserted in an oil tank, fitted with a feed hand pump and vapour pipe that forms the burner in the fire box; this boiler boils the oil, and its vapour is used to generate steam in the larger boiler. Patented in the year 1868.

the perforations in the tube, where it is also ignited upon or under the grate of the furnace of the steam boiler.

In the event of there being too high a pressure a condenser may be used, as shown; but, under ordinary circumstances, it is only necessary to check the issue of the vapour from the boiler to the tube that serves to heat the said boiler or generator.

The next liquid fuel boiler and preparing

LIQUID FUEL BOILERS.

apparatus that we notice is illustrated by Fig. 642, being the invention of a Mr. Stevens, an American, the arrangement of which is, that over the fire grate is arranged a steam boiler with flues passing through it for the flame from the fire to pass through. Over the boiler is a mixing chamber separated from the boiler by a flue space, for the purpose of dry-heating the vapours that are to be mixed in the chamber; and through

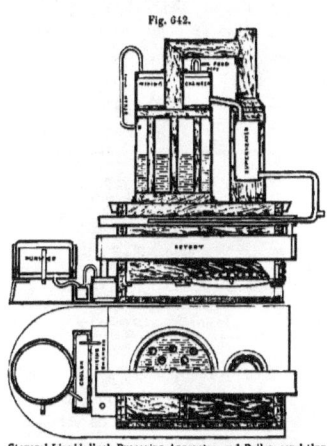

Fig. 642.

Stevens' Liquid Fuel Preparing Apparatus and Boiler, consisting of a retort steam boiler and various oil chambers. Patented in the year 1868.

the chamber passes the flue pipe, which passes from thence to the chimney. A pipe leads from the top or steam space in the boiler up into the mixing chamber for the passage of the steam into the chamber, and through another pipe a hydrocarbon vapour is forced.

These gases are then mixed, and to some extent dried by contact with the fire-heated surfaces in the mixing chamber, from which the mingled vapours of the hydrocarbon and hydro-oxygen—steam—are taken through a pipe into a superheater, which is a closed vessel placed in the smoke box.

From the superheater the mixed and dry vapours may be taken off through a pipe to the place where they are to be afterwards used or further treated as follows:—If the object be to distil the hydrocarbon in order to render it pure as a commercial article, then it may be taken to a condensing apparatus of any suitable well-known kind and construction through a pipe and condensed, and the water of condensation separated from the hydrocarbon, when the latter will be found to be perfectly pure and clear of all crude matter.

If the mixed vapours are to be burned as a fuel for generating steam, or for heating purposes of any kind, they may be taken directly from the pipe to the point of combustion; and it will be found highly advantageous and economical to mix with the vapours at or near the point or place of combustion, atmospheric air, which intensifies the burning and heating properties.

Following Stevens, came a Mr. Lafone, with two oil burners for liquid fuel: the first burner is shown by Fig. 643, on the next page, which is a box with recesses on the top to contain the oil; and the second burner is shown by Fig. 644, on the next page, which is a perforated plate for the oil to pass through.

A Mr. Weir next introduced a simple apparatus for burning liquid fuel, as illustrated by Fig. 645, on the next page also.

LIQUID FUEL BOILERS.

In practically carrying out this invention, a shallow tray of sheet iron is fixed horizontally in the fire box, and is covered or filled to a depth of a few inches with iron borings.

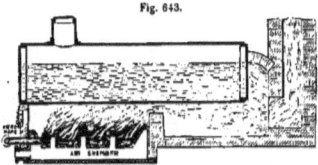

Fig. 643.

Lafone's Liquid Fuel Boiler, the oil burner being a box with recesses on the top and air holes, and the oil being contained in the recesses. Patented in the year 1868.

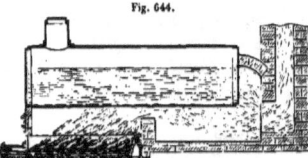

Fig. 644.

Lafone's Liquid Fuel Boiler, the oil burner being a perforated plate for the oil to rise through. Patented in the year 1868.

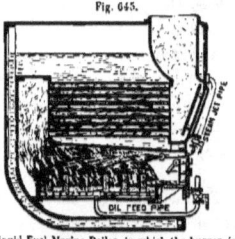

Fig. 645.

Weir's Liquid Fuel Marine Boiler, in which the burner is a cast-iron tray containing iron borings and brick slabs, under which is the oil feed pipe. Patented in the year 1868.

The oil is led into the furnace from a tank placed by preference at a slightly higher level than the tray by a pipe, which passing up through the tray bottom is jointed to the middle of a horizontal pipe placed slightly above the tray bottom, and having holes perforated along its under side. This position of the pipe, with the perforations on its under side, is adopted as the least liable to have the flow of oil interrupted by the borings.

A series of fire-brick slabs are placed across the tray, being partly embedded in the borings, and when they become intensely heated, serve to insure the thorough ignition of the vapours rising from the oil; these slabs may be made with a number of holes through them, to promote the intermixing of the gases, vapours, and air.

The oil entering by the pipe rises to the

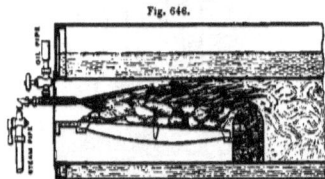

Fig. 646.

Smith's Liquid Fuel Boiler, in which the fuel is injected by steam on to brick slabs or coals. Patented in the year 1868.

surface of the borings, which act as a kind of wick for it, and it burns at and above that surface, the air necessary for its combustion entering by an aperture or apertures provided in the fire door. If required, the inward flow of the air may be increased by jets of steam or otherwise, and water in minute jets or spray may be thrown on the flames to improve the combustion.

Next came a Mr. Smith, full of the idea of injecting by steam, liquid fuel on to burning slabs of brick or coals, as illustrated by Fig. 646, which, by the way, is illustrated

LIQUID FUEL BOILERS.

also on page 242 of this work by Fig. 638, being another example of the hard, careful study and attention devoted to the interests of patentees by their legal guardians at the Patent Office.

After which a Mr. Sauvage's opinion on liquid fuel boilers was published as a patent, as shown by Fig. 647.

The arrangement is adapted to a locomotive boiler, and the oil fuel is contained in a tank placed either on the engine or tender, and descends by the oil feed pipe. This pipe forms two branches, each of which is provided with a cock to regulate the arrival of the oil at both ends of the long distributing pipe, which itself fulfils the function of a feeder. By means of a screw handle a uniform flow is obtained into the small pipes which distribute the oil to the fire bars through the orifices. The fire bars are vertical, with the top face of each having a gutter formed along it to contain and direct the oil.

The air necessary for the combustion passes through the spaces between the bars, its admission being regulated by means of a valve or damper, the handle for actuating the same being within reach of the driver.

A casing and two brick bridges supported by iron stays direct the first course of the flame. The door of the fire box is opened to facilitate the repair of the tubes, but it is closed when the engine is running.

The lighting is effected by the aid of a few lighted shavings—the cocks being slightly opened—assisted by the draught of the ordinary chimney. If it is required to get up the pressure very quickly, a supplementary draught is employed, produced by a jet of steam from any boiler near, or by a jet of air from a fan. In running the engine, the

Fig. 647.

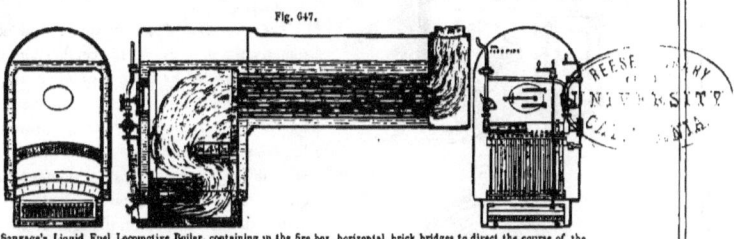

Sauvage's Liquid Fuel Locomotive Boiler, containing in the fire box, horizontal brick bridges to direct the course of the flame arising from a grate composed of a series of vertical bars, on which the oil burns. Patented in the year 1868.

natural admission of the air may also be replaced by an air blast.

A French engineer, named Ravel, next came into notice, in England, for his patent of an arrangement of automatical liquid fuel burning and feed water apparatus, as illustrated by Fig. 648, on the next page.

The boiler is composed of a single series of coil pipes connected at the lower extremity with a water casing, and at the upper end to a steam chamber formed by the shell. The water enters at the lower part of the casing by the water feed pipe, and, after filling it, flows into the lower end of the bottom coil, in which it rises, and then passes in the form of steam by the central top tube into the steam chamber.

The oil burner is placed directly beneath the lowest coil, and the gases there evolved are compelled to ascend through the cylindrical coil space, and then to descend between the exterior of the coils and the interior

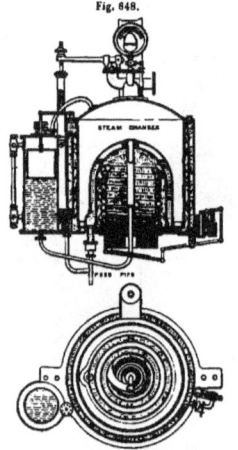

Fig. 648.

Ravel's Liquid Fuel Coil Tube Boiler, in connection with a dome-shaped fire box enclosed by a shell that is partially surrounded by a flame space. The feed water is regulated by a weight and lever gear, and the liquid fuel supply is regulated by plungers, springs, and levers actuated by the pressure of the steam. Patented in the year 1868.

of the water casing, to pass through the flame space which surrounds the shell and up the chimney, not shown.

The water is regulated by an apparatus that consists of a cylindrical vessel communicating at its upper part with the steam chamber, and at its lower part with the bottom coiled pipe. In the vessel is a float which rises and falls with the level of the water—which is the same in the coil as in the vessel—and is held in position by means of the rod and lever, which opens or closes a valve communicating by a tube with the feed pipe.

The oil burner shown in section in the elevation is composed of concentric annular rings placed in a cylindrical box constantly filled with petroleum or other mineral oil; and in these rings are suitable porous wicks, which dip into the oil, and between these spaces the air can pass as well as through the central opening.

The quantity of air supplied and consequently the intensity of the combustion of the oil are regulated by the extent of opening the lamp damper, which is actuated automatically by the increasing or decreasing pressure of the steam, by means of an apparatus consisting of a lever plunger, a spring plunger, and a system of levers acting on the damper. The under side of the lever plunger is in communication with the steam chamber. The spring plunger works in a box, to the side of which is connected the oil supply pipe. When the pressure in the boiler is normal, the lever plunger is lowered, and the spring plunger allows the petroleum to flow; but if the pressure increases, the flowing is reduced.

The year 1868 came to an end, and with it was introduced a Mr. Spartali's arrangement of pipes and chambers for injecting and burning liquid fuel in marine boilers, in which was contained an auxiliary boiler under the main fire box to commence the process of combustion, as illustrated by Fig. 649.

Next came from Canada an appeal for our

LIQUID FUEL BOILERS.

patent law protection by a Mr. Taylor, because he had thought out an arrangement of liquid fuel burning apparatus, as shown by Fig. 650,

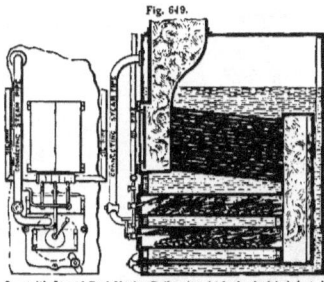

Fig. 649.

Spartali's Liquid Fuel Marine Boiler, in which the fuel is injected by steam on to porous slabs of brick or other suitable material, and an auxiliary boiler, to commence the process of combustion, is fitted below the main fire box. Patented in the year 1868.

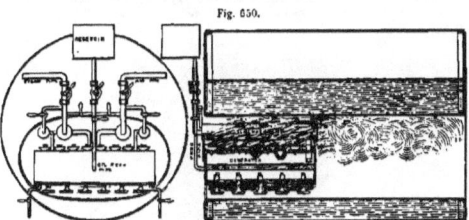

Fig. 650.

Taylor's Liquid Fuel Boiler, in which the fuel is ignited over and under a series of burners, with guards and perforated seats, and air and steam introduced in the fire box to cause combustion. Patented in the year 1869.

in which gas is used to ignite the oil, and gas burners of a suitable form are introduced as shown by the complete end elevation.

A Mr. Robinson next appeared with his ideas of what a liquid fuel burning apparatus should be, which is illustrated by Fig. 651, on the next page, and also thereat explained.

We close this chapter by illustrating and describing a gas fuel boiler, as illustrated by Fig. 652, on the next page, which is the invention of a Mr. Cutler.

The boiler is cylindrical, having a dome-shaped top, and is set vertically upon a cast-iron frame, which supports it. The part of the boiler below the dome is covered with a jacket; the space between the jacket and the boiler being filled with fire clay.

From the main gas pipe a number of smaller pipes proceed, one of which is inserted into each of the vertical pipes, and through the upper part of each of those pipes small holes are drilled, through which the gas passes and mingles with the atmospheric air drawn in through the lower holes. The gas and air thus combined pass up the vertical pipes and escape through ring-shaped caps, each formed by two metallic plates, which are kept at a suitable distance apart by pins.

At those points the gas and air are ignited, and the flame therefrom ascends in consequence of the draught caused by the rush of air through openings in the jacket; spaces being made for its admission.

The boiler is divided into two series of horizontal chambers, one series containing

2 K

water, and the other series as combustion chambers; the inventor states that both are so arranged that in section they appear, when the apparatus is in use, like "alternate layers of fire and water," each compartment of the same series having free communication with the adjoining one, and the top chamber leads to the chimney.

Fig. 651.

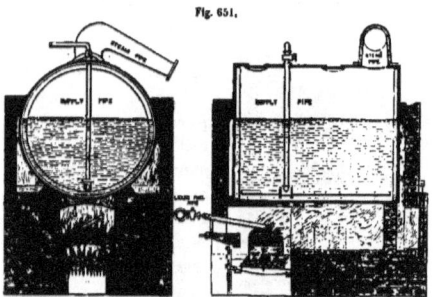

Robinson's Liquid Fuel Boiler and Apparatus, which consists of an arched brick bridge to receive the liquid fuel above an ordinary coal fire. Patented in the year 1860.

Fig. 652.

Cutler's Gas Fuel Boiler, in which the flame passes through openings alternately arranged in horizontal flat water spaces. Patented in the year 1868.

CHAPTER VII.

LOCOMOTIVE BOILERS.

THE first locomotive* was invented by Mr. R. Trevithick, a Cornish engineer, in the year 1803. It was constructed and worked at

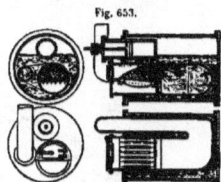

Fig. 653.

The First Locomotive Boiler, invented by R. Trevithick in the year 1803; constructed at Penydarran, South Wales.

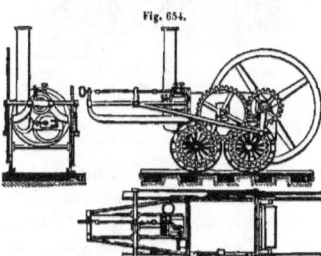

Fig. 654.

General Arrangement of the First Locomotive Engine and Boiler, invented by R. Trevithick in the year 1803.

Penydarran, South Wales, and the boiler was a cylindrical cast-iron shell with a wrought-

* 'Life of Richard Trevithick,' by Francis Trevithick, C.E. Spon. 1872.

iron return tube, as illustrated by Fig. 653. The arrangement of the locomotive is illustrated by Fig. 654. The steam from the engine exhaust pipe passed *through* the chimney, so Trevithick invented the "blast pipe" in the year 1803 also; and, in the year 1804, he improved on that locomotive, as illustrated by Fig. 655.

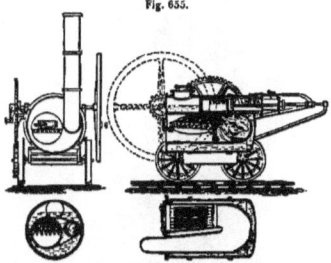

Fig. 655.

Trevithick's Improved Locomotive Engine and Boiler, constructed and worked at Newcastle-on-Tyne in the year 1804. Diameter of cylinder, 7 inches; stroke of piston, 3 feet. Diameter of boiler, 4 feet; length, 6 feet 6 inches. Diameter of fire box, 2 feet 3 inches. Diameter of flue at chimney, 1 foot.

In the year 1826, a Mr. Neville invented and patented a vertical tubular boiler, as shown in sectional elevation and plan by Fig. 656. The shell is cylindrical, with a dome top, and the crown of the fire box is fitted with vertical up-take flame tubes in connection with a semi-globular combustion chamber, to which are connected return down-take flame tubes that lead to a smoke box under the ashes box and the water space which form the base of the boiler.

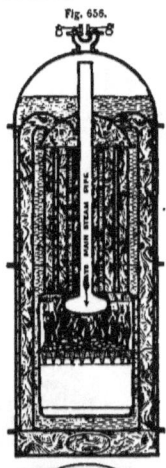

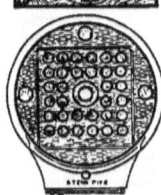

Neville's Vertical Return Tube Tubular Boiler, composed of a cylindrical dome-top shell, fitted internally with flame tubes over and around the fire box. Patented in the year 1826.

It is worthy of remark that Mr. Neville proposed that the steam should be superheated in a tube—suspended from the combustion chamber—over the fire grate. He stated also that his arrangement of the tubes was equally applicable for horizontal boilers; and from that statement we conclude that he invented, in purpose, the tubular locomotive boiler which George Stephenson introduced in the year 1829, via the "Rocket," as illustrated in sectional elevations by Fig. 657.

Stephenson was followed by Napier, in the year 1831, with a cylindrical shell containing a flue tube and return tubes, as illustrated by Fig. 658.

In the year 1833, a Mr. Fraser ignored tubes for the passage of the flame in a loco-

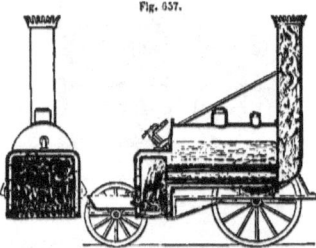

* Sectional elevations of the boiler of the "Rocket" Locomotive, constructed by George Stephenson, who won the prize of 500l. in the year 1829. Diameter of boiler, 3 feet 4 inches. Length of cylindrical part, 6 feet. Length of tubes, 6 feet. Diameter of tubes outside, 3 inches. Width of fire grate, 3 feet; length, 2 feet. Diameter of chimney, 14 inches.

motive boiler, and considered flat flues preferable, as shown by Fig. 659; and in the same year Mr. Joshua Field ignored flues and flame tubes also, and invented a water tube locomotive boiler, composed entirely of tubes enclosed in a casing, as illustrated by Fig. 660.

The locomotive boiler in actual use at that period, 1833, was an improvement on the "Rocket" boiler, and much credit is due to Stephenson for developing that boiler into a

LOCOMOTIVE BOILERS.

better arrangement, as shown by Fig. 661. This boiler was more adaptable for locomotion, and the entire arrangement of the details thorn appeared with a return tube tubular locomotive boiler, as illustrated by Fig. 662. The flame in this case traverses the boiler

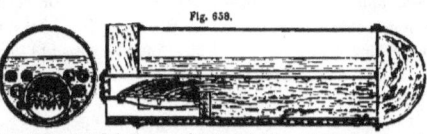

Fig. 658.

Napier's Locomotive Boiler, being a cylindrical shell with a cylindrical flue tube, semi-globular combustion chamber, and return flame tubes on each side of the flue. Patented in the year 1831.

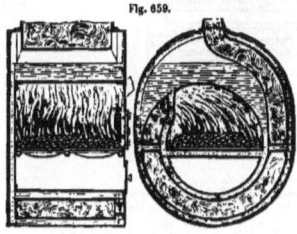

Fig. 659.

Fraser's Locomotive Boiler, being a curved short shell, containing a fire box in connection with flat flues and water spaces. Patented in the year 1833.

forth and back through two separate sets of tubes; the engine supply steam pipe is perfo-

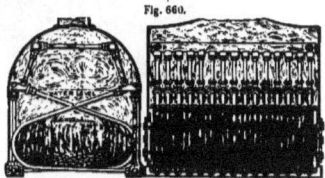

Fig. 660.

Field's Water Tube Locomotive Boiler, composed of tubes enclosed in a flame box. Patented in the year 1833.

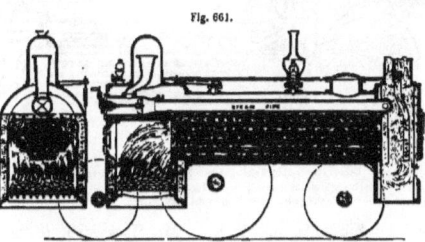

Fig. 661.

Stephenson's Locomotive Tubular Boiler, as constructed and used in the year 1833.

show an evidence of care resulting from experience in practice.

Those matters, however, were not permitted to rest long; for, in the year 1839, Mr. Haw-rated with slots on its upper side, and extends near the roof of the steam space in the boiler; the chimney is at the front end, and the engine exhaust steam pipe passes from the smoke box over the boiler to the chimney; therefore the internal arrangement was entirely different from Stephenson's.

Hawthorn proposed also the use of twin fire boxes, and the engine's exhaust pipe inside the boiler, as shown by Fig. 663.

Nothing worthy of our notice here was then done in locomotive boilers until the year 1842, when a Mr. Lewthwaite patented the

The speed of the locomotive about that period, 1842, was much increased, and the diameter of the driving wheel was also demanding attention in relation to the height of the boiler above the rails; accordingly Mr. Crampton brought forth his locomotive boiler to suit those relations, as shown by Fig. 665, and he also introduced a water bridge in the fire box.

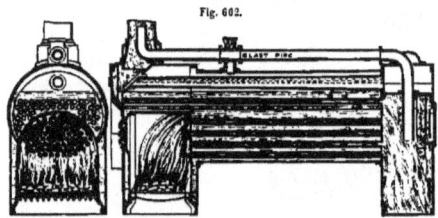

Fig. 662.

Hawthorn's Locomotive Return Tube Tubular Boiler, in which the flame passes from the fire box through tubes to the smoke box, and from there returns through a second set of tubes over the crown of the fire box to the chimney at the front end. Patented in the year 1830.

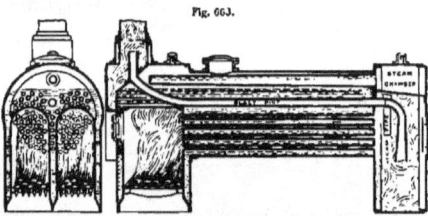

Fig. 663.

Hawthorn's Locomotive Return Tube Tubular Boiler, in which there are two fire boxes —and the passage of the flames is as in Fig. 662—but the engine blast pipe inside instead of outside the boiler. Patented in the year 1839.

arrangement shown by Fig. 664, in which the invention consists of a square shell containing flat water and flame spaces instead of tubes; and to cause a draught a fan is placed at the bottom of the chimney to "draw up" the products of combustion.

Mr. Hackworth, in the year 1847, patented the idea of arranging the tubes in vertical rows over each other, so that a brush or scraper could be passed down between them, as illustrated by Fig. 666.

Very soon after Hackworth, came the ideas

of a Mr. Johnson, claiming legal protection for the arrangement of the fire boxes and

Fig. 664.

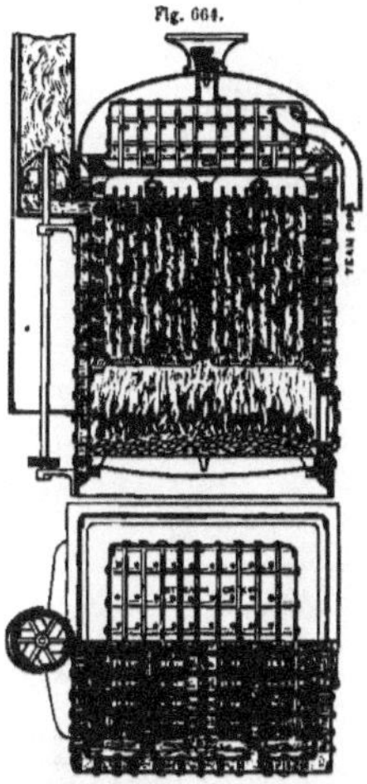

Lewthwaite's Locomotive Boiler, being a square shell, containing flat water and flame spaces, and a blast fan at the bottom of the chimney. Patented in the year 1842.

carried out in the ensuing year, 1848, as shown by Figs. 670 to 672, on the next page.

In the year 1850, Mr. Paul Rapsey Hodge patented the locomotive boiler as illustrated, on the next page, by Fig. 673, the main feature of which is the fixing of a tube superheater in the smoke box; and should the boiler be used as a portable boiler, the engine cylinder would be on the roof of the boiler, as shown.

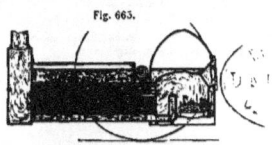

Fig. 665.

Crampton's Locomotive Boiler, having a recess behind the dome for the large driving wheel shaft to work in, and thereby suspending the boiler "low" over the rails, and a water bridge in the fire box. Patented in the year 1842.

Whilst Hodge was in the humour, he thought of another arrangement, as illustrated by Fig. 674, on page 257, which is that the boiler is fitted with return tubes, and the final smoke box is situated over the fire box, and the chimney is surrounded by the steam chest;

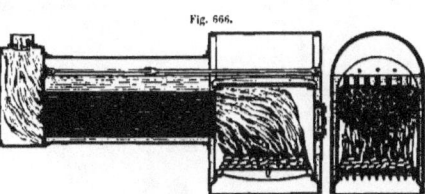

Fig. 666.

Hackworth's Locomotive Boiler, in which the tubes are arranged directly over each other in vertical rows. Patented in the year 1847.

central combustion chamber of locomotive boilers, as illustrated by Figs. 667 to 669, for the purpose of bringing the boiler "low" over the rails, which Mr. Johnson further

also the blast pipe passes through the boiler, most of which Hawthorn patented, and invented too, in the year 1839, as shown by Fig. 662, on page 254 of this work, again illustrating the arduous task fulfilled by our Patent Commissioners.

Our first International year, 1851, was water tubes in a locomotive boiler, as shown by Fig. 675.

A Mr. Barran began the year, 1852, of

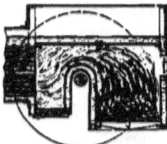

Fig. 667.

Johnson's Locomotive Fire Box, having a water-space-bridge in it, through which the driving wheel shaft works. Patented in the year 1847.

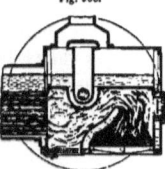

Fig. 668.

Johnson's Locomotive Fire Box, having a water bridge in it and a space, through the boiler over the crown of the fire box, for the driving wheel shaft to work in. Patented in the year 1847.

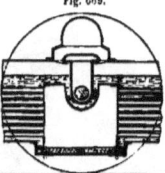

Fig. 669.

Johnson's Central Combustion Chamber, fitted with an open space through the boiler for the driving wheel shaft to work in. Patented in the year 1847.

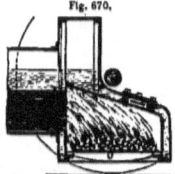

Fig. 670.

Johnson's Locomotive Fire Box, having the roof low down for the driving wheel shaft to work over it in front of the steam chest. Patented in the year 1848.

Fig. 671.

Johnson's Locomotive Fire Box, having a space through the water and steam spaces for the driving wheel shaft to work in above the fire box. Patented in the year 1848.

Fig. 672.

Johnson's Locomotive Fire Box, having a space between the fire bars for the driving wheel shaft to work in through the fire box. Patented in the year 1848.

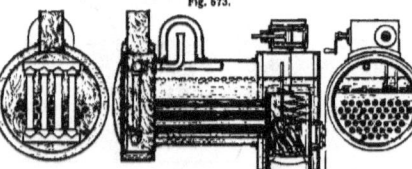

Fig. 673.

Hodge's Locomotive Boiler, having a tube superheater in the smoke box, and a steam jacketed engine cylinder on the roof of the boiler. Patented in the year 1850.

graced with the invention of a Mr. Stenson, which was that he formed the roof of the fire box with a projected flame space fitted with inventions by inventing the increase of heating surface by fixing in the fire box sides "cups," as illustrated by Fig. 676.

Next a French engineer, Bresson by name, informed the English community that, by pumping air into a chamber, and passing it in gusts through a fire, and from there into

Fig. 674.

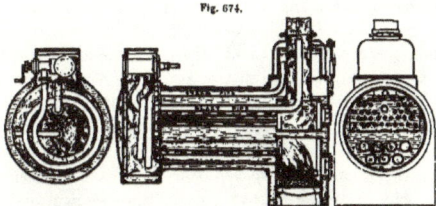

Hodge's Locomotive Return Tube Tubular Boiler, in which the flame from the fire box passes through large tubes to the smoke box, and from there returns through small tubes to the smoke box over the fire box. The back end smoke box is fitted with a coil tube superheater and the blast pipe is in the boiler. Patented in the year 1850.

Fig. 675.

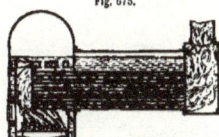

Stenson's Locomotive Tubular Boiler, in which the roof of the fire box is fitted with water tubes and a narrow flame space projecting over the horizontal flame tubes. Patented in the year 1851.

Fig. 676.

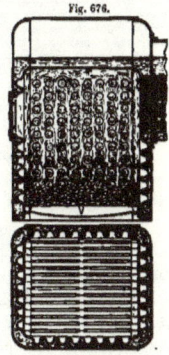

Darran's Locomotive Fire Box, fitted with flame "cups." Patented in the year 1852.

the atmosphere, the recoil became a motive power for propulsion; and here is his arrangement of the same for a locomotive, shown by Fig. 677, which illustrates also what misplaced confidence leads people to think and how they act after.

It will, of course, have been remarked that

Fig. 677.

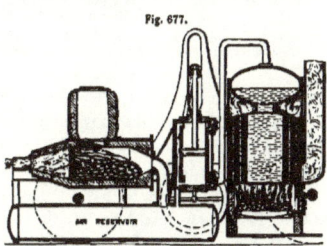

Bresson's Locomotive Vertical Boiler and Recoil Engine, the motion of the carriage being obtained by the discharge of rarified air in gusts into the atmosphere. Patented in the year 1852.

we, at every opportunity, point out the glaring advantages or defects in all the arrangements we have noticed in this work, but now we come to an improvement worthy of special notice, because it is a locomotive boiler constructed without angle iron, in the year 1852, by Mr. Adamson, as shown by Fig. 678; and, considering the period of the invention, and the general use of angle iron then, Mr. Adamson made a bold step of advance in boiler construction, inasmuch that the bending of curved plates meant "splitting them" in those days generally.

In the year 1853, Mr. Beattie conceived the idea that chimneys for locomotive boilers could be dispensed with, provided the combustion was perfect, or nearly so; and the mechanical question was answered, as shown by Fig. 679, which consists in the arrangement and combination of two furnaces and

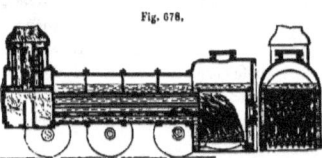

Fig. 678.

Adamson's Locomotive Tubular Boiler, being the first that was constructed entirely without any angle iron. Patented in the year 1852.

three combustion chambers. The first chamber is situated between the large and small furnaces, and there are tubes leading from the first furnace through the water space into the chamber; and tubes also through the water space of that chamber, leading into the large furnace, thus connecting the combustion chamber and the furnaces together.

There are two more combustion chambers attached to and situated between the short cylindrical tubular portions, and connected with the large furnace by tubes leading therefrom. Those two chambers are in connection with each other by the short tubular portions, which will allow the flame and heated gases to pass through into the last chamber, and from thence through the long tubes into the smoke box, in which is a chamber at the top with a perforated bottom plate, through which the gases ascend and are drawn off by the smoke pipe, into which the air pipe is inserted, having a trumpet mouth; and the other end of the pipe terminates in the smoke chamber—under the first set of tubes—which communicates with the air-tight ash pan by a long slot or opening running the entire length along the bottom, and which can be closed or opened, or partially

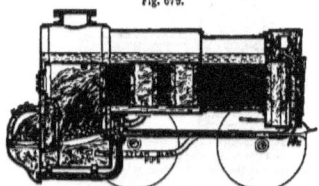

Fig. 679.

Beattie's non-Chimney Locomotive Boiler, composed of two furnaces divided by water spaces and a small combustion chamber. From the large furnace are three sets of tubes and two combustion chambers, and the gases from the smoke box passes back through a pipe into the large furnace, so that no chimney is fitted. Patented in the year 1853.

opened, by a flap valve. And another arrangement is shown by Fig. 680, in which water bridges in the combustion chambers formed the main feature.

Mr. Beattie was a bold man to propose even that a locomotive boiler could work without a chimney, and bolder by showing how he thought it could be done, which, by the way, has yet to be accomplished.

In the beginning of the year 1853, a Mr. Scott proposed to divide the fire grate with a vertical water bridge, and above that, at an angle, to divide the fire box with another water bridge—the bridges in both cases being perforated with flame tubes, as shown by Fig. 681.

Scott was immediately followed by a Mr. Shaw, who was desirous of improving on return tube locomotive boilers by the use of tubular flues and tubes, as illustrated by Fig. 682.

Fig. 680.

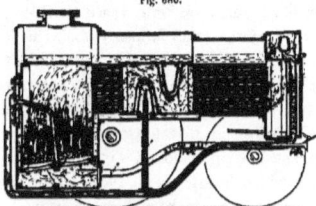

Beattie's non-Chimney Locomotive Boiler, in which are coal and coke furnaces, and two combustion chambers fitted with water bridges, and the gases returning to the furnaces, as in the previous example. Patented in the year 1853.

Fig. 681.

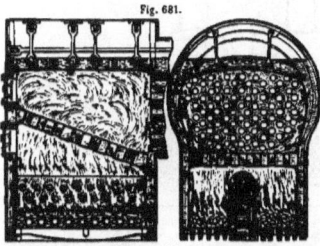

Scott's Locomotive Fire Box, containing vertical and angular water bridges perforated with flame tubes. Patented in the year 1853.

Directly after Shaw's boiler, came the idea for heating feed water around the fire box, and also a trumpet mouth combustion chamber in front of the tubes of a locomotive boiler, as shown by Fig. 683.

Mr. Dunn's fire box for a locomotive boiler, as illustrated by Fig. 684, in which the crown and sides are corrugated to increase the

Fig. 682.

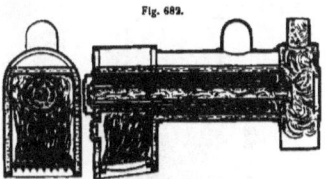

Shaw's Locomotive Boiler, in which the shell is fitted with two flame flues, between which are a series of small flame tubes for the flame to return in, and then proceed through the central flue to the smoke box. Patented in the year 1853.

Fig. 683.

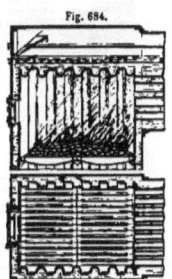

Newton's Locomotive Fire Box and Combustion Chamber, also a feed water heating space around and above the fire grate. Patented in the year 1853.

Fig. 684.

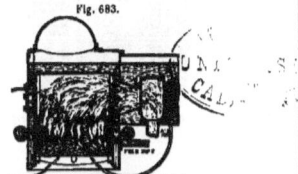

Dunn's Locomotive Fire Box, having a corrugated crown and sides. Patented in the year 1853.

surface, appeared next; and it was followed by Kendrick's, with deeper corrugations, as shown by Fig. 685. But Kendrick extended that idea, as shown by Fig. 686, where the crown and the base of the fire box are connected by narrow flame and water spaces, and a combustion chamber beyond, in front of the tubes; while a further belief in the corrugation principle is illustrated by Fig. 687, in which the tubes are even displaced by corrugations beyond the fire box.

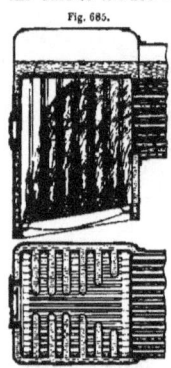

Fig. 685.

Kendrick's Locomotive Fire Box, having deep corrugated sides. Patented in the year 1853.

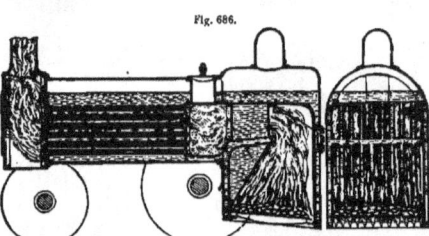

Fig. 686.

Kendrick's Locomotive Boiler, in which the fire box is constructed with narrow flame and water spaces in a line with the fire bars, connecting the base with the crown, and a combustion chamber beyond in front of the tubes. Patented in the year 1853.

The next inventor was Mr. Allan, who distinguished himself, as shown by Fig. 688.

The year 1854 made its appearance, and very soon after Mr. Beattie's ideas on locomotive boilers again appeared, which were more extensive than previously. In this case he began with the arrangement shown by Fig. 689, which is, that there are two fire boxes—or furnaces, as he called them—having an inclined transverse water space partition; the crown of the small furnace is roofed with perforated fire tile segments having sufficient openings to form a communication between the two furnaces. In the large furnace are placed vertical fire tiles, resting on a fire clay lump, and bearing against the inverted water bridge, which is fitted with tubes; the tiles are so formed as to have a space between them, whilst their extreme side surfaces come together, and a combustion chamber is formed by their ends, next to the tube plate of the short tubes, being made circular or concave, to allow the flame and gases to be fully developed before entering the combustion chamber through the short tubes.

LOCOMOTIVE BOILERS.   261

The action of the combined furnaces is as follows:—The large furnace is charged with carbonaceous coal or anthracite coal, and the small furnace with bituminous or gaseous surface of the incandescent fuel in the large furnace before entering the interstices between the fire tiles and the tile chamber leading into the short tubes and combustion chamber,

Fig. 687.

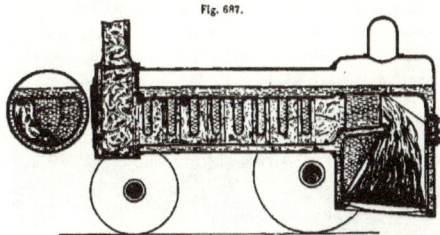

Kendrick's Locomotive Boiler, in which the fire box is constructed as in the previous example, but beyond the fire box vertical flame and water spaces are constructed in the place of tubes. Patented in the year 1853.

Fig. 688.

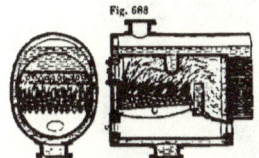

Allan's Locomotive Fire Box, containing an inverted water bridge and a vertical brick bridge. Patented in the year 1853.

where time and space are given to the gases for final combustion; and this object is aided by the inverted water space bridge in the combustion chamber, which is also fitted with tubes, and which checks the current of flame and gases before entering into the long tubes leading from the combustion chamber to the smoke box.

Fig. 689.

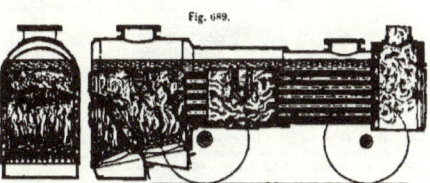

Beattie's Locomotive Boiler, in which there are two fire boxes and two grates; leading from the large fire box is a set of short flame tubes, in connection with a combustion chamber having an inverted water bridge in the roof, and a set of long flame tubes leading to the smoke box. Patented in the year 1854.

coal. The flame and combustible gases pass through the openings in the fire tile crown of the small furnace, and are deflected over the

Another method for effecting combustion was also proposed by Mr. Beattie, as shown by Fig. 690, on the next page.

A third, and further modification by Mr. Beattie, is illustrated by Fig. 691, showing another combination of furnaces, which consists of two furnaces and a preparatory combustion chamber formed by a longitudinal water space partition and a transverse partition, which is provided with an aperture, and an opening is formed also in the longitudinal partition.

and communicate with air passages in the bottom of the chamber, and the air thus admitted is brought into immediate contact and admixture with the gases evolved from the fuel in the furnaces; the quantity of air to be admitted is adjusted by a flap valve.

The final combustion chamber between the tubes is provided with a double series of

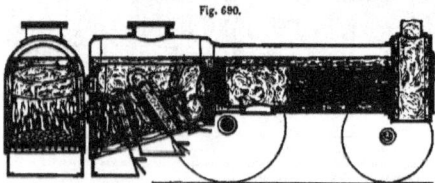

Beattie's Locomotive Boiler, in which there are three water bridges in the fire box, with two grates, but the remainder of the arrangement very much as in Fig. 689. Patented in the year 1854.

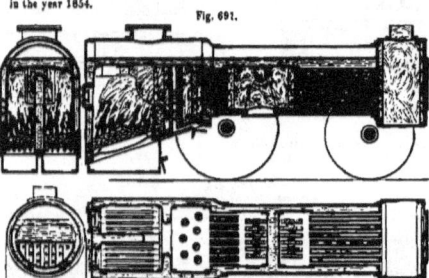

Beattie's Locomotive Boiler, in which there are twin fire boxes side by side and a water bridge at the end of the grates, beyond which are fire tile columns and air pipes, in front of the short tubes, and the intermediate combustion chamber is fitted with fire tiles and an inverted water bridge also. Patented in the year 1854.

The furnaces are provided with separate fire doors, ash boxes, and dampers; the preparatory combustion chamber is fitted with a fire tile bottom, on which are placed a number of fire tile columns, some of which are for air pipes, being hollow and perforated,

vertical fire tiles placed about two and a half or three inches apart, so shaped and formed as to leave an open mixing space at each end in the centre of the chamber.

The operation of the furnaces and combustion chambers will be as follows:—When the furnaces are charged alternately with fresh coal, the products of combustion will be well mixed in passing through the openings, and are further mixed and mingled together in passing between the fire tile columns before entering by the short tubes the final chamber, where the combustion of the gases is completely effected in passing between the vertical fire tiles before entering the long tubes leading from the combustion chamber into the smoke box.

A fourth, and entirely different arrangement, by Beattie, is shown by Fig. 692, in which there are two furnaces formed by the water space partition; the flame and gases passing through the flues are delivered into the top combustion chamber, which extends the entire length of the boiler and passes into the return chamber, and through the tubes into the chamber—which is formed by the water space partition in the furnace—and from there into a longitudinal flue which leads to the smoke box and chimney.

A fifth modification, by the same inventor, is illustrated by Fig. 693; in this case, he, to a great extent, combined his former examples, but beyond the intermediate combustion chamber was a set of return flame tubes, causing the chimney to be in the centre of the length of the boiler.

Mr. Beattie's sixth modification is illustrated by Fig. 694, showing an arrangement and combination of furnaces and cylindrical retorts where coal entirely, or coal in combi-

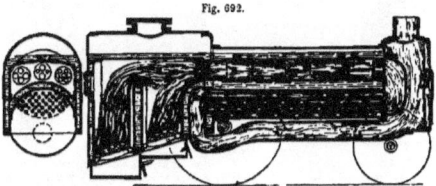

Fig. 692.

Beattie's Locomotive Boiler, in which there are two fire grates divided by a vertical water bridge in the fire box; the combustion chamber is over the tubes, and below them is a long return-back flue leading to the smoke box. Patented in the year 1854.

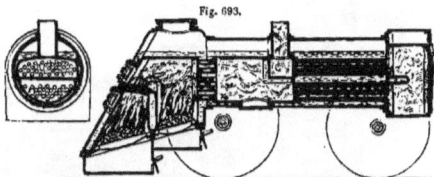

Fig. 693.

Beattie's Locomotive Boiler, in which there are two fire boxes and grates, and the front of the boiler constructed at an angle; there are short tubes, and beyond them a combustion chamber, and beyond it a set of return tubes, causing the chimney to be in the centre of the length of the boiler. Patented in the year 1854.

nation with coke, can be used, and the gases more perfectly consumed. The two furnaces are separated by the transverse water space partition, and the inside furnace is charged with coal at the outer end of the cylindrical retorts, which are fitted with doors, as shown in the drawing. The internal end, which rests on the partition, is left open or partially

In the space between the retorts another similar retort may be placed, or vertical fire tiles may be introduced instead, situated in such a way as to have openings or spaces between them to allow the flame thrown off the coal in the outer furnace to ascend between them.

In the intermediate combustion chamber is

Fig. 694.

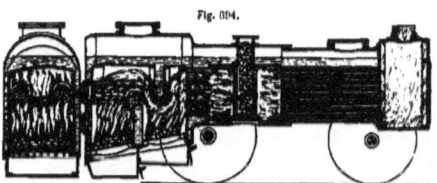

Beattie's Locomotive Boiler, in which there are two fire grates; above the outer fire grate are two horizontal cylindrical brick retorts, through which the fuel for the inner fire grate is passed; between and above the fire grates are two water bridges, and in front of the inverted bridge are short tubes leading into a combustion chamber fitted with a vertical brick retort filled with coke, and from this chamber are the long tubes leading to the smoke box. Patented in the year 1854.

Fig. 695.   Fig. 696.   Fig. 697.   Fig. 698.

  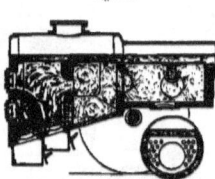 

Beattie's Locomotive Fire Box, containing two grates, two water bridges, and brick tiles in front of the tubes. Patented in the year 1854.

Beattie's Locomotive Fire Box, containing two grates, two water bridges, and brick tiles above and on each side of the water bridges. Patented in the year 1854.

Beattie's Locomotive Fire Box, containing three water bridges, brick tiles, and, beyond the box, a combustion chamber fitted with an inverted water bridge and brick tiles also. Patented in the year 1854.

Beattie's Locomotive Fire Box, fitted with two grates divided by a vertical water bridge. Patented in the year 1854.

open to allow the gas produced by the distillation of the coal to escape, and the coke produced from the coal after distillation to be delivered into the inner furnace by the fireman pushing it forward with a rake made for that purpose.

placed a fire clay retort filled with coke, and placed on the perforated fire tile stand. This retort has holes or slots to allow the gases and flame escaping from the furnaces to pass through it and through the incandescent fuel contained in it before entering into the long

LOCOMOTIVE BOILERS.

tubes leading from the combustion chamber through the boiler into the smoke box.

There is an opening, having a close fitting cover leading into the top of the combustion chamber through the water space over it, through which the retort can be charged with the coke.

In the year 1855, another French locomotive boiler invention appeared, as shown by Fig. 700, and this also is sufficiently explained under the illustration.

Mr. Beattie next showed to the world that his ideas on locomotive boilers were not exhausted, which is evident from the facts of

Fig. 699.

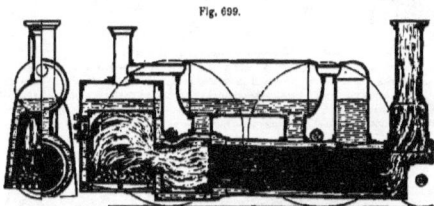

Blavier's Locomotive Boiler, arranged to have the base of the shell near the level of the rails while the upper portion is higher than generally. Patented in the year 1854.

Fig. 700.

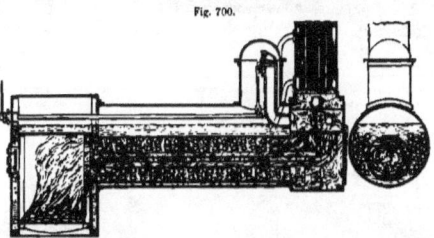

Montely's Locomotive Boiler, containing a tube flue fitted with a superheating steam coil, and on each side of the flue are flame tubes passing through the water space. In the chimney is fitted a tubular superheater or feed water heater, as may be required. Patented in the year 1855.

At the same period, appeared four different examples of locomotive fire boxes, as illustrated by Figs. 695 to 698.

Directly after Beattie, came a Frenchman with a most contradictory design—which nevertheless was allowed a patent by our obliging law—as illustrated by Fig. 699, and equally well explained under it also.

the nine modifications we now illustrate and explain.

Fig. 701 illustrates a locomotive boiler, in which there are four fire boxes, a prolonged combustion chamber fitted with sheet water spaces, and short tubes leading to the smoke box. Fig. 702 illustrates a locomotive boiler, in which U-shaped water spaces, with tubes in

2 M

the barrel, form the main feature. Fig. 703 illustrates curvous water spaces in connection with those shaped as before in the barrel combustion chamber, and the remaining five

Fig. 701.

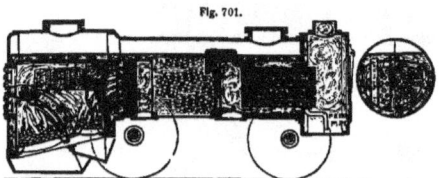

Beattie's Locomotive Boiler, in which there are four fire boxes; the two outer fire boxes are arched with curved water spaces and perforated brick slabs, and the inner fire boxes are fitted with brick bridges. In the combustion chamber beyond the fire boxes is a perforated brick slab, and beyond it are three vertical narrow water spaces that divide the flame before it enters the short tubes leading to the smoke box. Patented in the year 1855.

Fig. 702.

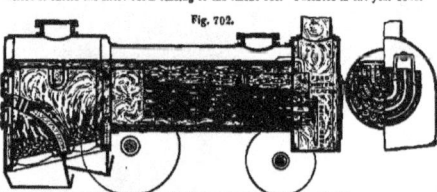

Beattie's Locomotive Boiler, in which there are two fire boxes; the outer box is arched by a water bridge and a perforated brick slab; the crown of the inner box has an inverted water bridge in it, and beyond is a prolonged combustion chamber; the remainder of the barrel of the boiler being fitted with U-shaped water spaces, between which are tubes. Patented in the year 1855.

Fig. 703.

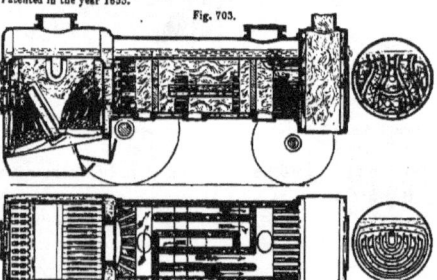

Beattie's Locomotive Boiler, in which there are two fire boxes divided by an angular water space; the barrel combustion chamber is fitted with a perforated slab and curvous narrow water spaces, blocked up between at the ends to cause the flame to make a return circuit before passing through the concentric U spaces beyond. Patented in the year 1855.

examples refer to transverse sections of various chambers, while the ninth example illustrates

Fig. 704.

Transverse Section of Beattie's Locomotive Barrel Combustion Chamber, fitted with vertical water spaces and horizontal flame tubes. Patented in the year 1855.

Fig. 705.

Transverse Section of Beattie's Locomotive Barrel Combustion Chamber, fitted with corrugated water spaces. Patented in the year 1855.

Fig. 706.

Transverse Section of Beattie's Locomotive Barrel Combustion Chamber, fitted with curved and corrugated water spaces. Patented in the year 1855.

Fig. 707.

Transverse Section of Beattie's Locomotive Barrel Combustion Chamber, fitted with vertical water spaces projecting up and down. Patented in the year 1855.

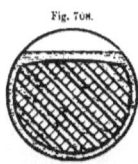

Fig. 708.

Transverse Section of Beattie's Locomotive Barrel Combustion Chamber, fitted with angular cross water tubes. Patented in the year 1855.

Fig. 709.

Transverse Section of Beattie's Locomotive Fire Box, fitted with central vertical and transverse horizontal water spaces, and also projecting water spaces supporting perforated brick retorts and slabs. Patented in the year 1855.

Beattie's extended ideas for a locomotive fire box.

In the year 1866, Mr. Crosland patented numerous arrangements of details for locomotive boilers, of which we have selected eleven examples.

Fig. 710 illustrates a locomotive boiler in which the fire box is fitted with three cross water tubes, with coal on them, situated above the ordinary grate; the chimney is fitted

with a tubular feed water heater. Fig. 711 differs only in the fire box, which is fitted

Fig. 710.

Crosland's Locomotive Boiler, in the fire box of which there are three cross water tubes and a horizontal moveable bridge, above the fire grate, and the fuel is placed on the tubes and grate. The chimney is fitted with a tubular feed water heater. Patented in the year 1856.

Fig. 711.

Crosland's Locomotive Boiler, in which the fire box is fitted with a perforated water space with fuel on it situated over the fire grate. Patented in the year 1856.

with a perforated water space instead of cross

tubes. Fig. 712 illustrates that the two horizontal moveable brick bridges at the front in the two previous examples were altered to suit the back of the fire box, as shown.

Mr. Crosland next ignored fire bars, and put tubes in their place, as shown by Fig. 713; and besides that, he put the feed water heater longitudinally on the shell.

Next appeared Crosland's six arrangements of fire boxes and moveable brick bridges, as illustrated by Figs. 714 to 719; and his eleventh arrangement is shown by Fig. 720.

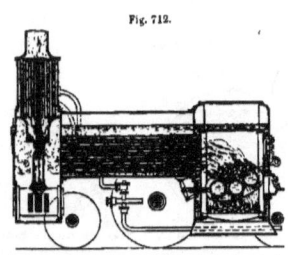

Fig. 712.

Crosland's Locomotive Boiler, in which the fire box has a horizontal moveable brick bridge at the back, next to the back cross tube. Patented in the year 1856.

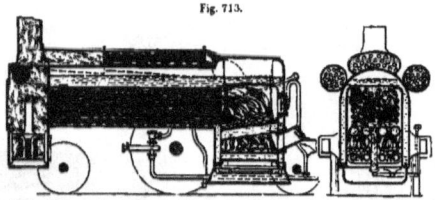

Fig. 713.

Crosland's Locomotive Boiler, in which the fire box has no fire bars, but longitudinal water tubes are used instead, and above the shell of the boiler are longitudinal tubular feed water heaters. Patented in the year 1856.

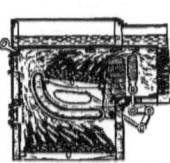

Fig. 714.

Crosland's Locomotive Fire Box, containing two fire grates, a curved water space above the bottom grate, and a vertically moveable brick bridge at the back of the fixed bridge. Patented in the year 1856.

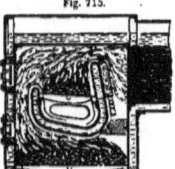

Fig. 715.

Crosland's Locomotive Fire Box, containing two fire grates, the top grate being contained in an air space chamber through the fire box, the back part of which is a water bridge, and behind it is a moveable brick bridge. Patented in the year 1856.

Fig. 716.

Crosland's Locomotive Fire Box, containing two fire grates, a curved water space over the bottom grate, and at the back end of the top grate an angular water bridge, and behind it a moveable brick bridge. Patented in the year 1856.

The next example of locomotive boiler occurred in the year 1857, and that we introduce merely as a novelty of idea rather than a practical improvement, as the illustration, Fig. 721, indicates.

The first day of the year 1858 gave birth

LOCOMOTIVE BOILERS.

Fig. 717.

Crosland's Locomotive Fire Box, containing two fire grates, an angular water space over the bottom grate, and at the back end of the top grate an angular water bridge, and behind it a moveable brick bridge. Patented in the year 1856.

Fig. 718.

Crosland's Locomotive Fire Box, containing two fire grates, an angular water space over the bottom grate, and at the back end of the top grate a vertical water bridge, and behind it a moveable brick bridge, and a combustion chamber beyond. Patented in the year 1856.

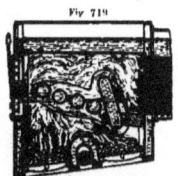

Fig. 719.

Crosland's Locomotive Fire Box, containing two fire grates divided by an arched water space for the wheel shaft to pass through, and above the grates are four cross water tubes and an angular water space, and behind it a moveable brick bridge. Patented in the year 1856.

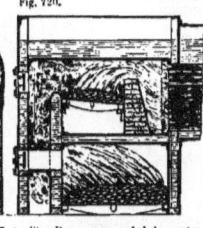

Fig. 720.

Crosland's Locomotive Twin Fire Boxes, surrounded by water spaced chambers, and above them is another fire grate situated opposite the tubes that lead to the smoke box. Patented in the year 1856.

practicable an arrangement as need be shown, but serves as a warning.

A Mr. Blinkhorn next informed the public that he also knew something of locomotive boilers, and the illustration, Fig. 723, on the next page, is an example; the arrangement being, that the ordinary flame tubes are made larger than general, and water tubes are passed through them and prolonged at each end through the fire and smoke boxes into the water spaces.

Another arrangement was also patented by

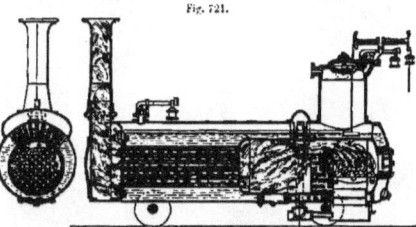

Fig. 721.

Malino's Locomotive Boiler, in which air pipes are situated in the crown of the combustion chamber beyond the fire box, and a blast used to cause a rapid combustion. Patented in the year 1857.

to the patent of a Mr. Clare, as illustrated on the next page, by Fig. 722, which is as im- | Blinkhorn, as illustrated by Fig. 724, in which the water tubes are divided longitudinally by a narrow water space chamber to cause a return action for the flame.

In the year 1859, appeared a flue locomotive boiler, by a Mr. Hunt, containing in the flue cross narrow water and flame spaces and short flame tubes leading into the smoke box,

Fig. 722.

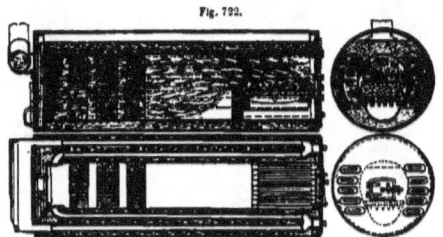

Clare's Locomotive Boiler, in which there is an "egg" section flue tube fitted with perforated brick slabs, and a twin set of return flame tubes on each side of the flue. Patented in the year 1858.

Fig. 723.

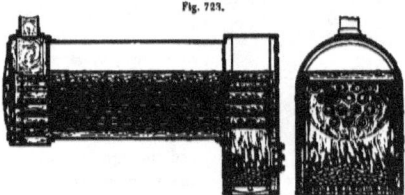

Blinkhorn's Locomotive Boiler, in which the barrel part of the shell is fitted with flame tubes, through which smaller water tubes pass that are connected to the water space plate of the fire box at the front ends, and the back ends are connected to the water box in the smoke box. Patented in the year 1859.

Fig. 724.

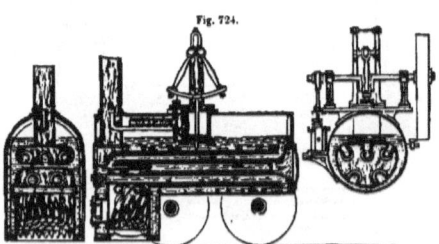

Blinkhorn's Locomotive Agricultural Boiler, in which there are longitudinal water tubes throughout the boiler, divided by a narrow water space to cause the return action of the flame. Patented in the year 1859.

LOCOMOTIVE BOILERS.

as shown in four sectional views, by Fig. 725. But Hunt did not originate the corrugated flue locomotive boiler, because it appeared in the year 1853, as shown on page 261 of this work.

After an interval of two years, a Mr. Martin introduced tubular superheaters in the smoke boxes of locomotives, as illustrated by Fig. 726. Two more years elapsed, and forth budded the fruits of wisdom of another Mr. Martin; but in this case the whole of the boiler was taken into consideration, as shown by Fig. 727, the principle of the arrangement being "the mixing of the gases to cause combustion in front of the tubes;" but other inventors had had that in their minds before Martin, as, for example, in the year 1852, a Mr. Selby evidently thought about it, because he patented the arrangement illustrated by Fig. 728, that consists of an angular crown fire box having a set of large tubes with curved ends, in connection with an intermediate combustion chamber, from which is a set of small tubes leading into the smoke box.

In the year 1864, also came forth the ideas of Mr. Fairlie on locomotive boilers; but he preferred the fire box to be in the centre of the arrangement, and two duplicate sets of

Fig. 725.

Hunt's Locomotive Boiler, in which the long tubes are omitted, and in their place cross water and flame spaces are constructed with short flame tubes leading to the smoke box. Patented in the year 1859.

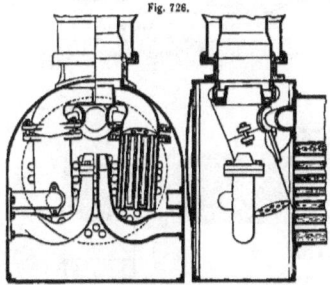

Fig. 726.

Martin's Locomotive Smoke Box, fitted with twin tubular superheaters. Patented in the year 1861.

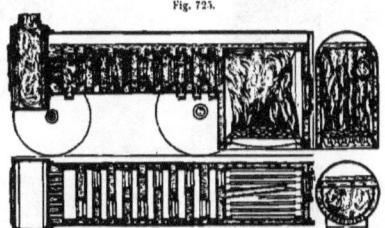

Fig. 727.

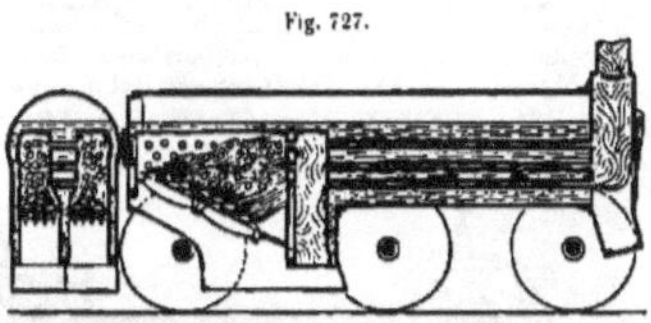

Martin's Locomotive Boiler, in which there are two fire boxes, divided by a longitudinal water space, perforated above the grates, at the back of which is a transverse water space, also perforated, and a combustion chamber in front of the usual flame tubes. Patented in the year 1864.

tubes on each side to cause a return action for the flame, with the smoke box and chimney above the fire box, while he also proposed a triple circuit for the flame with the chimneys at each extremity; and these three arrangements are shown by Figs. 729 to 731.

Fairlie could not fairly claim the return tube arrangement, because Hawthorn introduced it in the year 1839, as shown on page 254 of this work, and, apart from that, Stenson proposed return flame flue tubes, as shown by Fig. 732, in the year 1851.

Fig. 728.

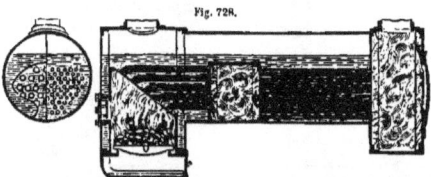

Selby's Locomotive Boiler, in which the fire box crown is angular, and fitted with large tubes having curved ends, and an intermediate combustion chamber, situated between those tubes and a set of small long tubes. Patented in the year 1852.

Fig. 729.

Fairlie's Locomotive Combined Boilers, in which there are return flame tubes in connection with the fire box and smoke boxes that are situated above each other, but the smoke boxes in this case are angular chambers passing through the water and steam over the fire box, which is in the centre of the arrangement, as also is the chimney. Patented in the year 1864.

Fig. 730.

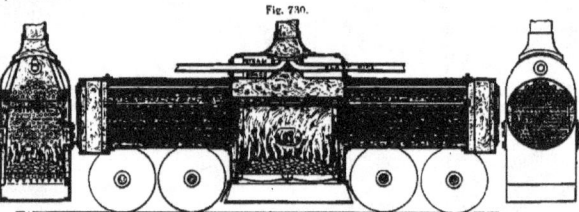

Fairlie's Locomotive Combined Boilers, in which there are return flame tubes in connection with the fire box and smoke box that are situated above each other in the centre of the arrangement, the chimney being central also, and separate steam chests in the smoke box. Patented in the year 1864.

The next example is a vertical tubular locomotive boiler, arranged as illustrated by Fig. 731.

cation from the top pipe to the bottom pipe with a forward and backward current.

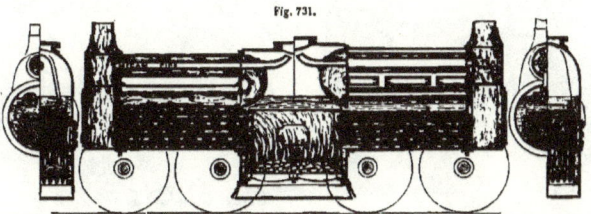

Fairlie's Locomotive Combined Boilers, in which there are return flue tubes over the ordinary tubes, and above them a return back flue tube, or a set of small flame tubes, to act as a superheater in either case, the chimnies being situated one at each end. Patented in the year 1864.

Fig. 733, the patent of a French "householder," as he styled himself.

Mr. Fairlie next put forth his ideas on locomotives again; but in this case they referred to the fire boxes only, which he persisted in placing between a twin set of tubes. He also proposed to burn either coal and oil

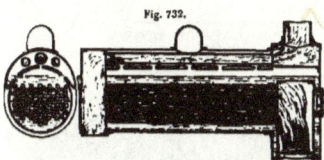

Stenson's Locomotive Boiler, in which there are return flame flue tubes over the ordinary flame tubes, and the smoke box and chimney over the fire box. Patented in the year 1851.

separately or both together, if better; in fact, the margin for discretion was so large in the mechanical arrangement that the original feature could be dispensed with entirely, as shown by Fig. 734. And in the same year a French engineer patented a locomotive tube boiler, as illustrated by Fig. 735, the principle of which is, that there is a perfect communi-

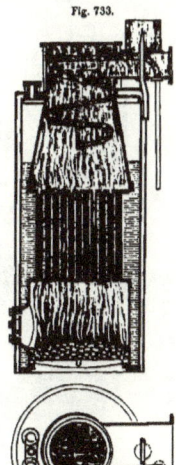

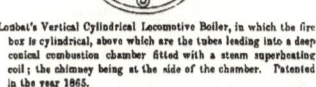

Loubat's Vertical Cylindrical Locomotive Boiler, in which the fire box is cylindrical, above which are the tubes leading into a deep conical combustion chamber fitted with a steam superheating coil; the chimney being at the side of the chamber. Patented in the year 1865.

LOCOMOTIVE BOILERS.

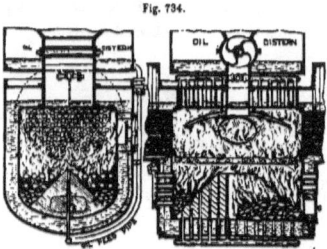

Fig. 734.

Fairlie's Locomotive Central Fire Box, fitted with apparatus to burn coal or oil, or both combined. Patented in the year 1865.

fire bars was patented in the ensuing year, 1866, which was perforations in the bottom of the fire box, as shown by Fig. 736.

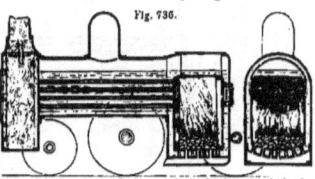

Fig. 736.

Woodward's Locomotive Boiler, in which the bottom of the fire box is perforated, instead of using fire bars. Patented in the year 1866.

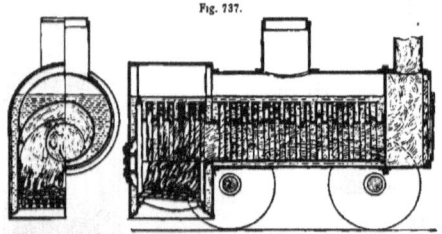

Fig. 737.

Holt's Locomotive Boiler, in which the long tubes are omitted, and in their place vertical water and flame spaces are fitted. Patented in the year 1866.

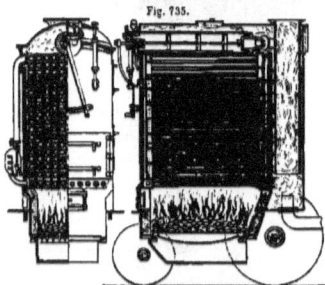

Fig. 735.

Belleville's Locomotive Tube Boiler, composed of horizontal water tubes connected at each end by twin nozzles, so as to cause an alternate end communication throughout. Patented in the year 1865.

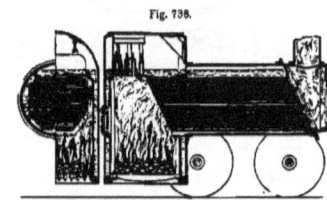

Fig. 738.

Holt's Locomotive Boiler, in which transverse flat sheet longitudinal flues are fitted in the place of ordinary tubes. Patented in the year 1867.

Mr Woodward's notion of a substitute for A Mr. Holt's locomotive boiler next appeared, in which a corrugated flue was proposed in the place of tubes, as shown by Fig. 737; and, in 1867, Mr. Holt proposed

LOCOMOTIVE BOILERS.

sheet flues for the same purpose, as illustrated by Fig. 738.

In the year 1868, a Mr. Bezy, a French locomotive boilers, in which vertical water spaces perforated with flame tubes were his improvement, as illustrated by Fig. 740.

After Miller, came a French engineer, Thirion by name, with the arrangement of vertical syphon tubes, U-shaped, in the place of the ordinary tubes, as illustrated by Fig. 741. And, following the Frenchman, came a Mr. Fox, who was filled with the notion of putting flame directly into water to raise steam, as illustrated by Fig. 742, and explained as well beneath.

The year 1871 began next, and immediately after that, a Mr. Norton invented a locomotive or portable boiler, as illustrated by Fig. 743, in which the tubes are water tubes fixed *across* the fire box and combustion

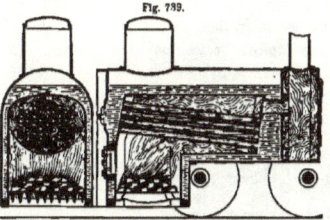

Fig. 739.

Bezy's Locomotive Boiler, in which there is a combustion chamber fitted with longitudinal water tubes. Patented in the year 1868.

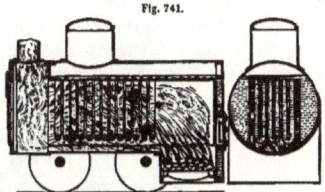

Fig. 741.

Thirion's Locomotive Boiler, in which the long tubes are omitted, and in their place vertical syphon tubes are fitted. Patented in the year 1869.

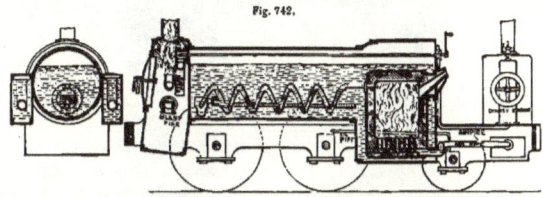

Fig. 742.

Fox's Locomotive Boiler, in which the fire box is lined with fire brick to burn liquid fuel that is forced by pumps into the box, from which the products of combustion are forced also by the pumps through a spiral coil, and from it forced into the water in the boiler to assist the generation of steam. Patented in the year 1869.

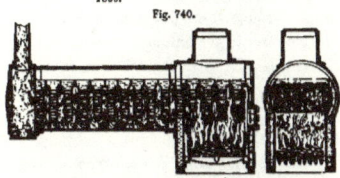

Fig. 740.

Miller's Locomotive Boiler, in which the long tubes are omitted, and in their place vertical water spaces, perforated with flame tubes, are fitted. Patented in the year 1869.

engineer, patented a water tube locomotive, as shown by Fig. 739.

Next came the ideas of a Mr. Miller on

chamber; and at one end of the tube is a communicative chamber in connection with each tube separately, that finally leads to the main end chamber; and at the other end is a single spaced main chamber only.

Following Norton's arrangement, came the ideas of a Mr. Girdwood on locomotive boilers, as shown by Fig. 744. The arrangement consists in having the main fire in only one of a set of two boilers, and a flue to lead the fire gases from the first boiler into a chamber in the second or chimney boiler. The fire box of the second boiler has two grates at the bottom, which is for the admission of air to complete the combustion of the fire gases should they not be completely burnt on leaving the first boiler. To insure ignition of the gases in the fire box a small quantity of incandescent coke may be kept on the grating.

A vertical up-take and chimney is fitted to the second boiler, for leading the fire gases away from the chamber, and the exhaust pipe of the engine is applied to increase the draught by directing the exhaust steam up the chimney, whilst to prevent the fire gases from proceeding from the inlet flue by too direct a course to the chimney, a vertical fire brick slab is placed in the fire box, so as to cause the gases to first pass down to the lower part of the box.

A Mr. Laharpe's notion of a locomotive boiler adapted for large driving wheels next appeared, as illustrated by Fig. 745. And at the same time came another arrangement by the same inventor for two pairs of large driving wheels, as shown by Fig. 746.

We conclude this chapter by illustrating and explaining a locomotive engine and boiler

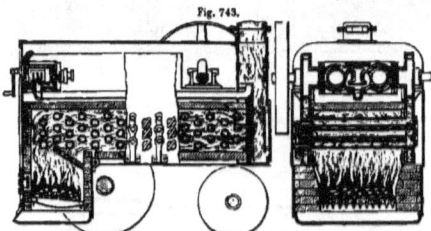

Fig. 743.

Norton's Locomotive or Portable Boiler, in which the water tubes are secured across the fire box and combustion chamber, which are formed by cast iron chambers, at the sides only, the end and the top being brickwork. Patented in the year 1871.

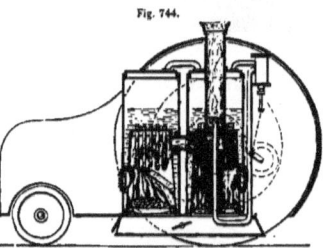

Fig. 744.

Girdwood's Vertical Locomotive Boilers, in which there are one fire grate in one boiler, and two fire grates in the other boiler, and hanging water tubes in both boilers, but only one chimney. Patented in the year 1871.

invented and patented by Mr. Perkins, in the year 1836. The engine is shown in sectional views, by Fig. 747. The steam acts only against one side of the piston; no cover or stuffing box is necessary to the other end of the cylinder, and, according to the arrangement shown, there only requires the guide bar, against which the spindle of the piston valve knocks at each stroke, in order to open the same for the passage of the steam into the atmosphere at the end of the stroke; and ment consists of generating steam through the medium of certain closed tubes containing confined and surcharged steam.

The boiler consists of a series of tubes, the one part of each tube projecting downwards into the flame, the other extending above the bottom of the boiler, and the tubes are consequently surrounded by the water in the boiler. The tubes are hermetically closed to prevent the escape of steam.

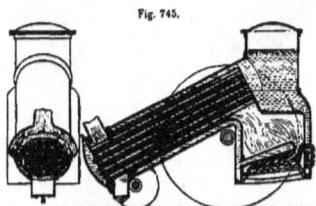

Fig. 745.

Laharpe's Angular Tube and Barrel Locomotive Boiler, adapted for a pair of large driving wheels. Patented in the year 1871.

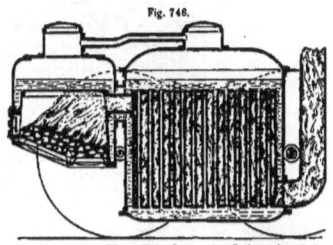

Fig. 746.

Laharpe's Vertical Water Tube Locomotive Boiler, adapted for two pairs of large driving wheels. Patented in the year 1871.

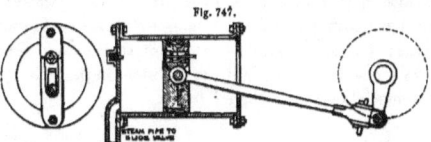

Fig. 747.

Perkins' Single Steam Action Locomotive Engine, requiring no piston or guide rods, or stuffing boxes, the steam acting on the back of the piston only. Patented in the year 1836.

Perkins recommended that high pressure steam should be employed and used expansively, say two hundred pounds to the square inch, and cut off at about one-eighth of the stroke, and expand down to atmospheric pressure.

The boiler is illustrated, on the next page, by Fig. 748; and the principle of the arrangement. By this arrangement, states Perkins, important results will be obtained; there will be no incrustation of the interior of the tubes, and the heat from the furnace will be quickly transmitted upwards; that the outer surfaces of the tubes will not be liable to scaleage or oxidation, which result will of course tend much to preserve the boilers so constructed.

The tubes are each to have a small quantity of water depending on the degree of pressure required to the engine; and in order for the working of this construction of boiler to the greatest advantage, the density of the steam in the tubes should be somewhat more than that intended to be produced in the boilers; one thousand eight hundredth parts, and so on, for greater or lesser degrees of pressure; by which means the tubes will, when the boiler is at work, be pervaded with steam, and any additional heat applied thereto will rise quickly to the upper parts of the tubes, and be given off to the surrounding water con-

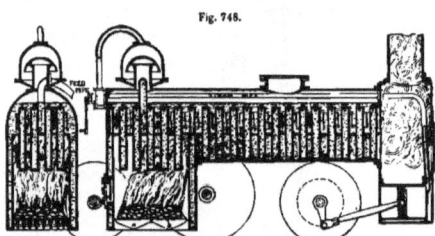

Fig. 748.

Perkins' Locomotive Boiler, in which vertical tubes containing surcharged steam assist the flame to raise the steam for working the engine at a pressure of 200 lbs. on the square inch. Patented in the year 1836.

and for steam and other boilers under atmospheric pressure, then the quantity of water to be applied in each tube is to be about one one thousand eight hundredth part of the capacity of the tube; for a pressure of two atmospheres, two one thousand eight hundredth parts; for three atmospheres, three tained in the boiler; because steam already saturated with heat requires no more heat to keep the atoms of water in their expanded state, consequently becomes a most useful means of transmitting heat from the furnace to the water in the boiler.

CHAPTER VIII.
BOILER STEAM SAFETY VALVES AND GEAR.

Fig. 749.

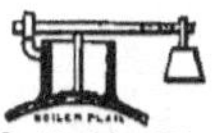

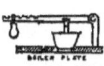

Papin's Safety Valve, consisting of a disc and rod seated in a recess formed with the boiler or digester, as it was termed; the disc was acted on by a weight lever. Invented in the year 1695.

Fig. 750.

Savery's Safety Valve, consisting of a solid cone seated on the boiler plate and acted on by a weight lever. Used in the year 1698.

Fig. 751.

Watt's Land Stationary Safety Valve, consisting of a disc acted on by series of direct weights, and a chain and handle connected to the valve rod above the weights to raise them and the valve. Invented in the year 1800.

Fig. 752.

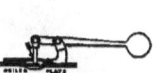

Vacuum or Atmospheric Safety Valve, for the purpose of permitting the atmosphere to fill the boiler when empty and prevent collapsing. Used since the year 1600.

Fig. 753.

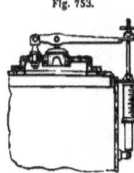

Locomotive Safety Valve, consisting of a disc acted on by a lever and box spring at its end, as used first in the year 1831.

Fig. 754.

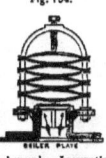

Stephenson's Locomotive Safety Valve, consisting of a disc acted on by a series of curved flat springs adjusted by a set screw. Invented in the year 1833.

Fig. 755.

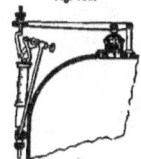

Locomotive Safety Valve, consisting of a half globe acted on by a lever fitted with adjustable lever gear in connection with the spring box. Used in the year 1840.

Fig. 756.

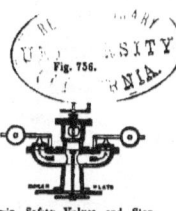

Twin Safety Valves and Stop Valve, used for land stationary boilers since the year 1850.

Fig. 757.

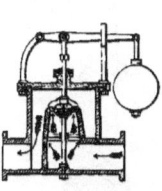

Spencer's Safety Valve, consisting of a double seated cone, the bottom seat being formed in the casing, and the top seat a disc secured by a vertical support rod and nut. Patented in the year 1852.

Fig. 758.

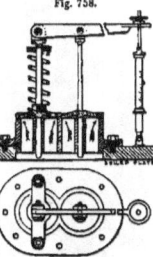

Scott's Twin Safety Valves, consisting of two discs that are acted on above by a spring lever and a spring direct, so that both valves lift the lever. Patented in the year 1853.

Fig. 759.

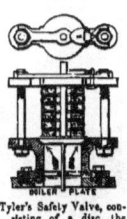

Tyler's Safety Valve, consisting of a disc, the seat for which is india-rubber, and the spring acting on the valve is formed by layers of india-rubber and metal. Patented in the year 1853.

Fig. 760.

Ramsbottom's Safety Valves, consisting of two cones acted on by a hand lever in connection below with a coil spring situated between the seat pipes of the valves. Patented in the year 1855.

280 BOILER STEAM SAFETY VALVES AND GEAR.

Fig. 761.

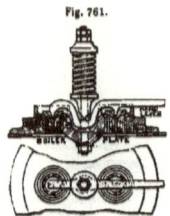

Ramsbottom's Safety Valves, consisting of two cones acted on by a hand lever in connection above with a coil spring that is adjusted with a cap and two nuts on the rod. Patented in the year 1855.

Fig. 762.

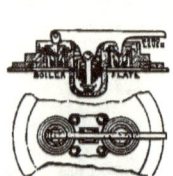

Ramsbottom's Safety Valves, consisting of two cones acted on by a hand lever in connection below with a volute spring that is secured in the recess between the valves. Patented in the year 1855.

Fig. 763.

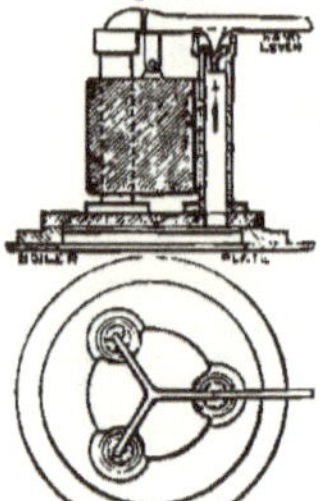

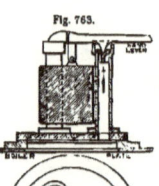

Ramsbottom's Safety Valves, consisting of three cones acted on by a hand lever in connection below with one direct weight situated between the seat pipes. Patented in the year 1855.

Fig. 764.

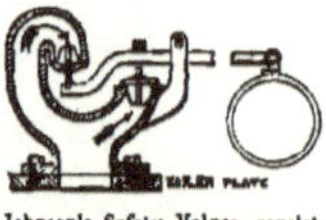

Johnson's Safety Valves, consisting of two valves opening up and down for the escape of the steam, and the "down" valve acts as the ordinary suspension pin for the lever. Patented in the year 1855.

Fig. 765.

Holt's Safety Valves, consisting of two discs that, when rising, permit the escape of the steam under and over the seats, the weight being direct on the top valve. Patented in the year 1856.

Fig. 766.

Holt's Safety Valves, consisting of two discs that, when rising, permit the escape of the steam under and over the seats, the weight being hung from the top valve. Patented in the year 1856.

Fig. 767.

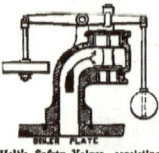

Holt's Safety Valves, consisting of two discs that, when rising, permit the escape of the steam under and over the seats; the lever is weighted at each end in proportion to the relative areas of the valves. Patented in the year 1856.

Fig. 768.

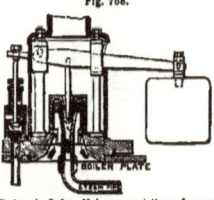

Bodmer's Safety Valves, consisting of a cap valve and weight lever in connection with a ball and socket regulating feed water valve. Patented in the year 1857.

Fig. 769.

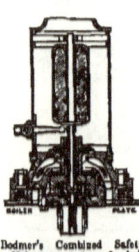

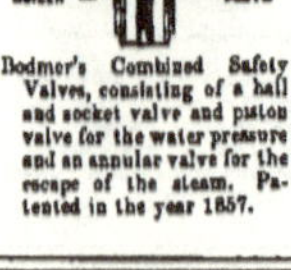

Bodmer's Combined Safety Valves, consisting of a ball and socket valve and piston valve for the water pressure and an annular valve for the escape of the steam. Patented in the year 1857.

Fig. 770.

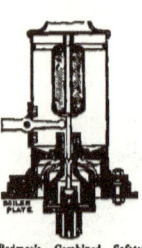

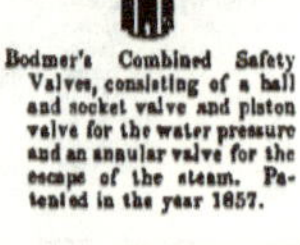

Bodmer's Combined Safety Valves, consisting of a ball and socket valve and piston valve for the water pressure and an annular valve for the escape of the steam. Patented in the year 1857.

Fig. 771.

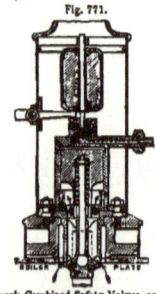

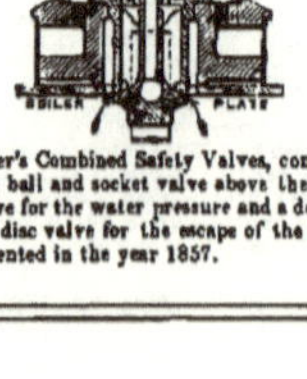

Bodmer's Combined Safety Valves, consisting of a ball and socket valve above the piston valve for the water pressure and a descending disc valve for the escape of the steam. Patented in the year 1857.

Fig. 772.

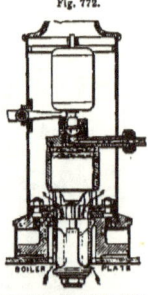

Bodmer's Combined Safety Valves, showing the steam escape valve opened. Patented in the year 1857.

BOILER STEAM SAFETY VALVES AND GEAR.

Fig. 773.

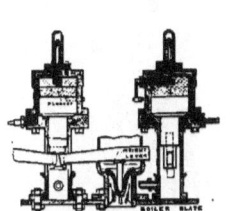

Bodmer's Regulating Feed Water Valves in connection with the weight levers of the steam escape valves. Patented in the year 1857.

Fig. 774.

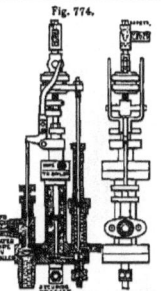

Bodmer's Regulating Feed Water Apparatus in connection with the spring levers of safety valves. Patented in the year 1857.

Fig. 775.

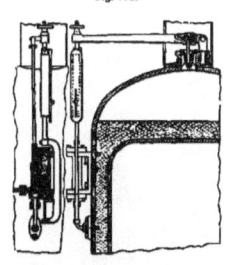

Bodmer's Safety Valve, consisting of a spring lever in connection with a feed water regulating piston valve and spring valve. Patented in the year 1857.

Fig. 776.

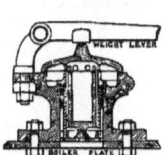

Haste's Safety Valves, consisting of a lever weight valve with two seats, and a cylinder formed within the smaller seat containing an independent direct weighted valve, so that the opening of the small valve will to some extent permit the steam to more readily open the large valve. Patented in the year 1858.

Fig. 777.

Illingworth's Safety Valve, consisting of a cylindrical valve with two seats, acted on by a direct weight; the steam escaping from the lower seat passes through the valve and weight. Patented in the year 1858.

Fig. 778.

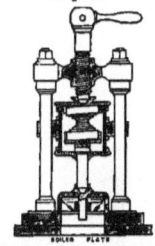

Smith's Safety Valve, consisting of a disc that is acted on by two volute spiral springs contained in a box, the adjustment being by a hand lever and screw stud, which raises or lowers the box. Patented in the year 1858.

Fig. 779.

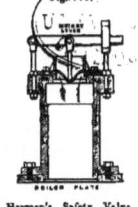

Harman's Safety Valve, consisting of a recessed cone portion below the seat for the lever rod to rest in, so that when the valve rises or falls it is free from the curve motion of the lever. Patented in the year 1859.

Fig. 780.

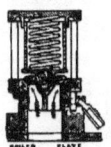

Clayton's Safety Valve, consisting of a recessed cone portion below the seat for the spring rod to rest in, so that when the valve rises or falls it is free from any side motion of the spring, which is contained in a box, and the adjusting casing padlocked. Patented in the year 1859.

Fig. 781.

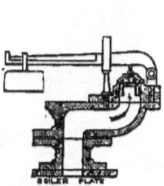

German Safety Valve, with a right angle branch for the passage of the steam. Used since the year 1860.

Fig. 782.

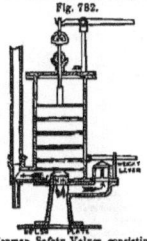

German Safety Valves, consisting of two separate discs acted on by a weight lever and a direct load. Used since the year 1860.

Fig. 783.

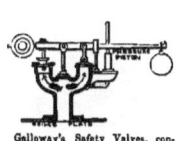

Galloway's Safety Valves, consisting of two valves closing and opening up and down, acted on by a lever in connection with a steam piston and rod which assist to open the valves as the pressure in the boiler increases. Patented in the year 1861.

2 o

Fig. 784.

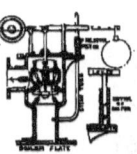

Galloway's Safety Valve, consisting of a double seat valve acted on by a lever in connection with a steam piston and rod which assist to raise the valve as the pressure in the boiler increases. Patented in the year 1861.

Fig. 785.

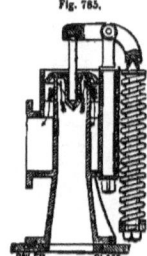

Naylor's Safety Valve, consisting of a disc with a deep recess for the lever rod to rest in, and the curved end of the lever is pressed up by the screw spring at the side of the casing. Patented in the year 1863.

Fig. 786.

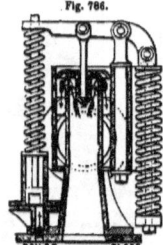

Naylor's Safety Valve, consisting of a disc with a deep recess for the lever rod to rest in; the curved end of the lever is pressed up by the screw spring, but the straight end is also pressed up by a screw spring acted on by a steam piston under it, so that the resistance is made more uniform. Patented in the year 1863.

Fig. 787.

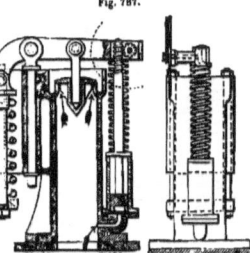

Naylor's Safety Valve, consisting of a disc with a deep recess for the lever rod to rest in; the curved end of the lever is pressed up by the screw spring, but the straight end is also pressed up by a screw spring acted on by a steam piston under it, so that the resistance is made more uniform; an indicator is also fitted. Patented in the year 1863.

Fig. 788.

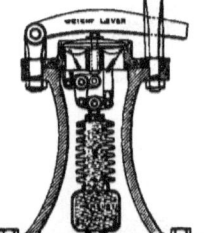

Maah's Safety Valve, consisting of a disc valve, over which is an ordinary weight lever, and under it a set of compound levers in connection with corrugated and drum cylinders filled with mercury, which expand and contract according to the surrounding temperature, and thus actuate the valve. Patented in the year 1863.

Fig. 789.

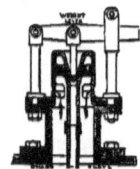

German Balance Piston Safety Valves, consisting of a piston seat for a cone that is formed with a cylinder in which the piston is fitted, the cylinder having a separate seat. Used since the year 1864.

BOILER STEAM SAFETY VALVES AND GEAR.

Fig. 790.

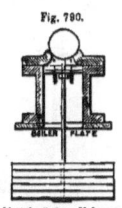

Baldwin's Safety Valve, consisting of a globe seated on an eccentric seat, so that when the valve rises the area of the opening is increased proportionately, the weight being suspended under the valve. Patented in the year 1866.

Fig. 791.

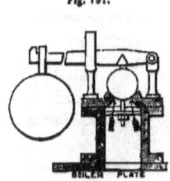

Baldwin's Safety Valve, consisting of a globe seated on an eccentric seat, so that when the valve rises the area of the opening is increased proportionately, the valve being acted on by a lever and weight. Patented in the year 1866.

Fig. 792.

Swann's Safety Valve, consisting of a disc acted on by a weight lever that is connected at the back end to a spring rod, the action of the spring being to hold the lever down, as is the weight at the opposite end. Patented in the year 1866.

Fig. 793.

Parson's Safety Valve, consisting of a mushroom-sectioned disc having a weight suspended below it in the boiler, so that the weight and valve may oscillate without the steam escaping. Patented in the year 1867.

Fig. 794.

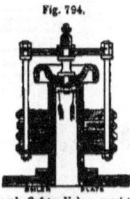

Parson's Safety Valve, consisting of a mushroom-sectioned disc having a cross bar above and two side rods suspending an annular weight below around the seating pipe above the boiler. Patented in the year 1867.

Fig. 795.

Parson's Safety Valve, consisting of a mushroom-sectioned disc acted on above by a bell-crank lever and suspended weight below. Patented in the year 1867.

Fig. 796.

Parson's Safety Valve, consisting of a mushroom-sectioned disc formed with a box above containing a spring in connection with a fixed rod passing through the valve. Patented in the year 1867.

Fig. 797.

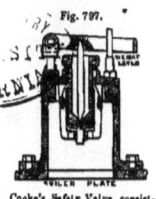

Cooke's Safety Valve, consisting of a deep cup valve with a knife edge seating and the lever rod pointed at each end, so that the action of the lever does not affect the valve. Patented in the year 1867.

Fig. 798.

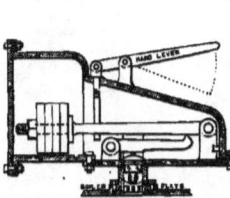

Cameron's Safety Valve, consisting of a disc acted on by two combined levers and a weight on the end of the upper and longer lever. Patented in the year 1867.

Fig. 799.

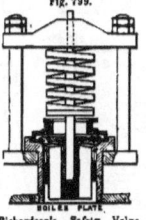

Richardson's Safety Valve, consisting of a disc fitted with an adjustable annular cap between the plate and the spring to cause a resistance to the escape of the steam independently of the action of the spring. Patented in the year 1867.

Fig. 800.

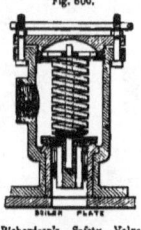

Richardson's Safety Valve, consisting of a disc formed with an inverted recess to cause a resistance to the escape of the steam independently of the action of the spring. Patented in the year 1867.

Fig. 801.

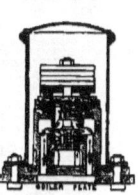

Sanders' Safety Valves, consisting of two discs of unequal diameters; and the top disc is fitted with a direct weight seated on four screw springs to counteract any vibration. Patented in the year 1868.

BOILER STEAM SAFETY VALVES AND GEAR.

Fig. 802.   Fig. 803.   Fig. 804.

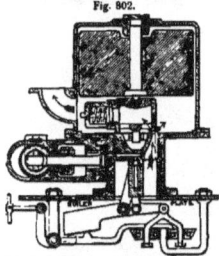 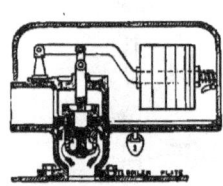 

Church's Safety Valve, consisting of discs and piston valves having vertical and horizontal motions, and acted on by springs, weights, and levers. Patented in the year 1868.

Ashcroft's Safety Valves, consisting of a cylinder with a top seat and two bottom seats, which are contained in a perforated box screwed in the casing, the valves being acted on by a weight lever above. Patented in the year 1868.

Ashcroft's Safety Valves, consisting of two discs above and below the seatings, the use of the lower disc being to prevent any water escaping with the steam and to close suddenly if the spring on the top disc breaks. Patented in the year 1868.

Fig. 805.   Fig. 806.   Fig. 807.   Fig. 808.

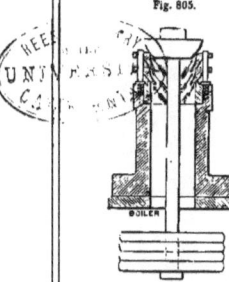 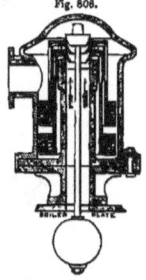 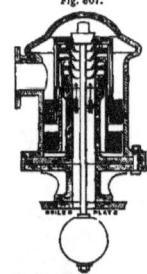 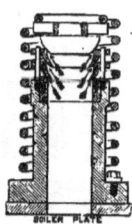

Hopkinson's Safety Valves, consisting of three cones, the upper cone being acted on directly by a heavy weight. Patented in the year 1870.

Hopkinson's Safety Valves, consisting of three cones; the top cone is acted on by a hung weight, and the other two cones by weights surrounding the cylinders of the cones. Patented in the year 1870.

Hopkinson's Safety Valves, consisting of five cones; the top cone is acted on by a hung weight, and the second and third cones by weights surrounding the cylinders of the cones. Patented in the year 1870.

Hopkinson's Safety Valves, consisting of three cones, the upper cone being acted on by a coil spring surrounding the seat pipe. Patented in the year 1870.

Fig. 809.   Fig. 810.   Fig. 811.

 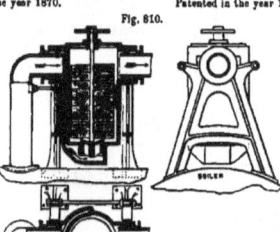

Hopkinson's Safety Valves, consisting of three cones separately acted on by separate coil springs surrounding the seat pipe and the cylinder of each valve. Patented in the year 1870.

Wilke's Marine Safety Valve, consisting of a disc with a cross bar and two side rods to suspend the direct weight below the casing, which is hung in two side frames to allow the weight and casing to oscillate. Patented in the year 1871.

Wilke's Marine Safety Valves, consisting of a disc acted on by a direct weight contained in the casing, which is hung in two side frames to allow the weight and casing to oscillate. Patented in the year 1871.

BOILER STEAM SAFETY VALVES AND GEAR.

Fig. 812.

Cowburn's Safety Valve, consisting of a semispherical disc in connection with a dome casing attached to an annular weight that surrounds the seat pipe. Patented in the year 1871.

Fig. 813.

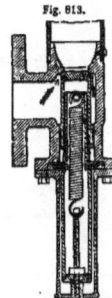

Mirchin's Safety Valve, consisting of a piston and cylinder seat valve in connection with a spring adjusted by a screw rod and hand wheel; the steam acting on the annular space of the piston causes the valve to open. Patented in the year 1871.

Fig. 814.

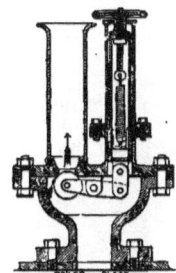

Mirchin's Safety Valve, consisting of a ball attached to a lever connected to a piston and spring above it; the steam acting on the annular space of the piston causes the valve to open. Patented in the year 1871.

Fig. 815.

Taylor's Safety Valve, consisting of a disc acted on by a series of flat curved springs. Patented in the year 1871.

Fig. 816.

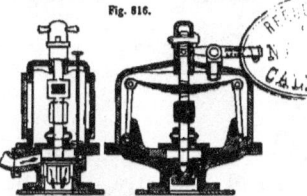

Taylor's Safety Valve, consisting of a disc fitted with a horizontal wheel and pinion for adjustment; and the valve is acted on by curved springs and a weight lever. Patented in the year 1871.

Fig. 817.

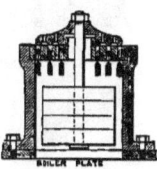

Lea's Safety Valve, consisting of a disc with three circular seats and three openings to correspond in it and the seating. Patented in the year 1871.

Fig. 818.

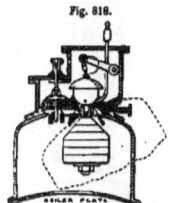

Watson's Marine Safety Valve, consisting of a hollow globe fitted with a cap and a seat having a groove in it to receive any lubricant; the small disc valve at the side forming no part of this invention. Patented in the year 1871.

BOILER STEAM SAFETY VALVES AND GEAR.

Fig. 819.

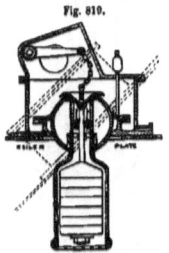

Watson's Marine Safety Valves, consisting of a hollow globe seated on a gland and packing; the top of the globe is formed with a seat, on which is a disc valve acted on by a hung weight contained in the casing attached to the globe. Patented in the year 1871.

Fig. 820.

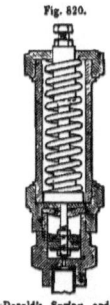

MacDonald's Spring and Piston Valve and Box, to assist the action of the lever of ordinary safety valves. Patented in the year 1872.

Fig. 821.

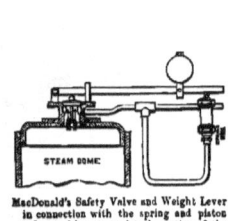

MacDonald's Safety Valve and Weight Lever in connection with the spring and piston valve and box to assist the action of the lever. Patented in the year 1872.

Fig. 822.

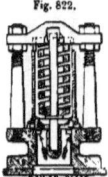

Giles' Land Safety Valve, consisting of a cylinder seated in a recessed seat, so that the steam, on the valve rising, is compelled to act on the outer edge, and thus compensate for the increasing resistance of the spring as it is compressed. Patented in the year 1872.

Fig. 823.

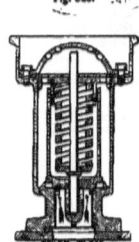

Giles' Locomotive Safety Valve, consisting of a cylinder seated in a recessed seat, so that the steam, on the valve rising, is compelled to act on the outer edge, and thus compensate for the increasing resistance of the spring as it is compressed. Patented in the year 1872.

Fig. 824.

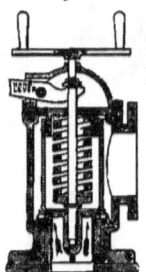

Giles' Marine Safety Valves, consisting of a cylinder seated in a recessed seating; the valve is turned on its seat by the twin handle lever, and a lift hand lever is also fitted to "ease" the valve. Patented in the year 1872.

Fig. 825.

Giles' Safety Valve, consisting of a disc seated on a cylinder with a recessed seating; and two coil springs are used with a cross bar to act on the valve rod. Patented in the year 1872.

Fig. 826.

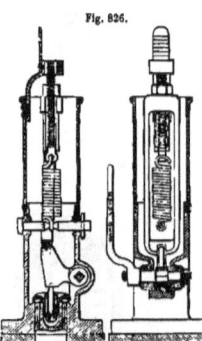

Field's Safety Valve, consisting of a disc having an annular seat and a recess for the hand lever rod to rest in, the upper part of the lever being in connection with the rods of a coil spring adjusted above by a screwed rod. Patented in the year 1971.

BOILER STEAM SAFETY VALVES AND GEAR.

Fig. 827.

Turton's Safety Valve Springs, consisting of two pairs of struts in connection with two cross bars having between them two spiral springs. Patented in the year 1872.

Fig. 828.

Turton's Safety Valve Springs, consisting of two struts in connection with a series of curved flat springs hinged and studded at each end, and the struts acting in the centre. Patented in the year 1872.

Fig. 829.

Turton's Safety Valve Springs, consisting of two struts in connection with a series of curved flat springs hinged on pins and adjusted by set studs and a bridle. Patented in the year 1872.

Fig. 830.

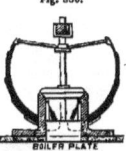

Turton's Safety Valve Springs, consisting of two struts in connection with a series of curved flat springs hinged at the centre line of the valve seat. Patented in the year 1872.

Fig. 831.

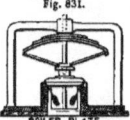

Turton's Safety Valve Springs, consisting of two struts in connection with a series of inverted curved flat springs suspended by the struts and bearing against the cross bar. Patented in the year 1872.

Fig. 832.

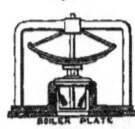

Turton's Safety Valve Springs, consisting of two struts in connection with a series of curved flat springs bearing on the collar above the valve. Patented in the year 1872.

Fig. 833.

Turton's Safety Valve Springs, consisting of two struts in connection with india-rubber springs. Patented in the year 1872.

Fig. 834.

Turton's Safety Valve Springs, consisting of two struts in connection with two levers that are attached to two coil springs. Patented in the year 1872.

Fig. 835.

Turton's Safety Valve Springs, consisting of two struts in connection with two single bar springs looped under pins below the struts and adjusted above by a cross bar and set screw, the bar being suspended. Patented in the year 1872.

Fig. 836.

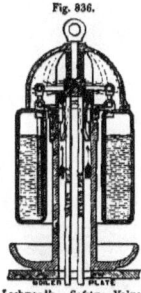

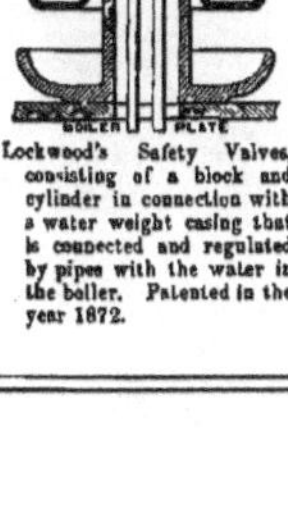

Lockwood's Safety Valves, consisting of a block and cylinder in connection with a water weight casing that is connected and regulated by pipes with the water in the boiler. Patented in the year 1872.

Fig. 837.

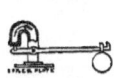

Casier's Safety Valve, consisting of a disc opening downwards and the weight lever under the valve. Patented in the year 1872.

Fig. 838.

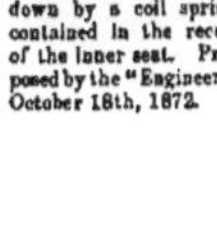

Safety Valve, consisting of an annular space for the steam to act up against a ring that bears on two seats; the ring is held down by a coil spring contained in the recess of the inner seat. Proposed by the "Engineer," October 18th, 1872.

BOILER ALARM SAFETY VALVES AND GEAR.

Fig. 839.

Annular Seat Safety Valve. Proposed in the "Engineer," November 8th, 1872.

Fig. 840.

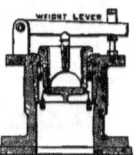

Double Seat Safety Valve. Proposed in the "Engineer," November 8th, 1872.

Fig. 841.

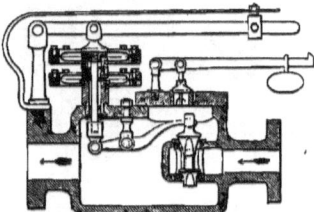

Pollard's Safety Valves in combination with valves, levers, and springs for the regulation of the steam pressure when passing from the boiler to the engine. Patented in the year 1867.

BOILER ALARM SAFETY VALVES AND GEAR.

Fig. 842.

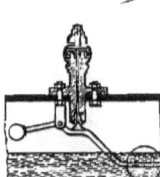

Alarm Whistle Valve, Float, Lever, and Balance. Used since the year 1840.

Fig. 843.

Alarm Float Chain-wheel Balance and Indicator. Used since the year 1840.

Fig. 844.

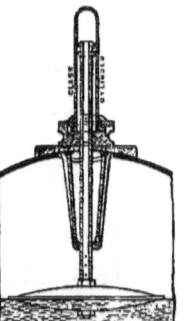

Johnson's Indicating Alarm Float, by means of a piston moving in a glass cylinder, secured on the top of the boiler. Patented in the year 1853.

Fig. 845.

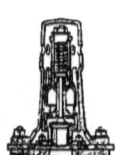

Taylor's Alarm Safety Valve, consisting of a disc and rod acted on by a coil spring contained in a whistle box; the whistle being situated between the valve and the spring. Patented in the year 1853.

BOILER ALARM SAFETY VALVES AND GEAR.

Fig. 846.

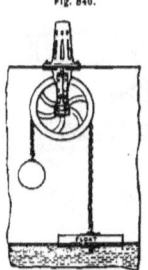

Tayler's Alarm Safety Valve Apparatus, inside the boiler, consisting of a disc float, chain, wheel, and counterbalance, in connection with the valve shown by Fig. 845. Patented in the year 1853.

Fig. 847.

Tayler's Alarm Safety Valve, Whistle and Spring. The valve opens downwards and is connected direct to the float. Patented in the year 1853.

Fig. 848.

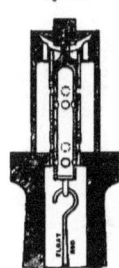

Tayler's Alarm Safety Valve, consisting of a sliding tube contained in a fixed tube, and both tubes perforated. Patented in the year 1853.

Fig. 849.

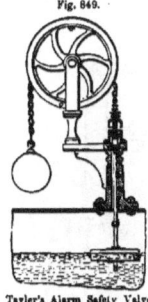

Tayler's Alarm Safety Valve Apparatus, outside the boiler, consisting of a disc float, chain, wheel, counterbalance, and whistle. Patented in the year 1853.

Fig. 850.

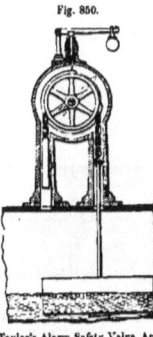

Tayler's Alarm Safety Valve Apparatus, outside the boiler, consisting of an eccentric rimmed wheel with a safety valve rod acting on it and a steam whistle to indicate when the valve is lifted. Patented in the year 1853.

Fig. 851.

Hall's Alarm Safety Valve Apparatus, consisting of a disc connected to a fusible plug on the fire box plate and a water pipe under and over the valve which, when lifting, admits water in the fire box besides through the plug casing. Patented in the year 1855.

Fig. 852.

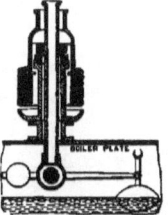

Cowburn's Water Level Float, Lever, and Counterbalance, in connection with a direct loaded safety valve, the hollow rod of which suspends the float lever. Patented in the year 1855.

Fig. 853.

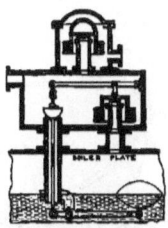

Cowburn's Water Level Float and Lever, in connection with a safety valve and lever, on the casing of which is a direct loaded safety valve. Patented in the year 1855.

2 P

BOILER ALARM SAFETY VALVES AND GEAR.

Fig. 854.

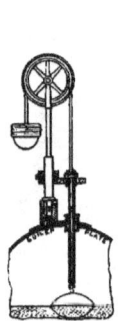

Cowburn's Water Level Float, Wheel, Chain, and Counterbalance, outside the boiler. Patented in the year 1855.

Fig. 855.

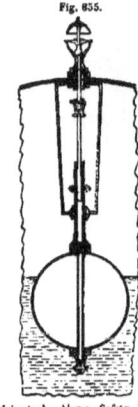

Johnston's Alarm Safety Valve Apparatus, consisting of a ball float that is directly connected to a spindle valve that descends to admit steam to the whistle above. Patented in the year 1856.

Fig. 856.

Routledge's Alarm Flame Tube, fitted with fusible plugs inside the boiler. Patented in the year 1856.

Fig. 857.

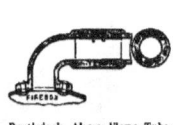

Routledge's Alarm Flame Tube, fitted with fusible plugs. Patented in the year 1856.

Fig. 858.

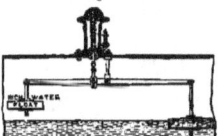

York's Alarm Safety Valve Apparatus, outside the boiler, consisting of two floats and a disc valve acted on by a spring; at the side of the casing is a whistle to indicate when the valve opens. Patented in the year 1856.

Fig. 859.

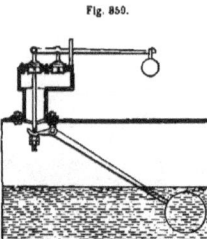

Knowelden's Alarm Safety Valve Apparatus, consisting of two safety valves; one being held down by a float lever inside the boiler, and the other acted on by the weight lever. Patented in the year 1856.

Fig. 860.

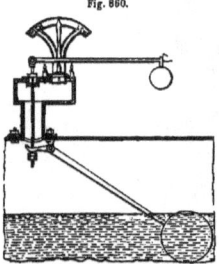

Knowelden's Alarm Safety Valve Apparatus, consisting of a float and lever that is connected to a weight lever which acts on a disc valve; the motion of the lever is indicated by a hand pointer and recording quadrant. Patented in the year 1856.

Fig. 861.

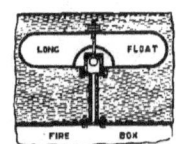

Knowelden's Alarm Low Level Safety Valve Apparatus, consisting of a ball connected to a long float that, when descending, admits the water in the boiler in the fire box. Patented in the year 1856.

BOILER ALARM SAFETY VALVES AND GEAR. 291

Fig. 862.

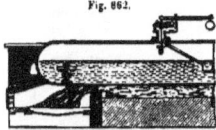

Knowelden's High and Low Level Alarm Safety Valve Apparatus, fitted to an egg-end boiler. Patented in the year 1856.

Fig. 863.

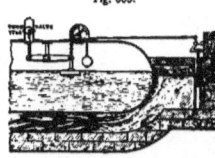

Walley's Alarm Safety Valve Apparatus, consisting of a float and lever in connection with the engine steam pipe throttle valve and a chain wheel that is connected to the flue damper. Patented in the year 1856.

Fig. 864.

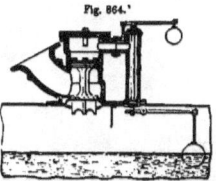

Horton's Safety Valves in connection with a water level float and lever, the action of the float and small valves being to regulate the pressure of the steam on the top of the piston of the large valve. Patented in the year 1857.

Fig. 865.

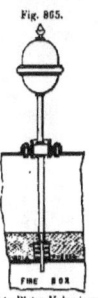

Parson's Safety Piston Valve in connection with a series of metal and india-rubber rings that are secured on the roof of the fire box; and the melting of the india-rubber permits the valve to fall and the steam to escape. Patented in the year 1858.

Fig. 866.

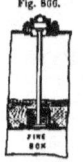

Parson's Safety Piston Valve in connection with a cylinder of india-rubber fixed in a casing on the roof of the fire box; and the melting of the india-rubber permits the valve to rise and the steam to escape. Patented in the year 1858.

Fig. 867.

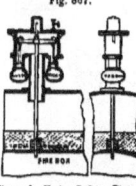

Parson's Twin Safety Piston Valves in connection with a series of metal and india-rubber rings that are secured on the roof of the fire box, for the same purpose as for Fig. 866. Patented in the year 1858.

Fig. 868.

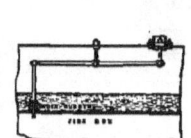

Parson's Safety Plug Valve and India-rubber Rings, for the same purpose as for Fig. 866, connected by a lever. Patented in the year 1858.

Fig. 869.

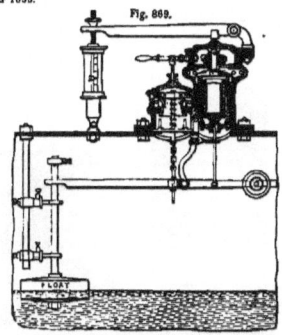

Haste's Safety Valves, acted on by a spring lever and a direct weight respectively; the weight valve is raised by the float lever, which also is in connection with a cylindrical safety valve that is lowered to open when the water is low in the boiler. Patented in the year 1858.

Fig. 870.

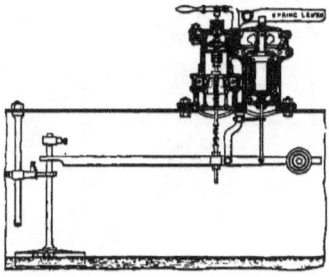

Haste's Safety Valves, acted on by a spring lever and a direct weight respectively; the weight valve is raised by the float lever, which is also in connection with a cylindrical safety valve that is lowered to open when the water is low in the boiler. Patented in the year 1858.

2 P 2

BOILER ALARM SAFETY VALVES AND GEAR.

Fig. 871.

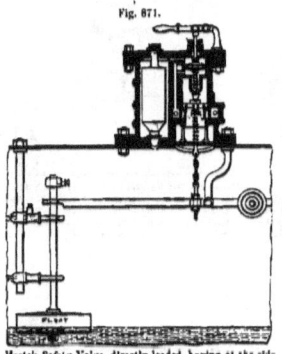

Hoste's Safety Valve, directly loaded, having at the side a cylindrical safety valve in connection with a float lever that lowers the valve when the water is low in the boiler. Patented in the year 1858.

Fig. 872.

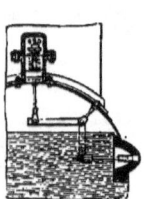

Parson's Alarm Valve in connection with lever gear in the boiler and a fusible plug at the side open to the flue. Patented in the year 1858.

Fig. 873.

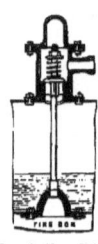

Parson's Alarm Valve in connection with a fusible plug on the roof of the fire box. Patented in the year 1858.

Fig. 874.

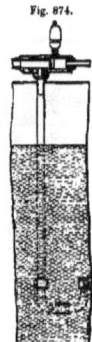

Parson's Feed Water Whistle Alarm, consisting of a piston working by the steam on one side and feed water on the other; and the whistle indicates when more feed is required. Patented in the year 1858.

Fig. 875.

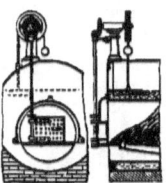

Wright's Water Float, Rod, Chain Wheel, Bevel Gearing, and Perforated Fire Door, for indicating the water level and regulating the draught. Patented in the year 1858.

Fig. 876.

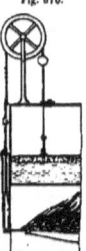

Wright's Water Float Rod and Chain Wheel in connection with the sliding furnace door. Patented in the year 1858.

Fig. 877.

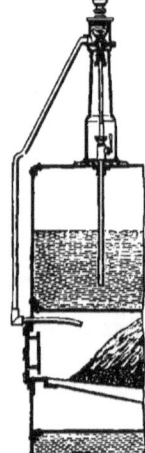

Bodmer's Alarm Valve and Pipes inside and outside the boiler, to indicate when the water is below the inside pipe. Patented in the year 1858.

Fig. 878.

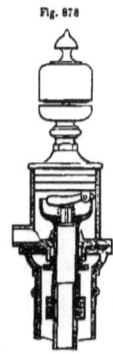

Bodmer's Alarm Valve and Whistle in connection with the apparatus shown by Fig. 877. Patented in the year 1858.

BOILER ALARM SAFETY VALVES AND GEAR.

Fig. 879.

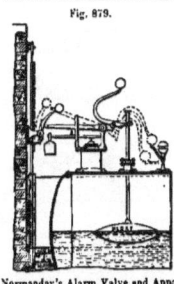

Normanday's Alarm Valve and Apparatus, consisting of a float and lever gear outside the boiler in connection with the damper and a whistle. Patented in the year 1858.

Fig. 880.

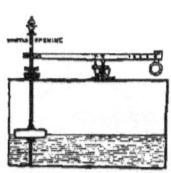

Archer's Alarm Valve Apparatus, consisting of a float, outside lever, safety valve, and steam whistle. Patented in the year 1858.

Fig. 881.

Archer's Reverse Seats Safety Valves in connection with the apparatus shown by Fig. 880. Patented in the year 1858.

Fig. 882.

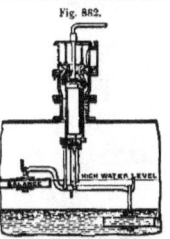

Illingworth's Alarm Safety Valve Apparatus, consisting of a cylindrical double seat valve with a direct weight under it, and below that a cone valve in connection with a lever float and balance weight; the central pipe over the disc valve being connected under a steam whistle. Patented in the year 1858.

Fig. 883.

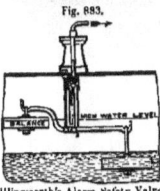

Illingworth's Alarm Safety Valve in connection with a lever, float, and balance weight inside the boiler. Patented in the year 1858.

Fig. 884.

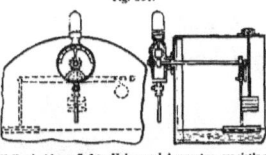

Walker's Alarm Safety Valve and Apparatus, consisting of a valve underhung with a weight, the rod being in connection with a float lever that, by a pinion and toothed quadrant, causes a whistle and hand pointer to indicate when the water is low in the boiler. Patented in the year 1859.

Fig. 885.

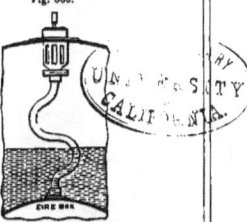

Davies' Alarm Safety Valve and Curved Pipe, connecting the valve casing to the fire box, so that on the valve rising the steam enters therein. Patented in the year 1860.

Fig. 886.

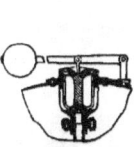

Davies' Double Seat Alarm Safety Valve and Casing, as shown with the curved pipe in Fig. 885. Patented in the year 1860.

Fig. 887.

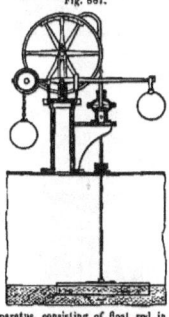

Galloway's Alarm Valve Apparatus, consisting of float rod in connection with a chain wheel shaft that is formed as a plug cock to admit steam under the diaphragm valve that acts on the lever of the safety valve. Patented in the year 1861.

Fig. 888.

McCarthy's Alarm Safety Valve and Apparatus, consisting of a direct weighted valve outside the boiler connected by a lever and rod to the float in the boiler; when the float lever raises the valve, the water in the boiler also flows through the inside and outside pipes to the fire box. Patented in the year 1862.

Fig. 889.

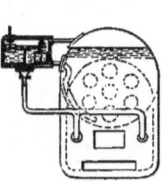

McCarthy's Alarm Safety Valve Float Lever Apparatus, contained in a box outside the boiler and connected to it by pipes. Patented in the year 1862.

BOILER ALARM SAFETY VALVES AND GEAR.

Fig. 890.

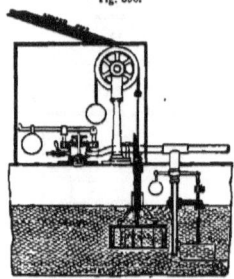

Turner's Alarm Safety Valve and Apparatus, consisting of a float, wheel, and counterbalance, also a float lever in connection with a throttle steam valve; and an ordinary safety valve. Patented in the year 1863.

Fig. 891.

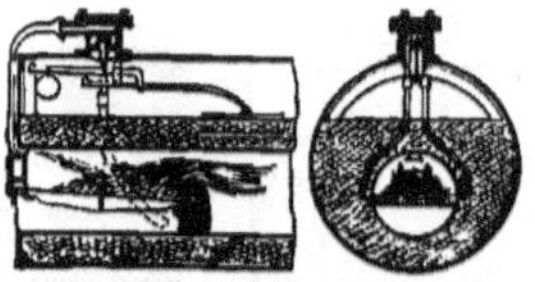

Hackett's Alarm Safety Valve and Apparatus, consisting of a valve in connection with a weight lever, float lever, and two water pipes, that, on the valve rising, conduct some of the water in the boiler to the fire box. Patented in the year 1866.

Fig. 892.

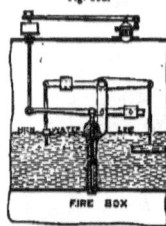

Bray's Alarm Safety Valve Apparatus, inside the boiler, consisting of a tube seating, on which is a cone acted on by a weight lever in connection with a weight connected to the lever of the safety valve outside the boiler; the tube also supports a float lever that, when in contact with the cone lever, admits the steam through the tube into the fire box. Patented in the year 1866.

Fig. 893.

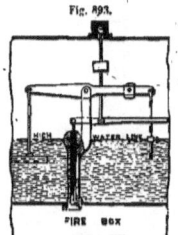

Bray's Alarm Safety Valve Apparatus, inside the boiler, consisting of a cone at the bottom of the tube secured on the roof of the fire box, and the action of the gear as in Fig. 892. Patented in the year 1866.

Fig. 894.

Bray's Alarm Safety Valve Apparatus, inside the boiler, consisting of the float and lever in connection with the lever and safety valve that admits steam into the fire box when the water is low. Patented in the year 1866.

Fig. 895.

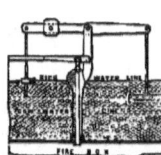

Bray's Alarm Safety Valve Apparatus, inside the boiler, showing the levers on the horizontal line in connection with Fig. 894. Patented in the year 1866.

Fig. 896.

Swann's Alarm Valve in connection with a float and chain direct. Patented in the year 1866.

Fig. 897.

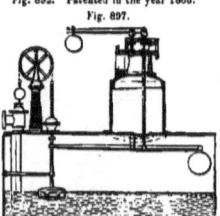

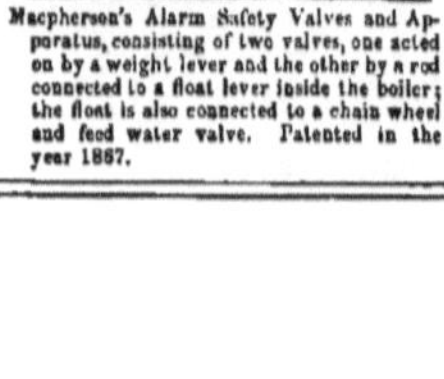

Macpherson's Alarm Safety Valves and Apparatus, consisting of two valves, one acted on by a weight lever and the other by a rod connected to a float lever inside the boiler; the float is also connected to a chain wheel and feed water valve. Patented in the year 1867.

Fig. 898.

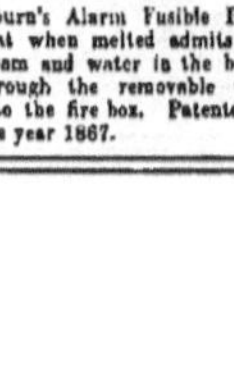

Cowburn's Alarm Fusible Plug, that when melted admits the steam and water in the boiler through the removable tube into the fire box. Patented in the year 1867.

Fig. 899.

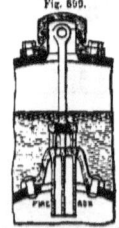

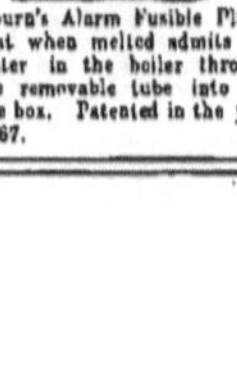

Cowburn's Alarm Fusible Plugs, that when melted admits the water in the boiler through the removable tube into the fire box. Patented in the year 1867.

Fig. 900.

Cowburn's Alarm Fusible Plugs, that when melted admits the water in the boiler in the fire box. Patented in the year 1867.

BOILER ALARM SAFETY VALVES AND GEAR.

Fig. 901.

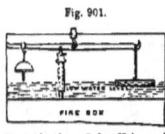

Fig. 902.

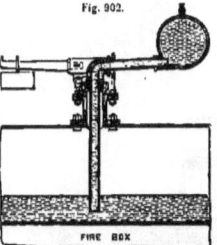

Fig. 903.

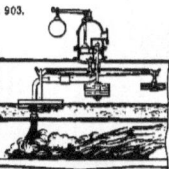

Kenyon's Alarm Safety Valve and Apparatus, inside the boiler, consisting of a tube secured to the fire box, having a valve on the top with a larger tube connected to a lever float and counterbalance; the lifting of the valve permits the water to enter the fire box. Patented in the year 1867.

Hugh's Alarm Safety Valve and Apparatus, consisting of a lever weight safety valve and also a direct underhung weight valve, the weight being in the boiler and the rod in connection with a lever float and counterbalance, and when that valve is lifted by the lever float the water in the boiler flows through the pipes into the fire box. Patented in the year 1869.

Benson's Alarm Safety Valve Water Pipe, consisting of a pipe vertical in the boiler and horizontal outside, with a hollow bell in connection with a lever weighted safety valve, which rises when the water is out of the pipe, to indicate the water is low in the boiler. Patented in the year 1868.

Fig. 904.

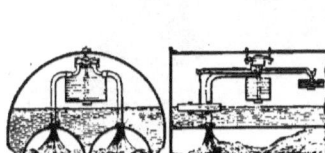

Fig. 905.

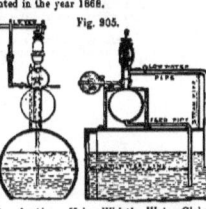

Fig. 906.

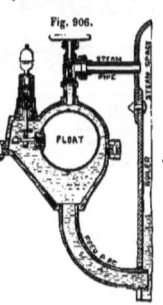

Hugh's Alarm Safety Valve and Apparatus, consisting of a valve and box, underhung weight float lever and counterbalance, and alarm water pipes. Patented in the year 1869.

Pratt's Alarm Valve Whistle, Water Globe, and Safety Valve combined; a quantity of water in the boiler, being in connection with the water in the globe, regulates the action of the safety valve and whistle. Patented in the year 1870.

Kimball's Alarm and Whistle Float Apparatus, situated outside the boiler front; the float rises and falls as the water in the boiler actuates when the steam valve is open. Patented in the year 1870.

Fig. 907.

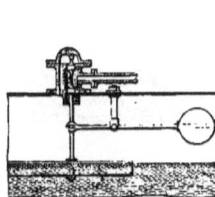

Fig. 908.

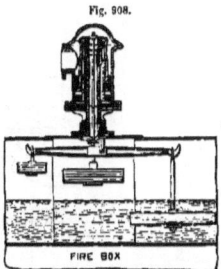

Fig. 909.

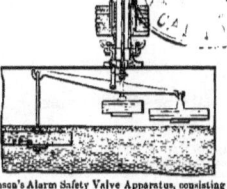

Langlet's Alarm Valve, Weight Lever, and Float, inside the boiler; the lowering of the float causes valve to descend and the steam to pass out through the casing pipe. Patented in the year 1870.

Hopkinson's Alarm Safety Valve Apparatus, consisting of two valve cylinders with weights surrounding the lower parts and a central valve with a hung weight in the boiler, there, in connection with a lever float and counterbalance, the central valve is lifted by the lever when the water is low in the boiler. Patented in the year 1870.

Adamson's Alarm Safety Valve Apparatus, consisting of one valve cylinder with weights surrounding the lower part and a central valve with a hung weight in the boiler, there, in connection with a lever float and counterbalance; the central valve is lifted by the float lever when the water is low in the boiler. Patented in the year 1871.

BOILER FEED PUMPS AND ENGINES.

Fig. 910.

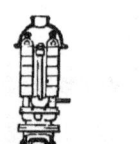

Cowburn's Alarm Safety Valve and Apparatus, inside the boiler, consisting of a direct weighted valve in connection with a float lever and balance; the lowering of the float assists to raise the valve. Patented in the year 1871.

Fig. 911.

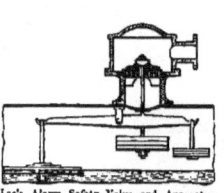

Lee's Alarm Safety Valve and Apparatus, consisting of a perforated valve and seat and a central curved valve underhung with a weight in connection with a lever, float, and counterbalance. Patented in the year 1871.

Fig. 912.

Kirk's Alarm Valve and Float in Mercury; the expansion of the mercury causes the float to lift the valve, and the steam blows the whistle above. Patented in the year 1871.

BOILER FEED PUMPS AND ENGINES.

Fig. 913.

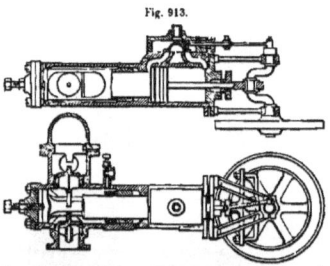

Cameron's Boiler Feed Pump and Engine, consisting of a piston rod-slotted head to drive the crank pin direct and the slide valve of the usual kind worked by a crank pin; the pump plunger is connected to the piston, and the pump valves are vertical, acting at the other extremity of the casing. Patented in the year 1852.

Fig. 914.

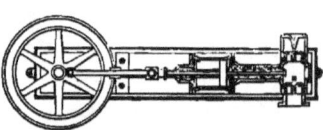

Johnson's Boiler Feed Pump and Engine, consisting of forming the pump barrel with the back end cover of the engine cylinder. Patented in the year 1852.

BOILER FEED PUMPS AND ENGINES.

Fig. 915.

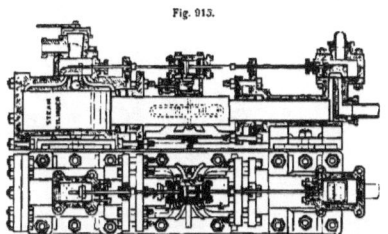

Newton's Boiler Feed Pump and Engine, consisting of a piston and plunger combined, the valves for the cylinder and pump being similar, and actuated by a steam piston contained in a small cylinder situated between the engine and pump; the valve of the small cylinder is worked by a rack and half pinion, the rod of which is moved by the plunger. Patented in the year 1853.

Fig. 916.

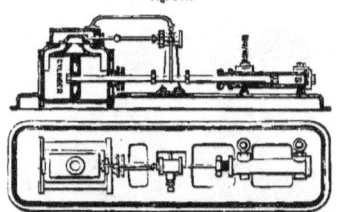

Knowelden's Boiler Feed Pump and Engine, consisting of an ordinary steam cylinder and slide valve that is actuated by a small engine supported on a standard; the piston is connected direct to the slide valve, and the steam actuates the piston by the lever and plug valve. Patented in the year 1861.

Fig. 917.

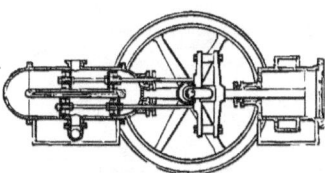

Cowan's Boiler Feed Pump and Engine, consisting of a twin barrel pump and four pistons fitted with India-rubber valves, the engine being the ordinary kind. Patented in the year 1862.

Fig. 918.

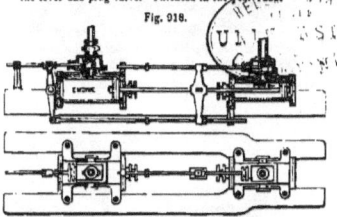

Duprey's Boiler Feed Pump and Engine, consisting of a direct acting engine and pump fitted with slide valves alike that are actuated by a cross bar on the piston rod and rods and a lever. Patented in the year 1864.

Fig. 919.

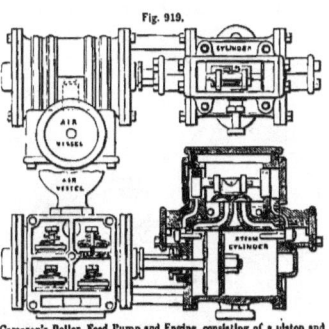

Cameron's Boiler Feed Pump and Engine, consisting of a piston and ported slide valve for the engine, actuated by the steam admitted by the tappet valves at the ends of the cylinder; the pump valves are metal discs with recesses for india-rubber and spiral springs at the backs. Patented in the year 1866.

Fig. 920.

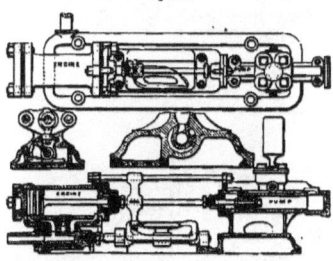

Tijou's Boiler Feed Pump and Engine, consisting of a steam cylinder fitted with a cylindrical valve that oscillates from the motion of a horizontal cam and a lever secured on the piston rod. Patented in the year 1868.

Fig. 921.

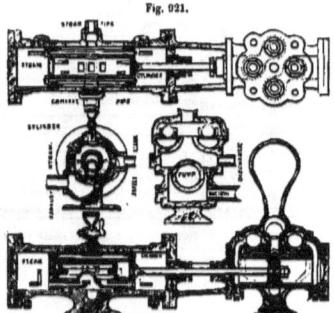

Maxwell's Boiler Feed Pump and Engine, consisting of a cylindrical piston containing a cylindrical valve with steam ports for supply and exhaust in the piston, that is forced by the steam backwards and forwards in the long cylinder. Patented in the year 1866.

Fig. 923.

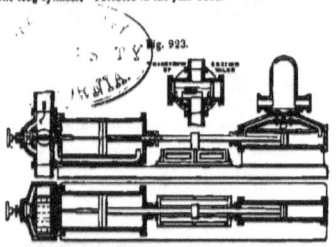

Ramsbottom's Boiler Feed Pump and Engine, consisting of a twin piston and ported sliding valve combined, situated in a casing at the back end of the cylinder, and the valve working at right angles to the motion of the piston, the valve's motion gear being a twisted bar and rod fitted in the piston rod. Patented in the year 1869.

Fig. 925.

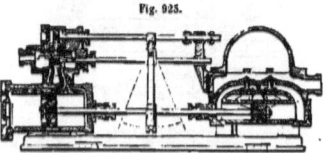

Davey's Boiler Feed Pump and Engine, consisting of a double ported slide valve, and above it a piston valve; and both valves are actuated by a lever that is actuated by the arm on the piston rod. Patented in the year 1871.

Fig. 922.

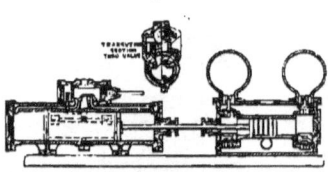

Walker's Boiler Feed Pump and Engine, consisting of a cylindrical piston which regulates the admission of the steam that actuates the cylindrical slide valve. Patented in the year 1872.

Fig. 924.

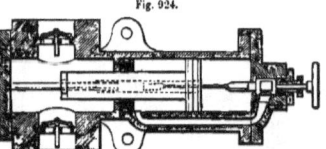

Ramsbottom's Boiler Feed Pump and Engine, consisting of a slide valve actuated by a twisted bar and rod passing through the engine piston and pump plunger, the motion of the valve being at right angles to the motion of the piston. Patented in the year 1869.

Fig. 926.

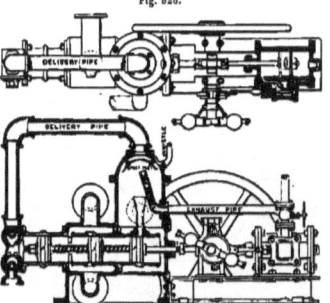

De Bergue's Boiler Feed Pump and Engine, consisting of a pump barrel fitted with three pistons, two of which act as valves to open and close apertures in the barrel that communicates with the injection condenser and the pipe leading to the boiler. Patented in the year 1871.

BOILER FEED PUMPS AND ENGINES.

Fig. 927.

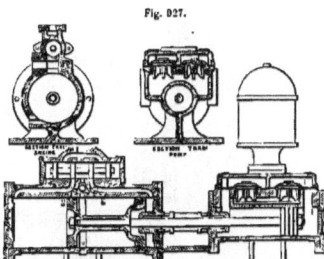

Clarkson's Boiler Feed Pump and Engine, consisting of a double disc piston that regulates the admission of steam to the piston valve casing above, by which means the valve is actuated, and therefore regulates the supply and exhaust steam to and from the cylinder. Patented in the year 1871.

Fig. 930.

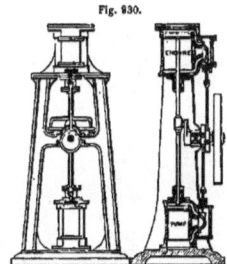

Johnson's Boiler Feed Pump and Engine, consisting of two cylinders with pistons, ports, and valves of similar arrangement and design; the crank pin is driven by a slotted crosshead, and the valves actuated by eccentrics. Patented in the year 1853.

Fig. 928.

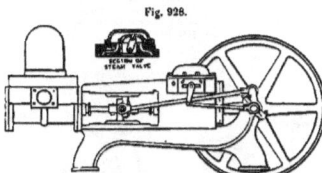

Cope's Boiler Feed Pump and Engine, consisting of the fly wheel bearing situated behind the back end cylinder cover, and a crank pin, rod, and lever to impart motion to the slide valve. Patented in the year 1872.

Fig. 931.

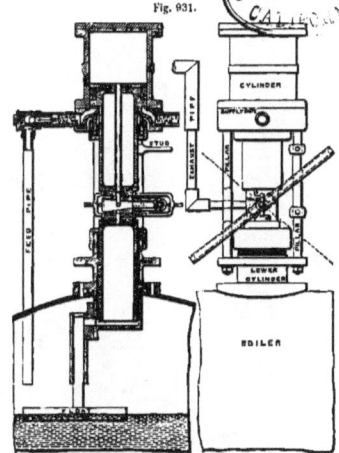

Fig. 929.

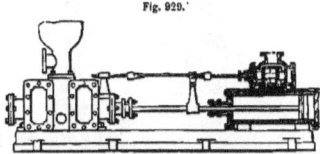

Wolstenholme's Boiler Feed Pump and Engine, consisting of a three-piston slide valve that also oscillates, the motion being derived from an arm secured on the piston rod with an angular grooved end that fits into toothed bosses secured on the valve rod. Patented in the year 1872.

Mellor's Boiler Feed Pump and Engine, consisting of a piston plunger, in the centre of which is a conical valve that is actuated by a mercury loaded balance lever; the steam from the boiler to the valve is regulated by the float and its plunger. Patented in the year 1855.

BOILER FEED PUMPS AND ENGINES.

Fig. 932.

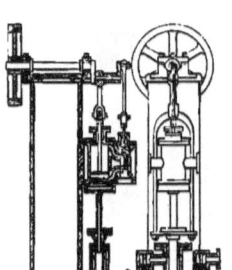

Cowburn's Boiler Feed Pump and Engine, consisting of an ordinary engine having two side rods secured to the piston rod and extended down to the cross bar of the pump plunger that is guided by the tubular projection on the bottom cover of the cylinder. Patented in the year 1855.

Fig. 933.

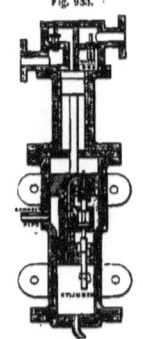

Mellor's Boiler Feed Pump and Engine, consisting of a piston plunger in the cylinder; the steam to actuate the piston is admitted and exhausted by tappet valves contained in the plunger; the pump is secured to the top cover of the cylinder. Patented in the year 1856.

Fig. 934.

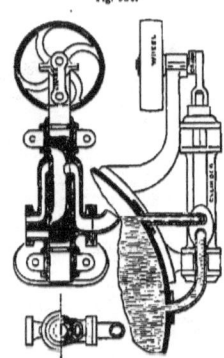

Gargan's Boiler Feed Pump, consisting of a plunger having openings on each side corresponding with similar openings in the casing instead of valves. Patented in the year 1860.

Fig. 935.

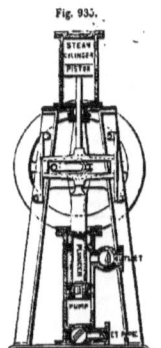

Knowelden's Boiler Feed Pump and Engine, consisting of an ordinary slotted crosshead to drive the crank pin; the pump plunger is hollow, with a valve in the end. Patented in the year 1861.

Fig. 936.

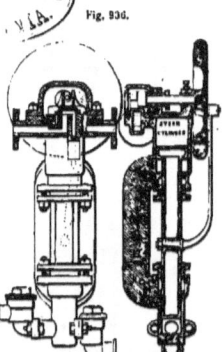

Brown's Boiler Feed Pump and Engine, consisting of a crank motion contained in the steam cylinder cover that moves the slide valve at right angles to the steam piston's movement. Patented in the year 1866.

Fig. 937.

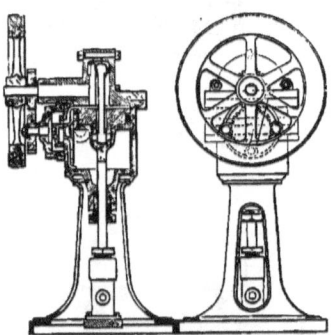

Kitloe's Boiler Feed Pump and Engine, consisting of the connecting rod being recessed in the piston, and the cylinder cover containing the crank pin and shaft bearing, beyond which is a spur tooth wheel that drives the wheel connected to the rotary steam valve. Patented in the year 1866.

BOILER FEED PUMPS AND ENGINES. 301

Fig. 938.

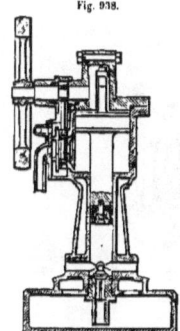

Kittoe's Boiler Feed Pump and Engine, consisting of the cylinder cover containing the connecting rod, crank, and pin; with the slide valve at the side, worked by an eccentric recess in the crank shaft bearing. Patented in the year 1866.

Fig. 939.

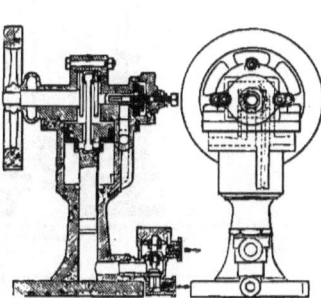

Kittoe's Boiler Feed Pump and Engine, consisting of two cranks contained in the cylinder cover and a rotary steam valve on the end of the shaft. Patented in the year 1866.

Fig. 940.

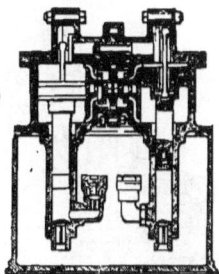

Kittoe's Boiler Feed Pumps and Engines, consisting of a twin arrangement of valves, pumps, and engines; the steam valves, with a tooth wheel between them, are rotary, and driven by a spur wheel secured on the left-hand crank shaft.—Patented in the year 1866.

Fig. 941.

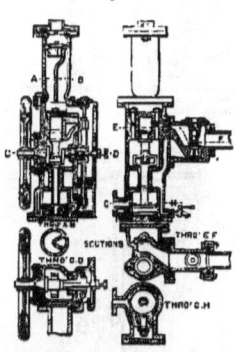

Samuel's Boiler Feed Pump and Engine, consisting of the steam cylinder being at the bottom and the pump at the top, with a space for the slotted crosshead between them contained in a casing with slide valves alike for the engine and pump. Patented in the year 1868.

Fig. 942.

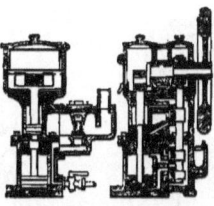

Samuel's Boiler Feed Pump and Engine, consisting of a steam cylinder and pump in a vertical line with each other, and above the pump is a space for the slotted crosshead, the slide valves for the pump and engine being alike and fitted to one rod that is worked by a slotted crosshead also. Patented in the year 1868.

Fig. 943.

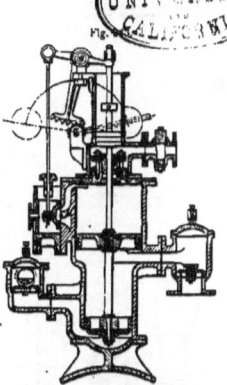

Macnbie's Boiler Feed Pump and Engine, consisting of three cylinders of unequal diameters, the top and middle cylinder being the engine and the bottom cylinder the pump; the steam slide valve is actuated by the tappet gear, and the pump valves are as usual. Patented in the year 1868.

BOILER FEED PUMPS AND ENGINES.

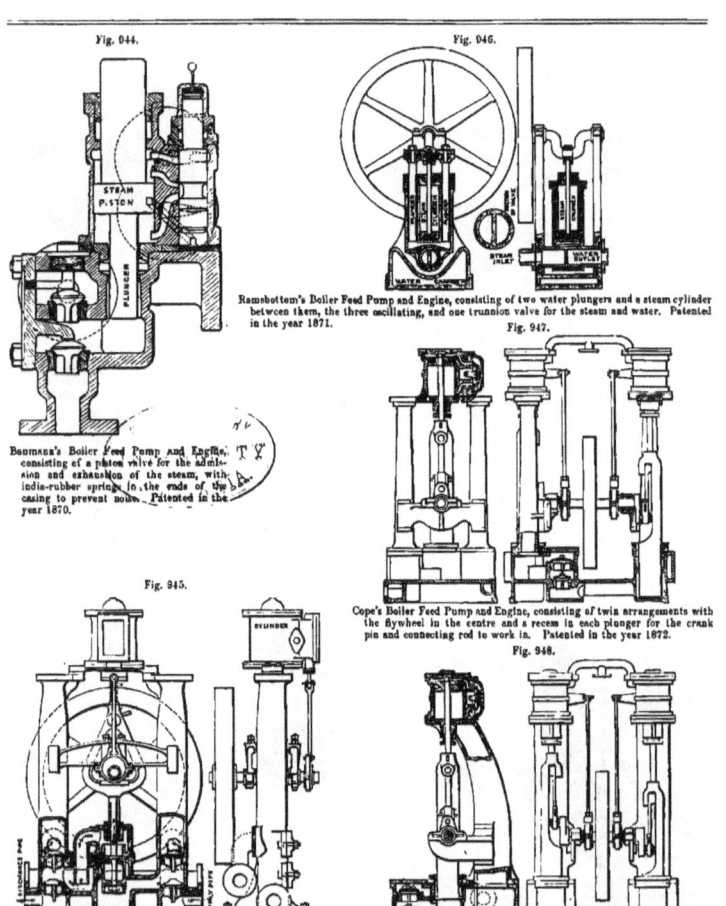

Fig. 944.

Baumann's Boiler Feed Pump and Engine, consisting of a piston valve for the admission and exhaustion of the steam, with india-rubber springs in the ends of the casing to prevent noise. Patented in the year 1870.

Fig. 946.

Ramsbottom's Boiler Feed Pump and Engine, consisting of two water plungers and a steam cylinder between them, the three oscillating, and one trunnion valve for the steam and water. Patented in the year 1871.

Fig. 947.

Cope's Boiler Feed Pump and Engine, consisting of twin arrangements with the flywheel in the centre and a recess in each plunger for the crank pin and connecting rod to work in. Patented in the year 1872.

Fig. 945.

Pearn's Boiler Feed Pump and Engine, consisting of a single plunger double acting pump with the valves on each side and a "bow" connection for the two piston rods, in which the crank and connecting rod work. Patented in the year 1870.

Fig. 948.

Cope's Boiler Feed Pump and Engine, consisting of twin arrangements, with the flywheel in the centre and a recess in each plunger for the crank pin and connecting rod to work in. Patented in the year 1872.

BOILER FEED PUMPS AND ENGINES.

Fig. 949.

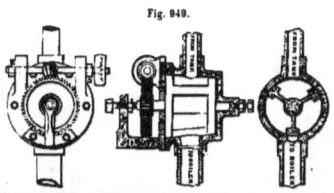

Whitaker's Boiler Feed Rotary Pump, consisting of a three-opening plug which, when revolving, passes water through a casing from the tank to the boiler by centrifugal action. Patented in the year 1856.

Fig. 950.

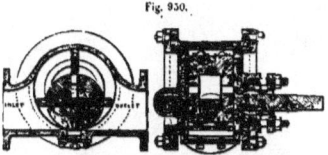

Wilson's Boiler Feed Rotary Pump, consisting of an eccentric disc fitted with four sliding slip pieces having packing at their outer ends, which are always in contact with the cylindrical casing; when the disc revolves, it passes the water from the inlet to the outlet. Patented in the year 1867.

Fig. 951.

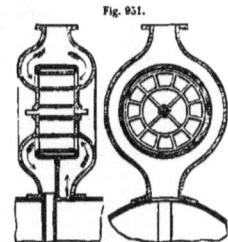

Davy's Boiler Feed Rotary Pump, consisting of a wheel with a hollow rim, through which water and steam passes—from centrifugal action—to the boiler. Patented in the year 1867.

Fig. 953.

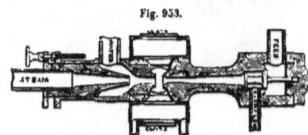

Giffard's Boiler Feed Injector, consisting of an arrangement of tubes and valve openings to admit steam and water simultaneously through them; and the condensation of the steam causes a partial vacuum, the effect of which, added to the pressure of the steam, forces the water into the boiler. Patented in the year 1858.

Fig. 952.

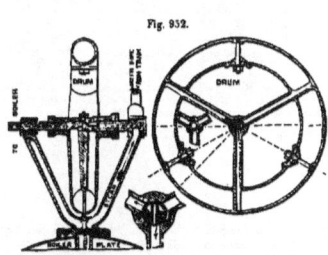

Faure's Boiler Feed Rotary Pump, consisting of a drum wheel composed of a hollow ring, hollow arms, and a hollow shaft; the steam and water enter separately through the same end of the shaft, and the rotation is caused by the water being admitted into and exhausted from one side of the drum only and the steam at the other side, and both meeting at the end of the bearing. Patented in the year 1870.

Fig. 954.

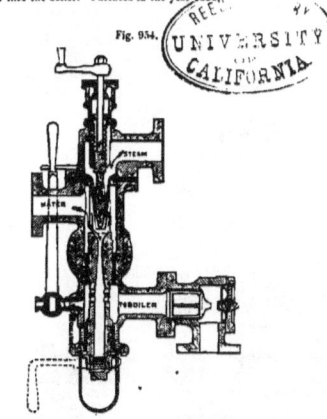

Giffard's Boiler Feed Injector, consisting of adjustable tubes and connections in proportion to the pressure of the steam. Patented in the year 1869.

304   BOILER FEED PUMPS AND ENGINES.

Fig. 955.

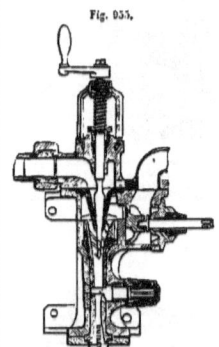

Fig. 956.

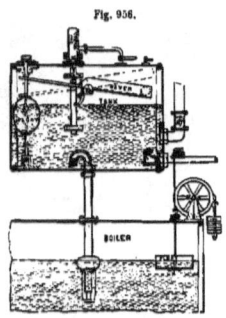

Fig. 957.

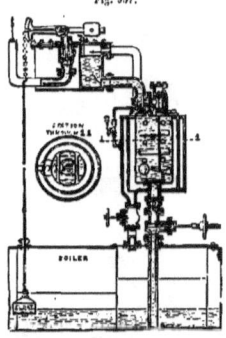

Friedman's Boiler Feed Injector, consisting of an addition of valves to Giffard's injector; also in spiral grooving the water nozzle to impart impetus to the steam and water. Patented in the year 1869.

Wagstaff's Boiler Feed Tank and Steam Lever Gear, consisting of a hollow lever with a ball in it to act as a counterbalance to the float connected at the opposite end; the lever valve admits steam into the tank which, pressing on the water, forces it through the syphon pipe into the boiler; the water entering the tank is regulated by the valve and wheel float. Patented in the year 1863.

Clark's Boiler Feed Apparatus, consisting of a tank filled with water heated by steam from the boiler, which also presses on the water in the tank above and forces it into the boiler through the steam casing tank. Patented in the year 1869.

Fig. 958.

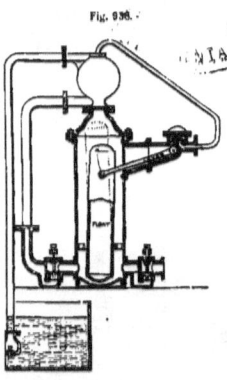

Fig. 959.

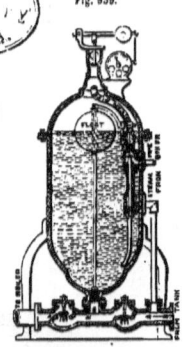

Fig. 960.

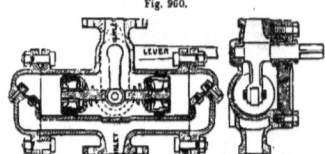

Holman's Boiler Feed Pump, consisting of two pistons on one rod that is actuated by a lever situated between the pistons that have disc valves and spiral springs on the rod for discharge valve, the suction valves being the usual loaded flap kind. Patented in the year 1867.

Fig. 961.

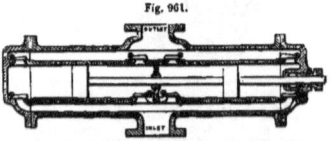

Macabie's Boiler Feed Water Apparatus, consisting of an arrangement of a tank, valves, float lever, and casing, so that steam can be admitted to press or force the water into the boiler as fast as it rises from the tank. Patented in the year 1870.

Grindrod's Boiler Feed Water Apparatus, consisting of a water casing containing a float, valve, and water injector; the steam from the boiler forces the water into the boiler. Patented in the year 1871.

Holman's Boiler Feed Pump, consisting of two pistons in one barrel and a corresponding set of valves for twin double action. Patented in the year 1867.

BOILER FEED PUMPS AND ENGINES.

Fig. 962.

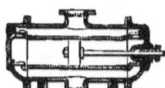

Holman's Boiler Feed Pump, consisting of a piston working in a barrel with a twin set of valves at each end. Patented in the year 1867.

Fig. 963.

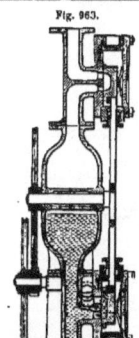

Macabie's Boiler Feed Apparatus, consisting of an arrangement of two slide valves—actuated by an eccentric driven by gearing—which admits water at the bottom and steam above to force water in the boiler. Patented in the year 1868.

Fig. 964.

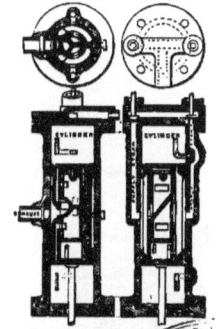

Maxwell's Boiler Feed Engine, consisting of cylindrical piston containing a cylindrical valve that is actuated by the steam that actuates the piston, as in Fig. 923 to Fig. 928. Patented in the year 1868.

Fig. 965.

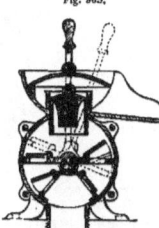

Bishop's Boiler Feed Oscillating Piston Pump, consisting of a flat piston fitted with discharge valves, and the suction valves, on seatings below, contained in a cylindrical casing. Patented in the year 1870.

Fig. 966.

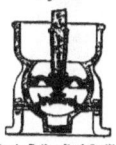

Bishop's Boiler Feed Oscillating Piston Pump, consisting of a flat piston fitted with discharge valves, and the suction valves, on seatings below, contained in a cylindrical casing. Patented in the year 1870.

Fig. 967.

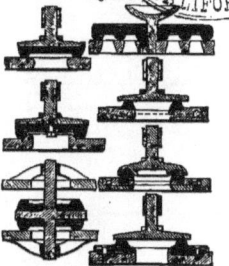

Kittoe's Boiler Feed Pump Valves and Seats, consisting of metal and India-rubber combined. Patented in the year 1869.

Fig. 968.

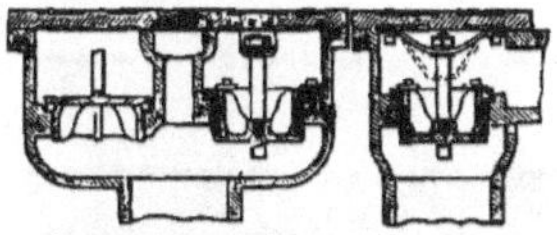

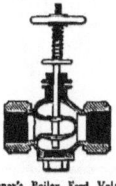

Bargh's Pump Valves, consisting of cast iron ribbed discs fitted with India-rubber rings to act as seats. Invented in the year 1871.

Fig. 969.

Turner's Boiler Feed Valves and Casing, consisting of two disc valves that are held up by the spring on the rod. Patented in the year 1863.

Fig. 970.

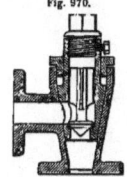

Johnson's Feed Water Supply and Non-return Valves combined. Patented in the year 1853.

2 R

306 SECURING AND CONNECTING TUBES.

CHAPTER IX.
SECURING AND CONNECTING TUBES.

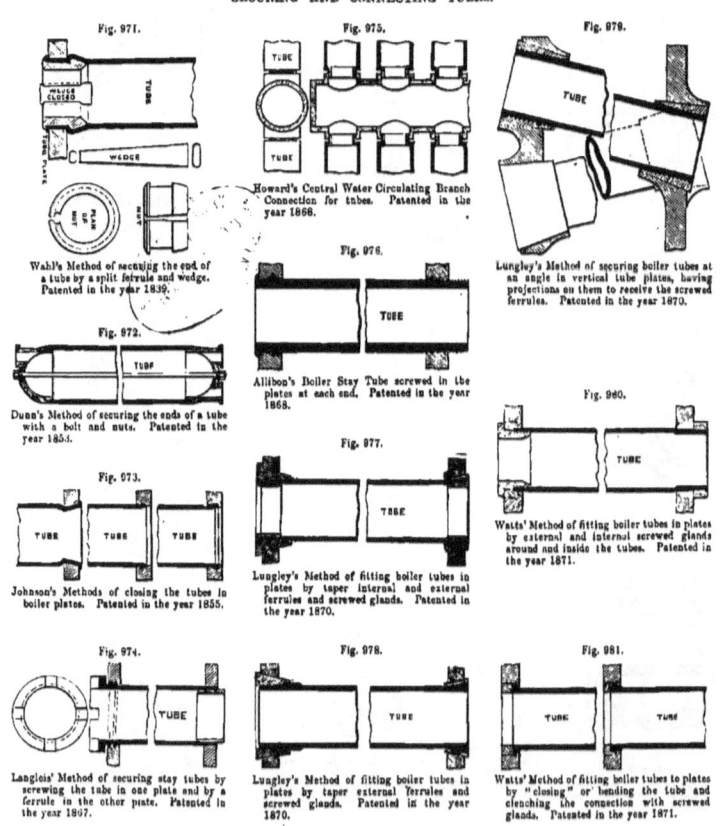

Fig. 971.
Wahl's Method of securing the end of a tube by a split ferrule and wedge. Patented in the year 1839.

Fig. 972.
Dunn's Method of securing the ends of a tube with a bolt and nuts. Patented in the year 1853.

Fig. 973.
Johnson's Methods of closing the tubes in boiler plates. Patented in the year 1855.

Fig. 974.
Langlois' Method of securing stay tubes by screwing the tube in one plate and by a ferrule in the other plate. Patented in the year 1867.

Fig. 975.
Howard's Central Water Circulating Branch Connection for tubes. Patented in the year 1866.

Fig. 976.
Allibon's Boiler Stay Tube screwed in the plates at each end. Patented in the year 1868.

Fig. 977.
Lungley's Method of fitting boiler tubes in plates by taper internal and external ferrules and screwed glands. Patented in the year 1870.

Fig. 978.
Lungley's Method of fitting boiler tubes in plates by taper external ferrules and screwed glands. Patented in the year 1870.

Fig. 979.
Lungley's Method of securing boiler tubes at an angle in vertical tube plates, having projections on them to receive the screwed ferrules. Patented in the year 1870.

Fig. 980.
Watts' Method of fitting boiler tubes in plates by external and internal screwed glands around and inside the tubes. Patented in the year 1871.

Fig. 981.
Watts' Method of fitting boiler tubes to plates by "closing" or bending the tube and clenching the connection with screwed glands. Patented in the year 1871.

SECURING WATER CIRCULATING TUBES.

307

Fig. 982.

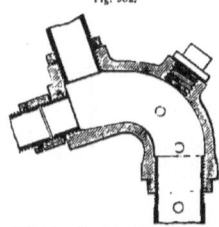

Field's Water Circulating Branch Piece for connecting tubes. Patented in the year 1833 (see Fig. 660 on page 253).

Fig. 983.

Hopkinson's Stay Tube Connection with inside and outside nuts. Patented in the year 1858.

Fig. 984.

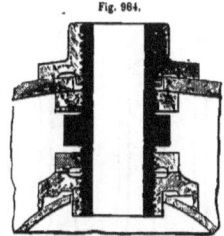

Harlow's Set Screw Connection for vertical water circulating tubes, which join horizontal tubes. Patented in the year 1801.

SECURING WATER CIRCULATING TUBES.

Fig. 985.

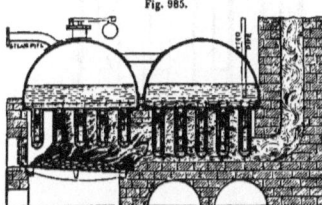

Perkins' Semi-globular Twin Land Boilers, fitted with vertical annular water circulating hanging tubes. Patented in the year 1831 (see Fig. 131 on page 52).

Fig. 986.

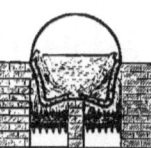

Perkins' Wagon Land Boiler, fitted with water circulating plates. Patented in the year 1831 (see Fig. 399, page 150).

Fig. 987.

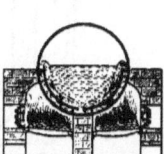

Perkins' Cylindrical Land Boiler, fitted with water circulating plates. Patented in the year 1831 (see Fig. 399, page 150).

Fig. 988.

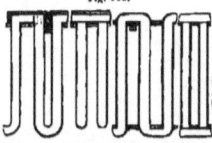

Joly's Arrangements of Water Circulating Vertical Tubes, as fitted in boiler tubes or formed as boilers in combination without a shell. Patented in the year 1857.

Fig. 989.

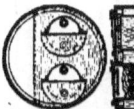 

Varley's Horizontal Cylindrical Boiler, in which the flue tubes are fitted with water pockets and stay water circulating tubes. Patented in the year 1859 (see also Fig. 419 on page 157).

2 R 2

SECURING WATER CIRCULATING TUBES.

Fig. 990.

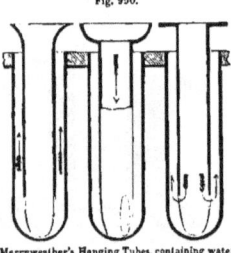

Merryweather's Hanging Tubes, containing water circulating tubes, perforated at the bottom end and capped at the top. Patented in the year 1862.

Fig. 991.

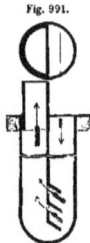

Marshall's Hanging Water Circulating Tube, fitted with a tube projecting above the tube plate, and below that is a plate having angular projections and apertures at the bottom end. Patented in the year 1864.

Fig. 992.

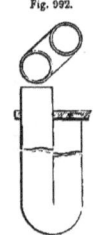

Marshall's Hanging Syphon Water Circulating Tube, with one leg projecting above the tube plate. Patented in the year 1864.

Fig. 993.

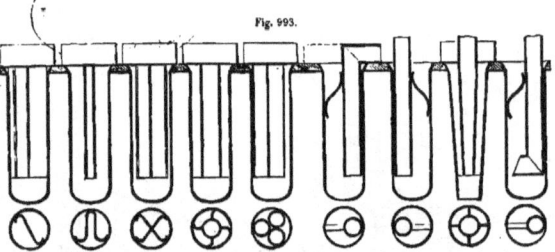

Smith's Hanging Tubes, containing water circulating tubes of various sections with caps of various forms. Patented in the year 1865.

Fig. 994.

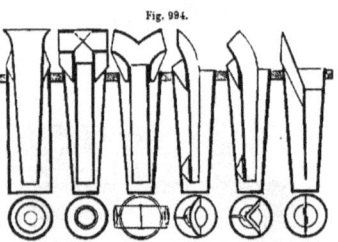

Wise's Hanging Tubes, containing water circulating tubes and plates of various sections, with bonnets acting as deflectors of various forms to promote the circulation. Patented in the year 1865.

Fig. 995.

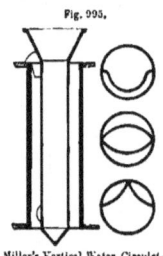

Miller's Vertical Water Circulating Tubes of various sections, contained in an ordinary tube. Patented in the year 1866.

Fig. 996.

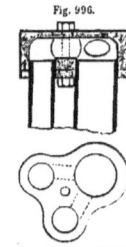

Feyh's Water Circulating Branch Piece, adaptable for three tubes. Patented in the year 1866.

SECURING WATER CIRCULATING TUBES.

Fig. 997.

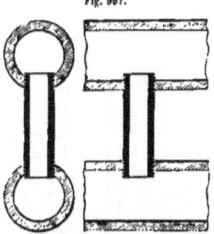

Perkins' Right Angle Tube Connection for Tube Boilers. Patented in the year 1868.

Fig. 998.

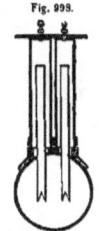

Howard's Vertical Water Circulating Tubes, in connection with a horizontal larger tube. Patented in the year 1868.

Fig. 999.

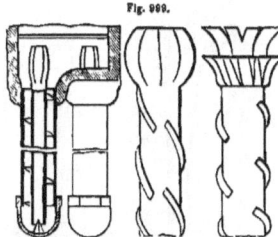

Wiegand's Vertical Water Circulating Tubes, being formed with screw projections and corrugated tops to assist the circulation. Patented in the year 1868.

Fig. 1000.

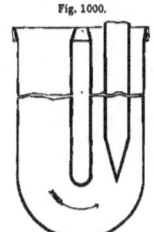

Thirion's Vertical Syphon Tube, in which one leg is fitted with a water circulating tube. Patented in the year 1869.

Fig. 1001.

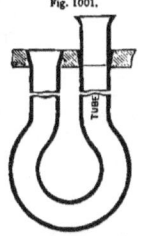

Desvignes's Vertical Syphon Tubes, having an enlarged ring connection at the lower ends. Patented in the year 1869.

Fig. 1002.

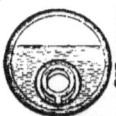

Lee's Horizontal Cylindrical Boiler, in which the flue tubes are fitted with a water circulating coil, and the fire boxes with "cups" and tubes for the same purpose. Patented in the year 1870.

Fig. 1003.

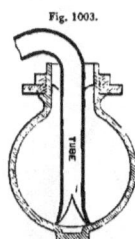

Lee's Globular Evaporating "Cup," fitted with a water circulating tube. Patented in the year 1870 (see Fig. 251 on page 96).

Fig. 1004.

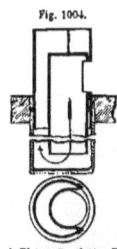

Nason's Water Circulating Tubes, having a side connection below the top of the outer tube; above the tube plate. Patented in the year 1868.

Fig. 1005.

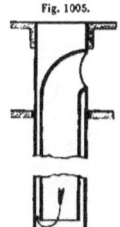

Smith's Water Circulating Tubes, having a side opening between the two plates. Patented in the year 1868.

Fig. 1006.

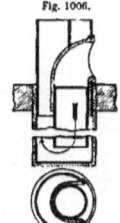

Nason's Water Circulating Tubes, having a side connection below the top of the outer tube; above the tube plate. Patented in the year 1868.

SECURING BRANCH PIECES TO CIRCULATING TUBES.

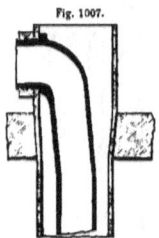

Fig. 1007.

Todd's Water Circulating Tubes, having a side connection below the top of the outer tube; above the tube plate. Patented in the year 1871.

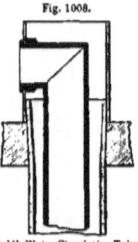

Fig. 1008.

Todd's Water Circulating Tubes, having a side connection below the top of the outer tube; above the tube plate. Patented in the year 1871.

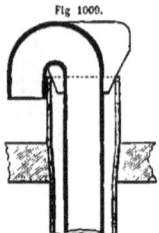

Fig 1009.

Todd's Water Circulating Tubes; the inner tube is syphoned over the tube plate, with short supports to suspend the tube. Patented in the year 1871.

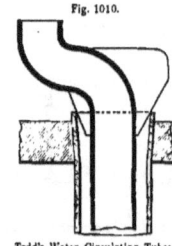

Fig. 1010.

Todd's Water Circulating Tubes; the inner tube being curved back over the plate, with short supports to suspend the tube. Patented in the year 1871.

SECURING BRANCH PIECES TO CIRCULATING TUBES.

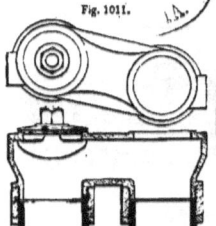

Fig. 1011.

Belleville's Water Circulating Branch Piece, for connecting tubes. Patented in the year 1865 (see Fig. 451 on page 174).

Fig. 1012.

Belleville's Perforated Water Circulating Tube, contained in a tube fitted with screwed branches, to connect on each side to twin tubes, and thus maintain a general circulation. Patented in the year 1865.

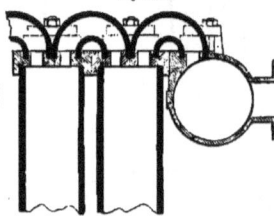

Fig. 1013.

Root's Water Circulating Branch Connections, as secured to a perforated tube plate. Patented in the year 1867 (see Fig. 455 on page 176).

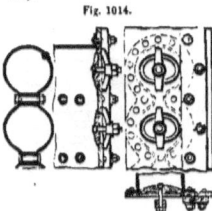

Fig. 1014.

Carville's Tubes, formed by corrugated plates bolted together. Patented in the year 1867.

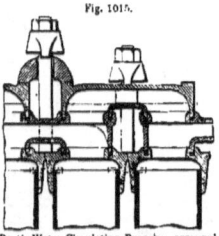

Fig. 1015.

Root's Water Circulating Branches, arranged to permit the flow from one tube to the other through right angle passages. Patented in the year 1870.

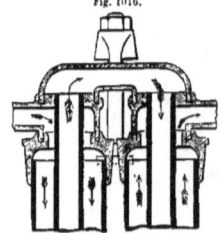

Fig. 1016.

Root's Water Circulating Branches in connection with internal and external tubes. Patented in the year 1870.

Fig. 1017.

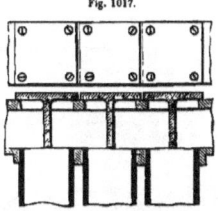

Root's Water Circulating Branches, fitted with covers having inside water guide projections. Patented in the year 1870.

Fig. 1018.

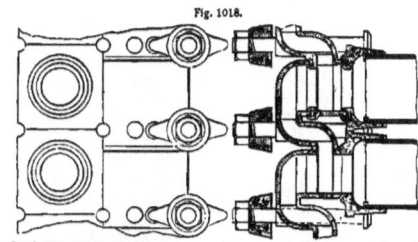

Root's Water Circulating Branches, arranged to permit the flow from one tube to the other through bend-passages. Patented in the year 1870.

Fig. 1019.

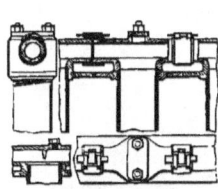

Westerman's Water Circulating Branches, that are connected by a square band and key. Patented in the year 1871.

Fig. 1020.

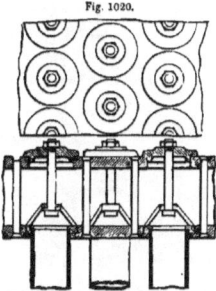

Watts' Water Circulating Branch, fitted with stay bolts and plates in a line with the tubes. Patented in the year 1871.

Fig. 1021.

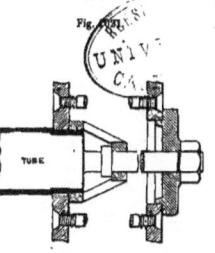

Watts' Flange Stay Bolts in connection with tubes secured by screwed rings and central bolts. Patented in the year 1871.

Fig. 1022.

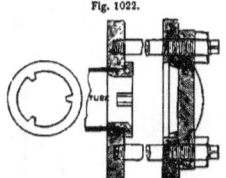

Watts' Flange Stay Bolts in connection with tubes secured by screwed glands. Patented in the year 1871.

Fig. 1023.

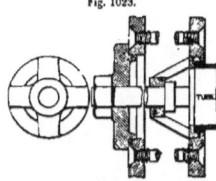

Watts' Flange Stay Bolts in connection with tubes secured by screwed glands and central bolts. Patented in the year 1871.

Fig. 1024.

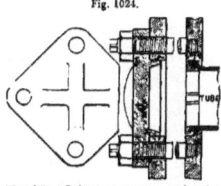

Watts' Stay Bolts in connection with tubes secured by screwed glands. Patented in the year 1871.

CONNECTION OF TUBES AT RIGHT ANGLES, ETC.

Fig. 1025.

Watts' Method of fitting boiler tubes to plates by flanged glands and cross bar connections. Patented in the year 1871.

Fig. 1026.

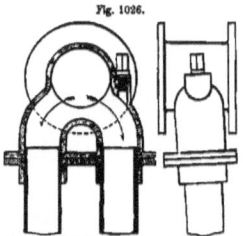

Mirchin's Water Circulating Branch Piece, to connect twin tubes. Patented in the year 1871.

Fig. 1027.

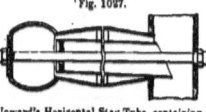

Howard's Horizontal Stay Tube, containing a water circulating tube, and both tubes communicating with two right angle tubes. Patented in the year 1871.

Fig. 1028.

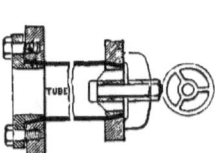

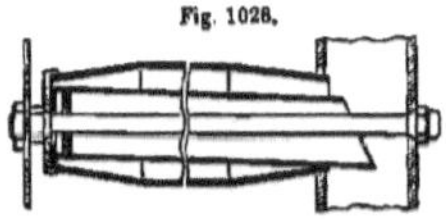

Howard's Horizontal Stay Tube, containing a water circulating tube, and both tubes communicating with a right angle tube. Patented in the year 1871.

Fig. 1029.

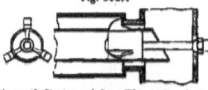

Howard's Horizontal Stay Water Circulating Tube, enclosed in a tube connected to a right angle tube. Patented in the year 1871.

Fig. 1030.

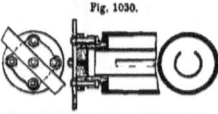

Howard's Horizontal Trough Water Circulating Tube, contained in a tube connected to a flange plate. Patented in the year 1871.

Fig. 1031.

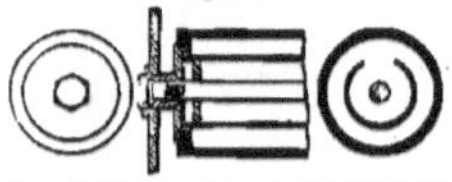

Howard's Horizontal Trough Water Circulating Tube, contained in a tube connected to a stay plate. Patented in the year 1871.

Fig. 1032.

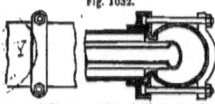

Howard's Horizontal Trough Water Circulating Tube, contained in a tube connected to a right angle tube by a band, bolts, and nuts. Patented in the year 1871.

Fig. 1033.

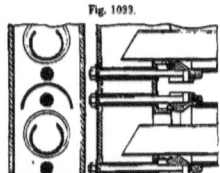

Howard's Horizontal Trough Water Circulating Tubes, contained in tubes at right angles secured by bolts and nuts. Patented in the year 1871.

CONNECTION OF TUBES AT RIGHT ANGLES, ETC.

Fig 1034.

Belleville's Water Circulating Tubes, in connection with a large tube closed at one end and flanged at the other, and screwed branches at the side. Patented in the year 1865.

Fig. 1035.

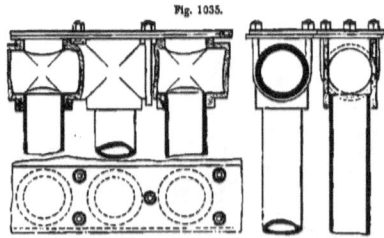

Turbill's Water Circulating Branches, to which the tubes are secured. Patented in the year 1865.

CONNECTION OF TUBES AT RIGHT ANGLES, ETC.   313

Fig. 1036.

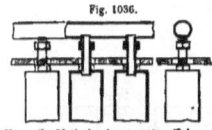

Howard's Method of connecting Tubes at right angles, to permit water circulation. Patented in the year 1866.

Fig. 1037.

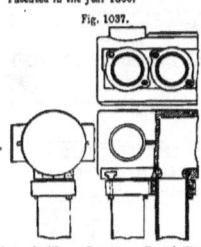

Howard's Water Circulating Branch Piece, connected to the tubes by a ring gland around the tube, that is attached to the piece by bolts and nuts. Patented in the year 1866.

Fig. 1038.

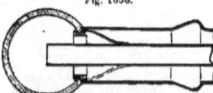

Howard's Horizontal Water Circulating Tubes in connection with a transverse tube. Patented in the year 1868.

Fig. 1039.

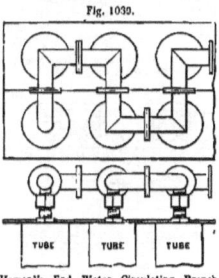

Howard's End Water Circulating Branch Connections for tubes. Patented in the year 1868.

Fig. 1040.

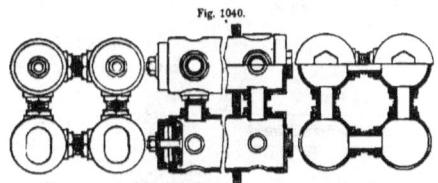

Bartlett's Method of connecting Tubes at right angles by screwed tubes, nuts, and packing washers, or lock nuts. Patented in the year 1871.

Fig. 1041.

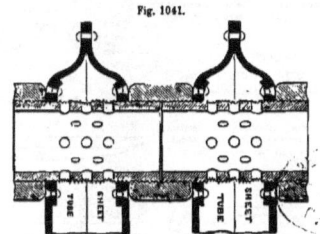

Lawthwaite's Water Circulating Sheet Tubes, connected by screwed tubes and nuts. Patented in the year 1842.

Fig. 1042.

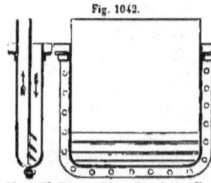

Marshall's Vertical Water Circulating Sheet-box, fitted with a central plate having apertures and angular projections at the lower extremity to assist the circulation. Patented in the year 1864.

Fig. 1043.

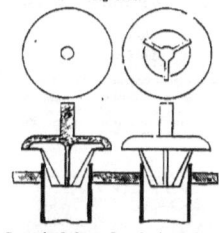

Paxman's "Deflectors," or "Mushroom Caps," consisting of discs having ribs that are forced into the tops of the tubes; the caps receive the impetus of the water rising out of the tubes, and thereby assist the circulation up the tube inside and down the tube outside. Patented in the year 1870.

Fig. 1044.

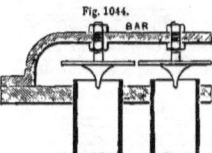

"Deflectors," or "Mushroom Caps," consisting of discs suspended over the tubes from a bar, and secured by bolts and nuts. Proposed in the year 1872 by Mr. Allison (see Plate 36A).

2 s

CHAPTER X.
PERFORATED FIRE BARS.

Fig. 1045.

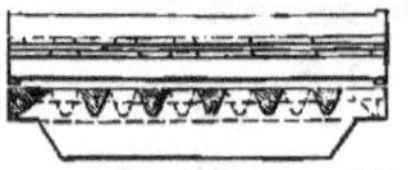

Moreau's Fire Bar, consisting of two shallow side bars connected by lugs to a deep narrow central bar, for the admission of air amongst the fuel. Patented in the year 1852.

Fig. 1048.

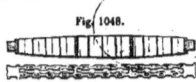

Harden's Fire Bar, consisting of vertical grooves, holes and channels, for the admission of air amongst the fuel. Patented in the year 1859.

Fig. 1051.

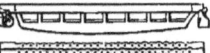

Jackson's Fire Bar, consisting of transverse and vertical holes, for the admission of air amongst the fuel. Patented in the year 1861.

Fig. 1054.

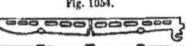

Lewis's Fire Bar, consisting of transverse perforations, for the admission of air amongst the bars to keep them cool. Patented in the year 1868.

Fig. 1057.

Brown's Fire Bar, consisting of twin sides connected by angular lugs, for the admission of air amongst the fuel. Patented in the year 1869.

Fig. 1060.

Raper's Fire Bar, consisting of two central ribs between two sides and vertical holes, for the admission of air amongst the fuel. Patented in the year 1871.

Fig. 1046.

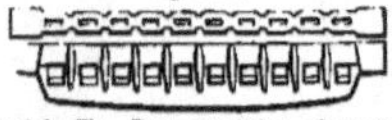

Martin's Fire Bars, consisting of vertical grooves between transverse and vertical holes, for the admission of air amongst the fuel. Patented in the year 1859.

Fig. 1049.

Stratford's Fire Bar, consisting of a double row of vertical perforations on each side of the rib, for the admission of air amongst the fuel. Patented in the year 1860.

Fig. 1052.

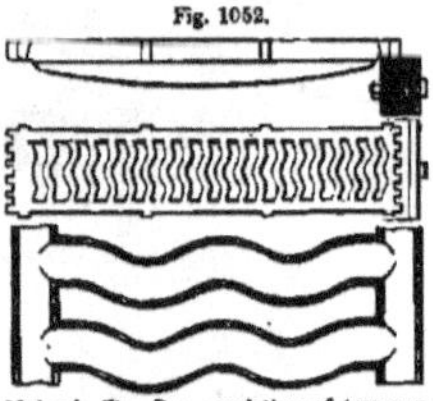

Myiren's Fire Bar, consisting of transverse openings between the sides, for the admission of air amongst the fuel; and in some cases the bar is cast hollow (as shown by the enlarged section) to receive the ashes, to prevent the bar from burning. Patented in the year 1863.

Fig. 1055.

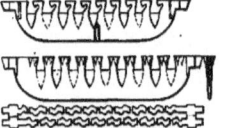

Fletcher's Fire Bar, consisting of grooves at the sides and top, for the admission of air amongst the fuel. Patented in the year 1866.

Fig. 1058.

Cone's Fire Bar, consisting of vertical openings through the bar on each side of the central rib, for the admission of air amongst the fuel. Patented in the year 1870.

Fig. 1047.

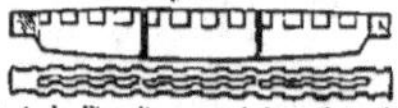

Harden's Fire Bars, consisting of vertical lugs and channels, for the admission of air amongst the fuel. Patented in the year 1859.

Fig. 1050.

Stratford's Fire Bar, consisting of a single row of vertical perforations on each side of the rib, for the admission of air amongst the fuel. Patented in the year 1860.

Fig. 1053.

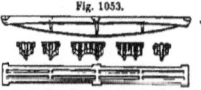

Harrison's Fire Bars, consisting of three or more ribs joined by five lugs, thus forming longitudinal spaces for the admission of air amongst the fuel. Patented in the year 1866.

Fig. 1056.

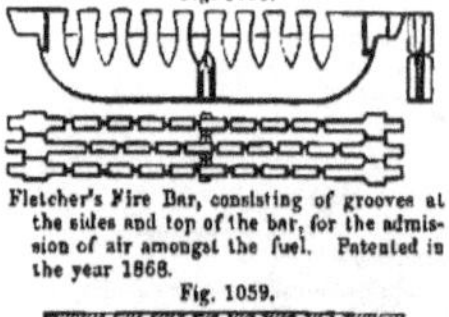

Fletcher's Fire Bar, consisting of grooves at the sides and top of the bar, for the admission of air amongst the fuel. Patented in the year 1868.

Fig. 1059.

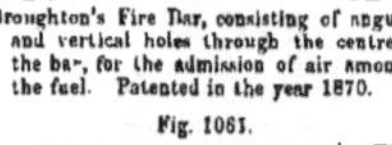

Broughton's Fire Bar, consisting of angular and vertical holes through the centre of the bar, for the admission of air amongst the fuel. Patented in the year 1870.

Fig. 1061.

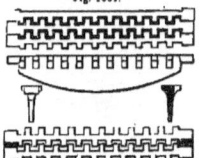

Dilnut's Fire Bars, consisting of vertical grooves at the sides, of unequal widths, for the admission of air amongst the fuel. Patented in the year 1872.

SOLID AND HOLLOW FIRE BARS.

Fig. 1062.

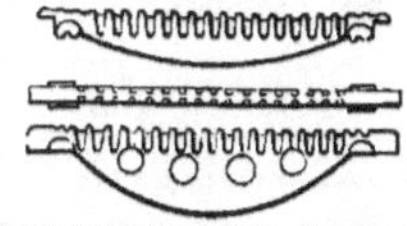

Chanter's Fire Bar, consisting of grooves at the sides and top of the bars and holes through the sides, for the admission of air amongst the fuel. Patented in the year 1844.

Fig. 1064.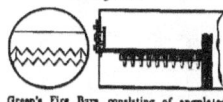

Green's Fire Bars, consisting of angulated bars, laid across the fire box. Patented in the year 1865.

Fig. 1066.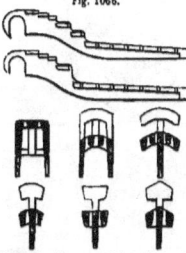

Jordan's Fire Bars, consisting of top steps and side lugs, for the admission of air amongst the fuel, as it is shaken on the steps by the rise-and-fall motion of the bar; the cross sections refer to bars for the same purpose, with steps also. Patented in the year 1868.

Fig. 1063.

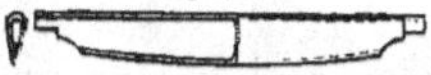

Blackwood's Fire Bar, consisting of a longitudinal passage through the bar, for air to pass through and prevent the bar from being burnt. Patented in the year 1860.

Fig. 1065.

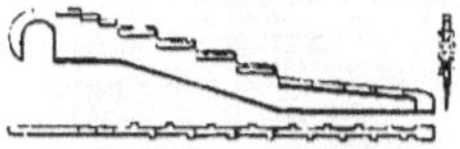

Jordan's Fire Bar, consisting of side lugs and top "steps," for the admission of air amongst the fuel, as it is shaken on the "steps" by the rise-and-fall motion of the bar. Patented in the year 1868.

Fig. 1067.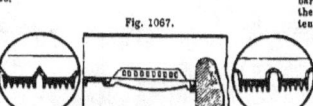

Fletcher's Fire Bars, consisting of an ordinary bar in connection with central and side deeper bars having lateral openings in them, for the admission of air amongst the fuel. Patented in the year 1869.

Fig. 1068.

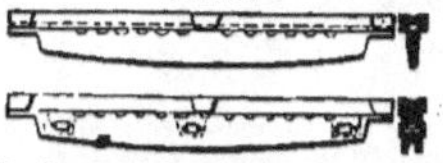

Batchelor's Fire Bars, consisting of combined bars, the top part being a plain bar, either dove-tailed or bolted to the bottom part, which has grooves across the joint surface, for the admission of air amongst the fuel. Patented in the year 1870.

Fig. 1069.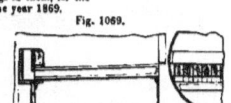

Whitelaw's Fire Bars, consisting of longitudinal grooves on each side of the bar, for the admission of air along the bar to the hollow bridge. Patented in the year 1871.

WATER FIRE BARS.

Fig. 1070.

Fig. 1071.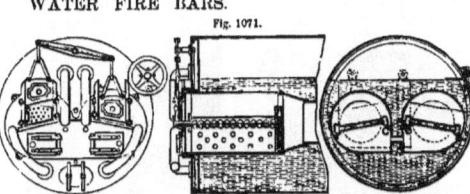

Haywood's Fire Bars, consisting of two side pipes connected by thirteen cross pipes placed in the fire box at an angle as shown; the boiler front is fitted with two vertical doors alternately raised and lowered, to regulate the admission of the air amongst the fuel. Patented in the year 1859.

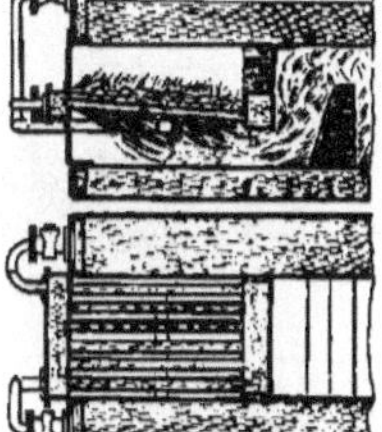

Miguet's Fire Bars, consisting of tubes secured longitudinally to a bridge box and a front box, communicating by water pipes and valves; above the bridge box are brick slabs to cause the air above the grate to force the flame down under the bridge box. Patented in the year 1864.

MOVABLE FIRE BARS.

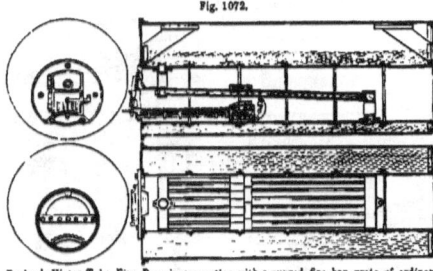

Fig. 1072.

Barlow's Water Tube Fire Bars in connection with a curved fire bar grate of ordinary form, so that the smoke from the lower grate may be consumed by the fire on the tube grate. Patented in the year 1867.

Fig. 1073.

Vicar's Fire Bars, consisting of two bars, one within the other longitudinally, the outer bar containing water to prevent the inner bar from being burnt, which is also moved forward and backward for the admission of air amongst the fuel. Patented in the year 1868.

Fig. 1074.

Vicar's Fire Bars, consisting of two fire bars, one within the other longitudinally, the outer bar containing water to prevent the inner bar from being burnt, which is also moved forward and backward for the admission of air amongst the fuel. Patented in the year 1868.

Fig. 1075.

Vicar's Fire Bars, consisting of two fire bars, one within the other longitudinally, the outer bar containing water to prevent the inner bar from being burnt, which is also moved forward and backward for the admission of air amongst the fuel. Patented in the year 1868.

Fig. 1076.

Gaze's Fire Bars, consisting of bars of an ordinary design, formed in sets by a feed water pipe passing through or under the thick part of the bar. Patented in the year 1869.

Fig. 1077.

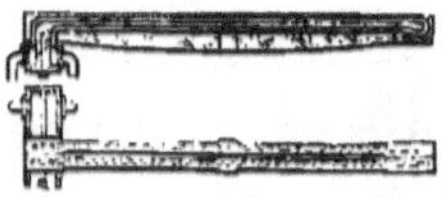

Ellis's Fire Bars, consisting of a bar of an ordinary design, formed with a return feed water pipe cast in the bar, but passing through one end only. Patented in the year 1872.

MOVABLE FIRE BARS.

Fig. 1078.

Drew's Fire Grate, that is raised and lowered by levers, cross bar, screw spindle, and bevel gearing, the object being to consume the smoke from the fixed grate by the fire on the movable grate. Patented in the year 1836.

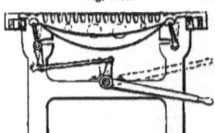

Fig. 1079.

Chanter's Fire Bar—Lever Motion, consisting of a series of levers connected at the ends of the bars, and by a hand crank lever the bars are raised and lowered, also moved forward and backward for the admission of air amongst the fuel. Patented in the year 1844.

MOVABLE FIRE BARS.

Fig. 1080.

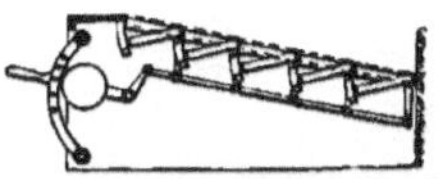

Mash's Fire Bar—Lever Motion, consisting of a series of slabs packed close together, and the shaft of each set fitted with a lever, each lever being connected by a rod attached to a hand crank lever, which, when moved, causes the bar slabs to rise and fall, for the admission of air amongst the fuel. Patented in the year 1856.

Fig. 1081.

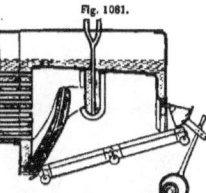

Johnson's Fire Bar—Eccentric Motion, consisting of eccentrics at the ends of the bars that rise and fall, due to the action of the eccentrics, which are actuated by a cam and ratchet gear (shown by Fig. 1082); the fuel is mechanically put on the bars by the hopper bin set in motion by the large eccentric. Patented in the year 1857.

Fig. 1082.

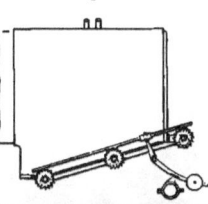

Johnson's Fire Bar—Catch and Pinion Motion, worked by a cam and counterbalance lever, to raise and lower the bars for the admission of air amongst the fuel. Patented in the year 1857.

Fig. 1083.

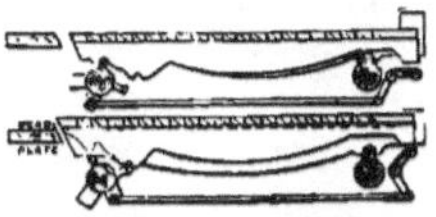

Annan's Fire Bar—Lever Motion, consisting of levers at each end that raise and lower the bars and also move them forward and backward at the same time; the bars have lugs on each side for the admission of air amongst the fuel. Patented in the year 1860.

Fig. 1084.

Colquhoun's Fire Bar—Lever Motion, which raises and lowers the bars, also moves them forward and backward at the same time, for the admission of air amongst the fuel. Patented in the year 1861.

Fig. 1085.

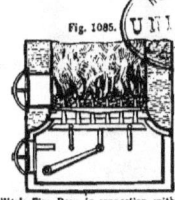

Shillito's Fire Bars, in connection with vertical rods secured on a frame that receives a vertical motion from a hand lever, for the admission of air amongst the fuel. Patented in the year 1863.

Fig. 1086.

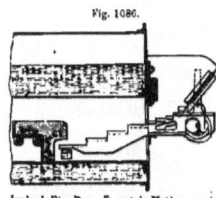

Jordan's Fire Bar—Eccentric Motion, consisting of an eccentric at the front end of each bar, that rise and fall during the motion of the eccentric, and thereby cause the fuel to advance on the "steps" formed with the bar, for the admission of air amongst the fuel. Patented in the year 1868.

Fig. 1087.

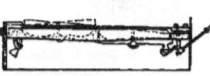

Lewis's Fire Bar—Lever Motion, consisting of levers and bearing rods that raise and lower the bars by rocking or tilting them for the admission of air amongst the fuel. Patented in the year 1868.

Fig. 1088.

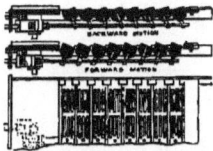

Holt's Fire Bars, consisting of perforated bars placed across the fire box, and therein rocked forward and backward for the admission of air amongst the fuel. Patented in the year 1871.

CHAPTER XI.

MECHANICAL FEED FUEL APPARATUS.

Fig. 1089.

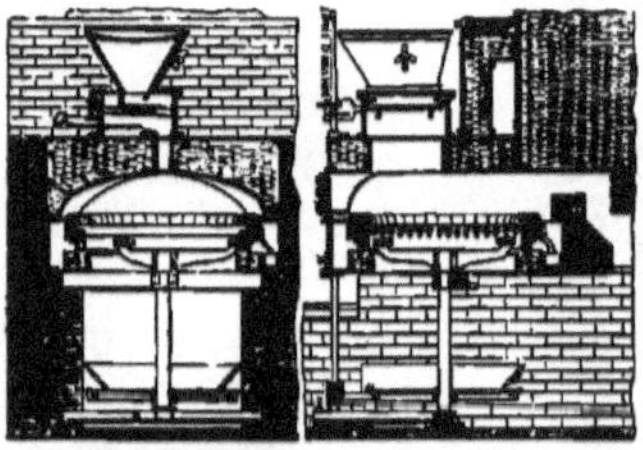

Brunton's Mechanical Feed Fuel Apparatus, consisting of a revolving circular fire grate that is actuated by a pinion and wheel driven by the steam engine; the coal is admitted on the fire grate through the hopper that is situated on the shell of the boiler. Patented in the year 1819.

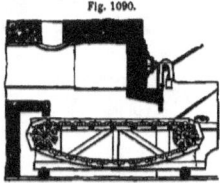

Fig. 1090.

Juckes's Mechanical Feed Fuel Apparatus, consisting of a series of solid links connected by pins, forming thereby endless chains, suspended at each end on driving rollers having flat sides corresponding with the length of each link; the fuel from the bin is admitted by raising the vertical door, and then conveyed by the motion of the links longitudinally while burning to the inner end, from which it falls into the ashpit below. Patented in the year 1848.

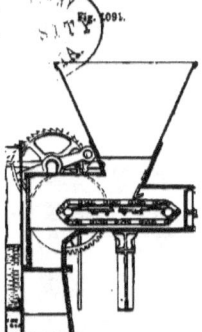

Fig. 1091.

Vicar's Mechanical Feed Fuel Apparatus, consisting of endless chains suspended at each end and worked by spur gearing; the fuel is thereby conveyed from the hopper above to the fire box below in such quantities as the speed of the chain regulates. Patented in the year 1867.

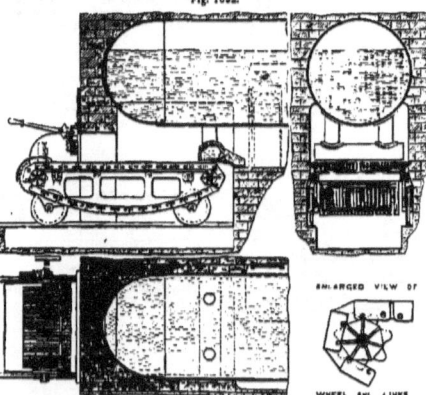

Fig. 1092.

Taylor's Mechanical Feed Fuel Apparatus, consisting of solid links pinned together to form endless chains suspended at each end on ribbed rollers that actuates the chain towards the cross bridge end, and thus convey the burning fuel in that direction also, the ashes falling below; the quantity of fuel is regulated by the vertical slide. Patented in the year 1868.

MECHANICAL FEED FUEL APPARATUS.

Fig. 1093.

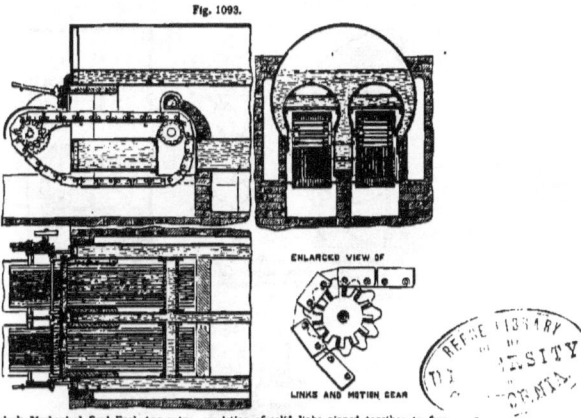

Taylor's Mechanical Feed Fuel Apparatus, consisting of solid links pinned together to form endless chains suspended at each end on ribbed rollers, the front roller being the "driver," and over the back roller is a cross-water bridge; the quantity of fuel is regulated by the vertical slide at boiler front. Patented in the year 1868.

Fig. 1094.

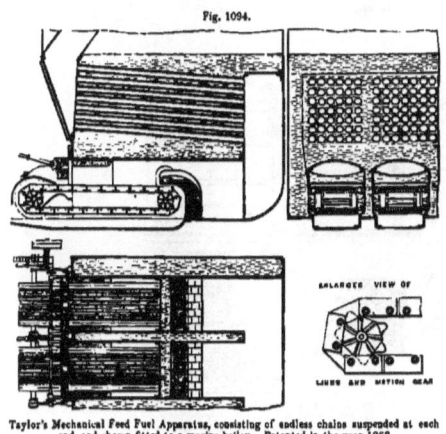

Fig. 1095.

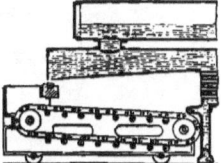

Crosland's Mechanical Feed Fuel Apparatus, consisting of solid links pinned together to form endless chains, suspended on large rollers at each end and small rollers between top and bottom, so that the chains do not drop from the straight line during working or at rest. Patented in the year 1869.

Taylor's Mechanical Feed Fuel Apparatus, consisting of endless chains suspended at each end, and shown fitted to a marine boiler. Patented in the year 1868.

MECHANICAL FEED FUEL APPARATUS.

Fig. 1096.

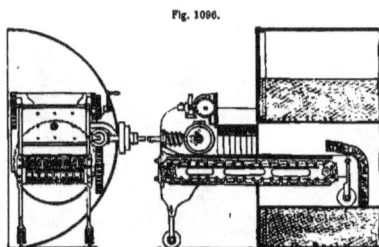

Burnley's Mechanical Feed Fuel Apparatus, consisting of a series of links forming endless chains, suspended at each end, and worked by angular rollers, receiving motion from a wheel and worm driven by a pulley. Patented in the year 1870.

Fig. 1097.

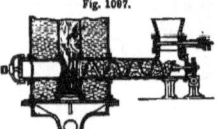

Shillito's Corkscrew Fuel Feeder, worked by a worm and pinion at the side of the boiler. Patented in the year 1863.

Fig. 1098.

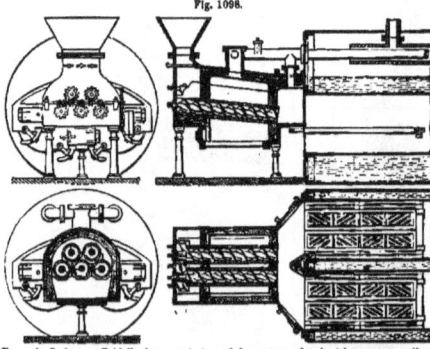

Turner's Corkscrew Fuel Feeders, consisting of fire retorts fitted with screw travellers, to convey the fuel from the bin to the fire grate, which is composed of perforated cast-iron slabs; motion to the screw travellers is imparted by worm and pinion gear. Patented in the year 1866.

MECHANICAL FEED FUEL APPARATUS.

Fig. 1099.

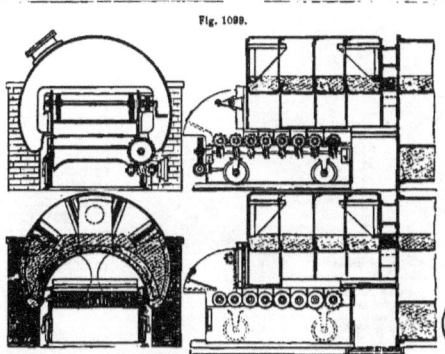

Ripley's Mechanical Feed Fuel Apparatus, consisting of rollers or drums of unequal diameters set in motion by worms and pinions, the main worm being driven by a pulley; the fuel is conveyed and burnt as the rollers revolve. Patented in the year 1866.

Fig. 1100.   Fig. 1101.

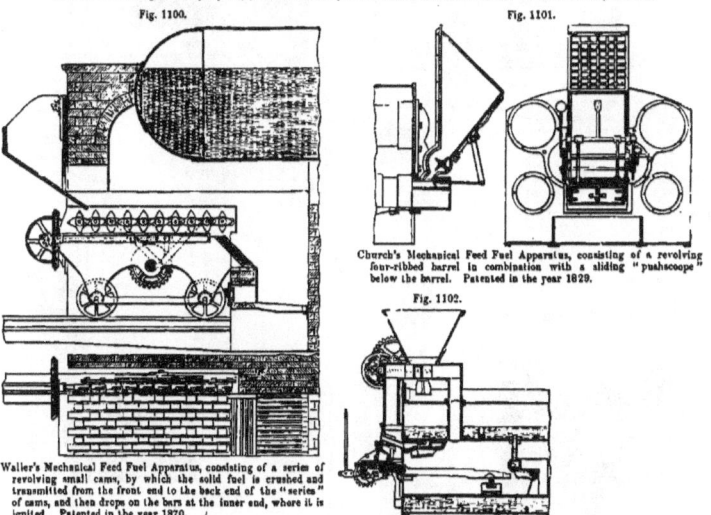

Church's Mechanical Feed Fuel Apparatus, consisting of a revolving four-ribbed barrel in combination with a sliding "pushscoope" below the barrel. Patented in the year 1829.

Fig. 1102.

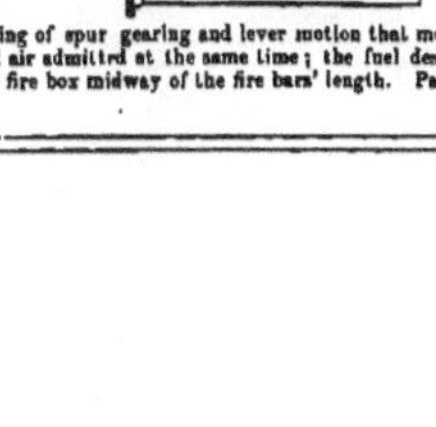

Waller's Mechanical Feed Fuel Apparatus, consisting of a series of revolving small cams, by which the solid fuel is crushed and transmitted from the front end to the back end of the "series" of cams, and then drops on the bars at the inner end, where it is ignited. Patented in the year 1870.

Vicar's Mechanical Feed Fuel Apparatus, consisting of spur gearing and lever motion that moves the fire bars forward and backward, by which the fuel is shaken on and air admitted at the same time; the fuel descends at the boiler front and through a tube, passing through the shell and fire box midway of the fire bars' length. Patented in the year 1867.

2 T

MECHANICAL FEED FUEL APPARATUS.

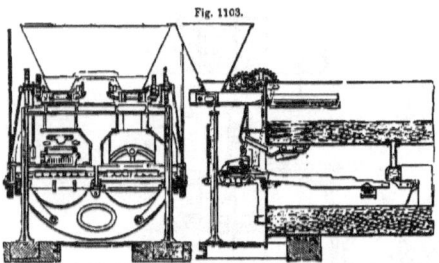

Fig. 1103.

Vicar's Mechanical Feed Fuel Apparatus, consisting of spur gearing and lever motion that moves the fire bars forward and backward in connection with a hopper and "push" plates at the fire boxes' front; the hopper and push plate convey the fuel on the end of the bars, and their motion shakes it further on and admits air at the same time. Patented in the year 1897.

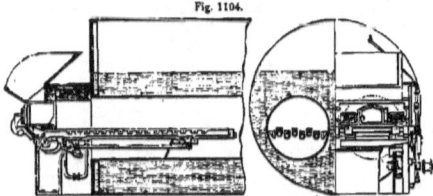

Fig. 1104.

Butterworth's Mechanical Feed Fuel Apparatus, consisting of tubular fire bars filled with water; at the front of the fire box is a "push" plate to push the fuel on the fire bars that are moved forward and backward by the cam and lever motion gear under the bars. Patented in the year 1870.

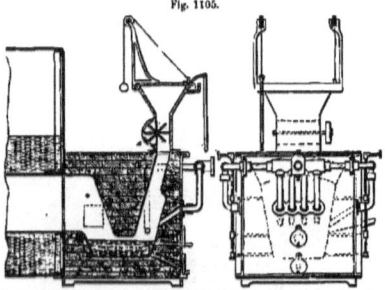

Fig. 1105.

Leigh's Mechanical Feed Fuel Apparatus, consisting of an armed revolving roller to permit the fuel to fall on molten iron contained in a fire-brick furnace situated in front of an ordinary land boiler, the ignition of the fuel and the molten state of the metal being "kept up" by blast through the air pipes, as shown. Patented in the year 1866.

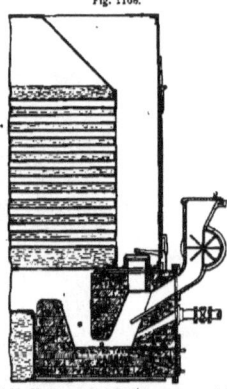

Fig. 1106.

Leigh's Mechanical Feed Fuel Apparatus, consisting of an armed revolving roller to permit the fuel to fall on molten iron contained in a fire-brick furnace situated within the front of a marine boiler, the ignition of the fuel and the molten state of the metal being "kept up" by blast through the air pipe, as shown. Patented in the year 1866.

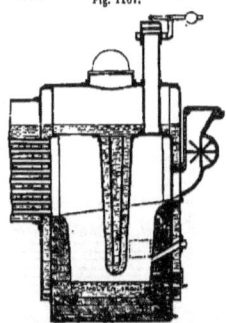

Fig. 1107.

Leigh's Mechanical Feed Fuel Apparatus, consisting of an armed revolving roller to permit the fuel to fall on molten iron contained in a fire-brick furnace situated in the fire box of a locomotive boiler, the ignition of the fuel and the molten state of the metal being "kept up" by blast through the air pipe, as shown. Patented in the year 1866.

FIRE-BOX DOORS.

Fig. 1108.

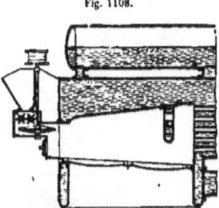

Crosland's Mechanical Feed Fuel Apparatus, consisting of two crushing rollers and a horizontal rotary spreading fan for hurling the small fuel on the fire grate. Patented in the year 1869.

Fig. 1110.

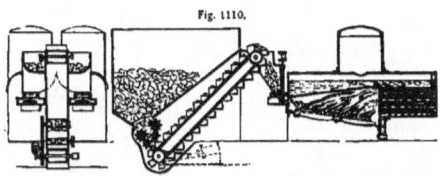

Crosland's Mechanical Feed Fuel Apparatus, consisting of a coal box fitted with four crushing rollers at the foot of a travelling bucket hoist in connection with rotary horizontal feed discs situated at the front of the fire boxes; the hoist and rollers are worked by a horizontal engine. Patented in the year 1869.

Fig. 1109.

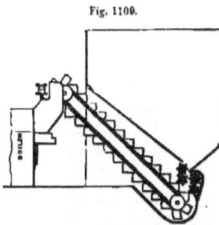

Crosland's Mechanical Feed Fuel Apparatus, consisting of four crushing rollers at the foot of a travelling bucket hoist in connection with rotary horizontal feed discs situated at the front of the fire box. Patented in the year 1869.

Fig. 1111.

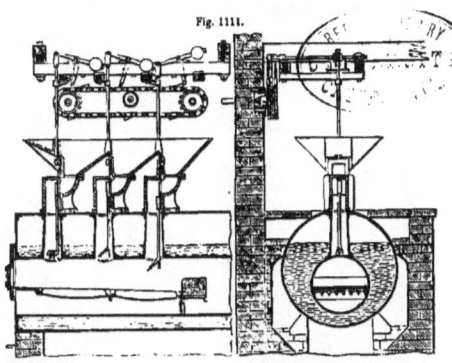

Heginbottom's Mechanical Feed Fuel Apparatus, consisting of vertical feed tubes through the boiler over the fire box for the fuel to pass through from hoppers fitted with vertical slides that are actuated by vertical rods that are moved up and down by levers receiving motion from an endless chain and "tappet" levers in connection; the chain is driven by toothed pinions. Patented in the year 1871.

FIRE-BOX DOORS.

Fig. 1112.

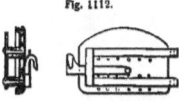

The Common Fire-box Door, that was used against Chas. Wye-Williams' door in the boiler experiments in the year 1857.

Fig. 1113.

Wye-Williams' Fire-box Doors and Frame: the top door admits the air over the grate through the perforated plate inside the door. Used at Newcastle in the year 1857.

Fig. 1114.

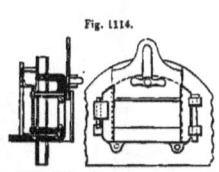

Wye-Williams' Fire-box Door and Frame, that was used in the boiler experiments at Newcastle in the year 1857.

FIRE-BOX DOORS.

Fig. 1115.

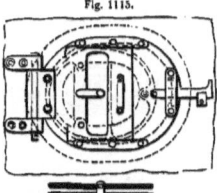

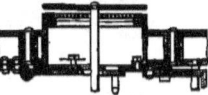

Beattie's Fire-box Door and Frame, containing a solid flame plate in front of a perforated plate and a horizontally sliding air plate, having an air hole in it corresponding with an air hole in the door plate. Patented in the year 1858.

Fig. 1116.

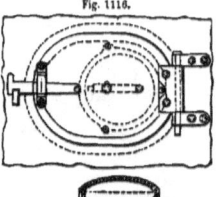

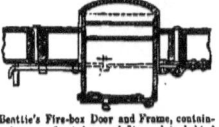

Beattie's Fire-box Door and Frame, containing a perforated curved flame plate, behind which are two perforated air plates contained in a box formed with the door, in which is an air door in connection with the box. Patented in the year 1858.

Fig. 1117.

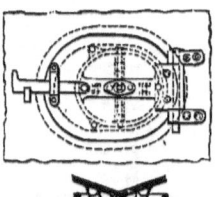

Beattie's Fire-box Door and Frame, containing movable flame plates that are regulated in position by a rod and curved levers. Patented in the year 1858.

Fig. 1118.

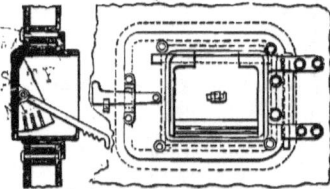

Beattie's Fire-Box Doors and Frame, in which the small door is fitted inside with air-distributing bars and a notched rod to keep the door open at various angles. Patented in the year 1858.

Fig. 1119.

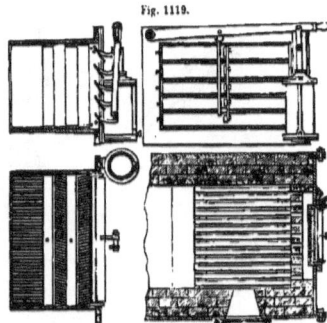

Prideaux's Fire-box Door and Frame: the door is horizontally apertured, and is fitted with metal strips that are shifted by levers and a rod in connection with a piston in a cylinder that is half filled with water which is compressed when the door is opened, and when the door is closed the expansion of the water causes the metal strips to cover the apertures in the door, and thus the draught is lessened as the ignition of the fuel increases; the multiple strips inside the flame split the air. Patented in the year 1855.

Fig. 1120.

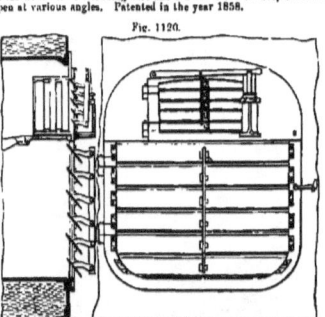

Prideaux's Fire-box Door and Frame, in which the air admitted through the door is regulated by metal strips that are raised or lowered by levers and rods, causing action as with a Venetian blind. Patented in the year 1855.

Fig. 1121.

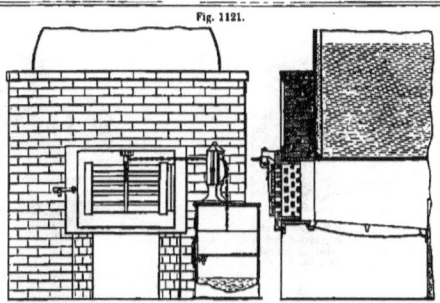

Cliff's Fire-box Door and Frame, in which the door is perforated and also fitted outside with a perforated movable sliding plate, and inside with three fixed perforated plates; the admission of air is regulated by the connection of the sliding plate to a piston—by a chain—contained in a cylinder, and the rise and fall of the piston causes the sliding plate to open and close; the piston is caused to rise and fall by the pressure of water. Patented in the year 1855.

Fig. 1122.   Fig. 1123.

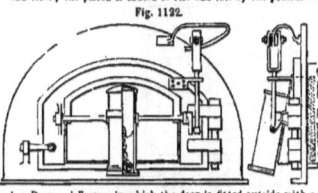 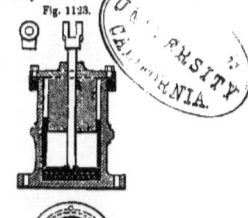

Auld's Fire-box Door and Frame, in which the door is fitted outside with a sand box, that by a lever and spring connection inverts the position of the box when the door is opened, so that when it is closed the falling of the sand regulates the amount of air gradually admitted over the grate for combustion. Patented in the year 1870.

Fig. 1124.

Prideaux's Fire-door Shutter Regulator, consisting of a cylinder fitted with a hollow piston and rod that are moved by the action of the mercury in the cylinder through an aperture in the bottom of the piston. Patented in the year 1870.

Fig. 1126.

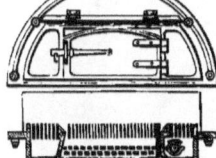

Fig. 1125.

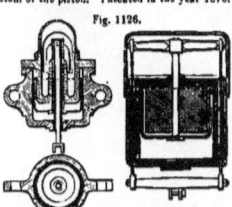

Sectional Plan of Prideaux's Fire Door and Frame, with the regulator contained in the centre of the door. Patented in the year 1870.

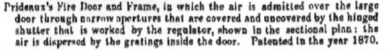

Prideaux's Fire Door and Frame, in which the air is admitted over the large door through narrow apertures that are covered and uncovered by the hinged shutter that is worked by the regulator, shown in the sectional plan; the air is dispersed by the gratings inside the door. Patented in the year 1870.

Prideaux's Inverted Regulators, for the actuation of fire-door shutters, the motion being obtained from the action of the mercury in the cylinder and the hollow piston; the weight on the rod—contained in the bell top of the cylinder—ensures the descent of the piston; also a double rod regulator is shown. Patented in the year 1872.

FIRE-BOX DOORS.

Fig. 1127.

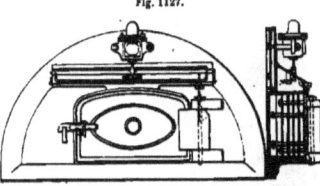

Prideaux's Fire Door and Frame, in which the air is admitted over the grate by the motion of the angular door or shutter that is opened and closed by the inverted regulator that is centrally situated over the shutter. Patented in the year 1872.

Fig. 1128.

Prideaux's Fire Door and Frame, actuated as in Fig. 1127, showing gratings formed in the door and frame in elevation and sectional plan. Patented in the year 1872.

Fig. 1129.

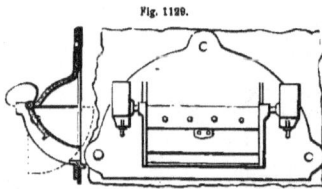

Martin's Patent Pendulous Fire Door, hung on hinges at the ends, and swung open by an inward or outward action. As used in the year 1873.

CHAPTER XII.
IGNITION OF COAL.

Fig. 1130.

NORMAL CONDITION.

An illustration of Coal, of average quality, raised in a midland county of England, 1873, and broken off to the proper size for putting it on the boiler fire grate.

Fig. 1131.

FIRST STAGE OF IGNITION.

An illustration of the piece of Coal in the first stage of ignition, producing flame, steam, and smoke by the release of the carbon from insufficient air to produce perfect combustion.

Fig. 1132.

SECOND STAGE OF IGNITION.

An illustration of the piece of Coal in the second stage of ignition, producing sharper flame, less steam, and but little smoke, because the draught was increased.

Fig. 1133.

THIRD STAGE OF IGNITION.

An illustration of the piece of Coal in the third stage of ignition, when the flame is fuller, the steam the same, and the smoke nearly nil, because the quantity of air over and under the coal was in better proportion to the release of the carbon.

Fig. 1134.

FOURTH STAGE OF IGNITION.

An illustration of the piece of Coal in the fourth stage of ignition, when the flame is at the maximum power of heat, the steam reducing and the smoke entirely nil, on account of the quantity of air admitted over and under the grate being in proper proportion to the heat in the fire box.

Fig. 1135.

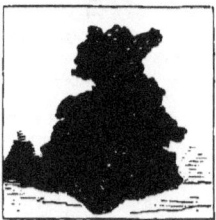

FIFTH STAGE.—FLAMING COKE.

An illustration of the piece of Coal when its constituents to produce flame are nearly exhausted; but it produces bright white-red heat of the most immediate power, and assists the process of ignition for the newer pieces of coal.

Fig. 1136.

SIXTH STAGE.—DEAD COKE.

An illustration of the piece of Coal converted into "Coke;" which means that the carbon is nearly burnt, and that the other constituents have been nearly exhausted during the process.

CHAPTER XIII.

THE CAUSE AND EFFECT OF COMBUSTION.

THE cause of combustion is the acquirement of sensible heat from any combustible that contains latent heat, which is released by the evolution of the chemical constituents of which the combustible is composed; as, for example, coal is the best solid combustible yet known for raising steam, because it contains so much more latent heat than any other equally available material. It is obvious, then, that the combustible properties of coal should be considered here, because it is most adaptable for marine and land boilers; and the table below shows the chemical constituents of coal now in every-day use.

AVERAGE PROPORTIONS OF THE CHEMICAL CONSTITUENTS OF COAL AS USED IN LAND AND MARINE BOILERS, 1873.

Carbon	86·32	
Oxygen	7·21	
Hydrogen	3·75	Water 1·
Nitrogen	0·41	Specific gravity 1·30.
Ashes	2·21	
Sulphur	0·10	
	100·00	

Carbon is the solid part that contains the most latent heat which, when liberated, becomes sensible heat.

Oxygen is the gas of the greatest density, and therefore heavier than the other gases, but is the chief gas in combustion, and causes flame.

Hydrogen is the most inflammable gas, but burns dull alone or in the absence of oxygen, which, when added, increases the brilliancy of the ignition.

Nitrogen is an invisible gas, which if used alone will not burn at all; but, nevertheless, is available in combustion as a diluent for the oxygen.

The two remaining constituents of coal, sulphur and ashes, contain but little latent heat, and therefore, in the production of sensible heat, are nearly useless, and, indeed, retard the action of the other constituents.

Being now conversant with the chemical constituents of the latent heat, we must dilate a little on the chemical action of the sensible heat. Let it be supposed a quantity of coals is put on a fire—the primary ignition of coals having been shown on the preceding page—which is composed of white red-hot fuel emitting no smoke, but say little pale yellow and light blue flames about two to two-and-a-half inches high, which are of course immediately quenched by the solid matter weighing them down. The portion of the coal nearest the hot fuel robs the heat therefrom, and the absorption must of course displace the carbon, which it not only does, but it also "sets on fire" the constituents of the coal, and causes what is known as "live" fuel; that is, coal in a state of motion caused by the release of its gases.

Most of us have seen this even in a domestic grate, where the coal is seen to expand at such a speed as to be credible to the naked eye; and when a split occurs, the gas escapes and ignites.

When the fuel has warmed the coal sufficiently to cause ignition at the points of contact, there is no more robbing of heat, but rather a congeniality or an affinity, and the new heat generating above ascends and ignites the remainder of the coal until the carbon is burnt out, when the affinity becomes more general. The chemical action of the constituents of coal is therefore set in motion by borrowing—or even so far as robbing, to commence with—a little heat to set the action in motion, which, once accomplished, needs no more recourse, but rather pays back than otherwise what it first required.

This action therefore relates specially to sensible heat, which is, as we have before explained, latent heat set in motion; and as that motion is a chemical action causing combustion, we next describe the constituents of combustion.

CHEMICAL CONSTITUENTS OF COMBUSTION.

Gases.	Proportions.	Weight.
Air	N 28 + O 8	36
Water	H 1 + O 8	9
Ammonia	H 3 + N 14	17
Carbonic oxide	C 6 + O 8	14
Carbonic acid	C 6 + O 16	22
Sulphurous acid	S 16 + O 16	32
Sulphuretted hydrogen	S 16 + H 1	17
Disulphuret of carbon	S 32 + C 6	38

These symbols and numerals also require a little explanation, which is, that

AIR is composed of nitrogen (N) of 28 parts in lbs. + oxygen (O) of 8 lbs. = 36 lbs.

WATER is composed of hydrogen (H) of 1 part in lbs. + oxygen (O) of 8 lbs. = 9 lbs.

AMMONIA is composed of hydrogen (H) of 3 parts in lbs. + nitrogen (N) of 14 lbs. = 17 lbs.

CARBONIC OXIDE is composed of carbon (C) of 6 parts in lbs. + oxygen (O) of 8 lbs. = 14 lbs.

CARBONIC ACID is composed of carbon (C) of 6 parts in lbs. + oxygen (O) of 16 lbs. = 22 lbs.

SULPHUROUS ACID is composed of sulphur (S) of 16 parts in lbs. + oxygen (O) of 16 lbs. = 32 lbs.

SULPHURETTED HYDROGEN is composed of sulphur (S) of 16 parts in lbs. + hydrogen (H) of 1 lb. = 17 lbs.

BISULPHURET OF CARBON is composed of sulphur (S) of 32 parts in lbs. + carbon (C) of 6 lbs. = 38 lbs.

Our next subject is the chemistry of those gaseous compounds, as they are arranged.

AIR.—Air is a compound of nitrogen and oxygen. The oxygen is the heavier of the two, bulk for bulk: the nitrogen is necessary to the oxygen as a diluent, but the nitrogen alone is a destroyer of heat and life, while the oxygen is a promoter of both. The action of air in combustion is, that the oxygen combines with the hydrogen and carbon, and accelerates the production of flame; as, for instance, in the case of applying a fan blast, the flame becomes brighter, longer, and hotter than with a natural draught.

WATER.—Water is a compound of oxygen and hydrogen, and when the hydrogen in the coal is emitting, the oxygen in the air admitted for combustion combines with it and forms water, which by the heat is converted into steam and accelerates combustion with a great deal of oxygen, but with not enough oxygen the water makes smoke by uniting the particles of carbon as they are released, which is the cause of black smoke and soot.

AMMONIA.—Ammonia is the result of putrefaction, or the combination of hydrogen and nitrogen, which occurs only with slow combustion, or what may be termed a "sweating fire."

CARBONIC OXIDE.—Carbonic oxide is the result of carbon being imperfectly burnt, and thus it may be said also to be the cause of smoke as well as a loss of heat.

CARBONIC ACID.—Carbonic acid is the result of there being sufficient oxygen introduced so as to effectually destroy the carbon and convert it into a gas, from which no smoke emanates.

SULPHUROUS ACID.—Sulphurous acid is the result of the combining of sulphur and oxygen, and forming a putrefied gas, which is poisonous as well as a non-conductor of heat and a preventive of combustion, but at the same time is formed in all cases during combustion according to the quantity of sulphur in the coal.

SULPHURETTED HYDROGEN.—Sulphuretted hydrogen is composed of sulphur and hydrogen, and the two are combined by the action of decomposition, or when the sulphur putrefies, it combines with the hydrogen, and thus the sulphuretted gas is formed.

BISULPHURET OF CARBON.—Bisulphuret of carbon is the result of the separation of the sulphur from the carbon after the sulphur combines with the hydrogen sufficiently to make sulphuretted hydrogen, and is therefore a final gas.

Having fully digested the "cause" of combustion, we must next devote attention to the "effect," or, what is often termed in science, the evaporative value. The authorities on this subject, whether English, Scotch, or French, put too much importance on the laboratorial experiments, and take them as the basis for calculations in practice, whereas we find from experience that the evaporative power of coal increases as the bulk is enlarged; as, for example, the late Professor Rankine, in a lecture "On the economy of fuel," at the United Service Institution, 1867, stated:—

"With regard to the total evaporative power of different sorts of fuel as distinguished from the available evaporative power, we already possess very full and accurate information. Numerous careful experiments on that subject have been made in scientific laboratories; they have consisted in completely burning small quantities of different combustible substances and measuring the heat with scientific apparatus of a very accurate kind, and taking special care that no heat escaped measurement. For example, in such experiments it has not always been practicable to prevent hot gases, the products of combustion, from passing away at an elevated temperature. But then the volume of those gases and their temperature could always be measured, and the heat that so passed away by the chimney calculated and allowed for. It is from experiments of that sort that our knowledge of the total or theoretical evaporative power of various combustible substances is derived.

"I will now make some remarks specially upon the total or theoretical evaporative power of certain sorts of fuel as ascertained by the experiments to which I have referred.

"We may distinguish the combustible substances to which those experiments relate into two classes—elementary substances and compound substances. I have stated the theoretical or total evaporative powers of the only elementary substances that are of any practical importance in this table:—

Element.	Oxygen per unit of weight.	Air per unit of weight.	Units of evaporation.
(1) Hydrogen gas	8	36	64·2
(2) Carbon, solid	2⅔	12	15·0
(3) Carbon, solid, with half supply of oxygen	1⅓	6	4·5 } 15·0
(4) Carbon, gaseous, in 2½ parts of carbonic oxide	1⅓	6	10·5 }
(5) Carbon, pure gaseous (inferred by theory)	2⅔	12	21 ?

"You will observe that there are three columns of figures following the name of the elementary substance. The first expresses the weight of oxygen that is required in order to burn completely an unit of weight of that elementary substance. For example, 1 lb. of hydrogen requires 8 lbs. of oxygen for its complete combustion. In order to supply that oxygen there are 36 lbs. of air required; and in the second column of figures is stated the weight of air. The third column gives the total or theoretical evaporative power, which in the case of hydrogen is 64·2 times its own weight. It has the greatest evaporative power of all known substances. Then follows carbon. A pound of carbon requires 2⅔ lbs. of oxygen to burn it completely, and to supply that oxygen 12 lbs. of air are wanted; and the complete combustion of carbon produces heat enough to evaporate 15 times its own weight of water.

"I may here observe that carbon exists in various states of aggregation, such as charcoal, coke, plumbago, and diamond, and that the total heat of combustion differs according to the state of aggregation. The more hard and dense the carbon is, the less is the heat we get by the combustion; the reason evidently being that a certain quantity of that heat is expended in overcoming the attraction of the particles of carbon for each other. For example, diamond does not evaporate so much as 15 times its weight of water, because of the heat required to overcome its own cohesion. That has a bearing upon some other phenomena to which I will presently refer. But the quantity I have here set down is the result obtained from carbon in the ordinary states of charcoal and coke.

"The third line in the table refers to the result produced by carbon when burned with only half the full quantity of oxygen. It is known to chemists that carbon combines in two different proportions with oxygen. One part of carbon by weight combined with one part and one-third of oxygen produces carbonic oxide. Carbonic oxide is itself a combustible gas, and in burning it combines with just as much additional oxygen as it already contains, so as to form carbonic acid. If we have a furnace ill supplied with air, so that the carbon only gets half the quantity of oxygen that it needs in order to form carbonic acid, then we have carbonic oxide as the product. Each pound of carbon takes up 1⅓ lb. of oxygen, and to supply that oxygen 6 lbs. of air are required. The total evaporative power is diminished not merely to one-half, but to a great deal less than one-half: it is only 4·5, or three-tenths of 15, the total evaporative power with a full supply of oxygen.

"If we next take that carbonic oxide, the weight of which will be 2⅓ lbs., namely, 1 of carbon, and 1⅓ of oxygen, and burn it, it takes an additional 1⅔ lb. of oxygen to burn it; and we get exactly the quantity of heat necessary to make up the deficiency, namely, 10½. The 4½ units of evaporation during the first process and the 10½ during the second give 15 in all, making up the whole evaporative power of carbon with a full supply of oxygen.

"A conclusion can be drawn from this, which I will now explain. It is to be observed that in both those stages of the combustion of carbon we have the very same thing happening chemically; we have 1⅓ lb. of oxygen combining with 1 lb. of carbon. But there is this difference in the two stages. In the first stage, where the carbonic oxide is produced from solid carbon, the solid carbon has to be converted from the solid state to a state of vapour or of gas. In the second stage we have the carbon already in the state of gas. Hence it appears that the cause of difference between the 4½ units of evaporation due to the first stage, and the 10½ due to the second stage of the combustion, must be, that during the first stage 6 units of evaporation disappear in transforming the carbon from the solid to the gaseous condition: in other words, the *latent heat of evaporation of carbon is six times that of water*. Thus we arrive at the conclusion, that the total evaporative power of pure gaseous carbon is 21, from which, if we subtract 6, the latent heat of evaporation of carbon, there remains 15, the total evaporative power of solid carbon.

"Now, as to the effect of the presence of oxygen in fuel. It was established some time ago by the researches of Dr. Joule, to whom science is immensely indebted for many discoveries of a similar kind, that a certain and definite quantity of heat is produced by the union of two chemical elements, and that precisely the same quantity of heat disappears in the separating of those elements. He verified that in a great number of cases, and established it as a universal law of nature. It holds in many other processes besides chemical combinations and decompositions, whether we are speaking of heat or of any other form of physical energy, that whatsoever quantity of energy we obtain by a given process, if we exactly reverse that process we have to expend precisely the same quantity of energy again.

"If 8 lbs. of oxygen combine with 1 lb. of hydrogen, they produce heat enough to evaporate 64 lbs. of water. If we decompose that water by any contrivance, whether by the use of a galvanic battery, by the superior affinity of carbon for oxygen, or by any other process, precisely the same quantity of heat disappears in overcoming the attraction of the hydrogen and oxygen for each other. Now, if oxygen enters into the chemical composition of any fuel, we have this result: calculate how much hydrogen that oxygen requires in order to form water—that is to say, one-eighth part of the weight of oxygen; then the oxygen present in the fuel renders just that quantity of hydrogen unavailable for the production of heat, and takes away from the total evaporative power 64 times the weight of the hydrogen so rendered unavailable. That being known, we have the following rule in the shape of a formula:—

$$E = 15\,C + 64\,H - 8\,O.$$

"This, then, is the rule for calculating the theoretical evaporative power of any sort of fuel whose combustible materials are carbon and hydrogen from its chemical analysis. Distinguish the constituents into carbon, hydrogen, oxygen, and refuse. The refuse does harm as being a useless weight, but it does not take away from the heating power of the constituent parts that remain. For every unit of carbon that remains in the compound we have 15 units of evaporation; and for every unit of hydrogen, in the absence of oxygen chemically combined with it, 64 units of evaporation; and if there is oxygen in the compound fuel, we have to subtract eight times the weight of that oxygen, because the oxygen renders inoperative one-eighth of its weight of hydrogen; and one-eighth of 64 is 8. The remainder gives the total evaporative power of that fuel.

"The next formula shows the amount of air necessary in order to supply oxygen for the complete combustion of a compound fuel. It consists of 12 units of weight of air for each unit of carbon, and 36 for each unit of hydrogen, deducting $4\frac{1}{2}$ for each unit of oxygen in the compound; or in symbols—

"$A = 12 \, C + 36 \, H - 4\frac{1}{2} \, O.$

"In an actual furnace it is seldom sufficient to supply just the quantity of air that contains the oxygen required for complete combustion. It is in general necessary to supply a surplus of air in order to dilute and sweep away the *burnt gas*, as we may call it, and to insure that every particle of fuel shall have air brought in contact with it. In furnaces with a draught produced by a chimney in the common way the practical result is, that we have to double the quantity of air, or nearly so. But when we produce a draught by a blast pipe, by steam jets, by a fan, or by other means, which thoroughly mix the air with the gaseous fuel, we reduce the surplus air required very much. In some cases we need only half; in some cases we need no surplus at all. Hence the quantity of air actually supplied ranges from once to twice the quantity necessary for the oxidation of the fuel.

"Here follows a table of examples of the total evaporative powers of some kinds of fuel, as calculated from their chemical composition:—

	C.	H.	O.	A.	E.	Evap. due to C.	H. — 0
Charcoal	·93	0	0	11·5	14·0	14·0	0
Coke	·88	0	0	10·6	13·2	13·2	0
$C_{10}H_{22}$	·84	·16	0	15·75	22·7	12·7	10·0
$C_{22}H_{22}$	·85	·15	0	15·65	22·5	12·66	9·84
Coal	·87	·05	·04	12·1	15·9	13·05	2·85
,,	·85	·05	·06	11·7	15·5	12·75	2·75
,,	·75	·05	·05	10·6	14·1	11·25	2·85
Peat, dry	·56	·06	·31	7·7	10·0	8·5	1·5
Wood, dry	·50	·05	·40	6·0	7·5	7·5	0

"The first line refers to charcoal; the second to coke of average quality; the third and fourth lines give two examples of hydrocarbons, which comprise between them the chief ingredients of rock-oil. In the third, fourth, and fifth columns are stated the proportions of the chemical constituents, ranging,

for the rock-oils, between $8\frac{1}{2}$ per cent. of carbon to 16 of hydrogen, and 85 per cent. of carbon to 15 of hydrogen. The fifth column gives the net weights of air required for the oxidation of the fuel. In the sixth column the theoretical evaporative powers are set down: for rock-oils they are from 22·7 to 22·5, or, in round numbers, we may say $22\frac{1}{2}$. In two additional columns are some figures to show how much evaporation is due to carbon and how much to hydrogen. In rock-oils, those quantities are $12\frac{1}{2}$ due to the carbon, and 10 to the hydrogen, out of the $22\frac{1}{2}$ units of evaporation. Lines 5, 6, and 7 give a few examples of coal. These might be immensely multiplied; but I have given only three specimens. They differ from the kinds of fuel previously mentioned, in having some oxygen in them, which somewhat lessens the evaporative power. Here, too, the units of evaporation due to carbon and to hydrogen are distinguished; for example, line 6 shows $12\frac{3}{4}$ units of evaporation due to carbon, and $2\frac{3}{4}$ due to hydrogen. Then follow some results for peat and for wood, on which I need not enlarge."

The late Professor, it will be noticed, not only treated of solid fuel, but estimated the evaporative properties of liquid fuel also, of which much more requires to be known; therefore for the present we shall proceed to the evaporative power of coal; but we must first dilate a little on the difference in Latent heat and Sensible heats.

Latent heat is a power given by nature to any body that is chemically formed to retain it, and also by a chemical change can give it out to produce Sensible heat; but it does not follow that one body must be hot and the other cold to produce the effect, because, on the contrary, cold water and cold lime mixed together produce sensible heat, and likewise cold water and cold sulphuric acid mixed will produce sensible heat. Ice also contains latent heat, but of such a small quantity that while melting it continues nearly at the same temperature. Latent heat is then practically Sensible heat confined, and in our present case we wish to know the best means of releasing it from coal, so that no smoke or hydrocarbon shall escape up the chimney. Our first purpose will be therefore to explain the theory, and afterwards the practice.

The theoretical amount of heat in a pound of coal is, according to the French authorities, MM. Favre and Silbermann, 14,500 units, *i. e.*, we must suppose that they extracted all the heat in a pound of coal they could, and, for the sake of convenience, termed it 14,500 parts, or units, the term unit being conventional, to express a certain amount of effect accomplished.

The authorities we have quoted put this matter in a tabulated form as follows:—

	British units of heat.	Evaporative power from 210°.
One pound of carbon imperfectly burned (C + O) produces	4,800	5·00
One pound of carbon perfectly burned (C + A) produces	14,500	15·00

The initial letters refer to carbon (C), oxygen (O), and acid (A). The pound of carbon refers to a pound of coal with the least possible amount of sulphur in it, and therefore we are to suppose that 14,500 units of heat is the maximum power that can be obtained from that quantity. But we know in practice that the carbon is more often imperfectly than perfectly burnt; and MM. Favre and Silbermann gave another table also of the compound ingredients of fuel as they more often occur in practice:—

Compound ingredient.	Lbs. of required air for combustion.	Lbs. of oxygen.	British units of heat.	Evaporative power from 210°.
One pound of carbon producing carbonic oxide by imperfect combustion, C + O	6	1·50	10,200	11·00
One pound of liquid fuel or hydrocarbon	10	2·5	20,000	21·00
One pound of olefiant gas	15	4·0	21,400	22·5

Alluding again to carbon, it will be remembered that its maximum heat per lb. is 14,500 units, and if we compare those with the others quoted we shall see the difference in practice—

	Units.
One lb. of carbon with perfect combustion	14,500
One lb. of carbon with practical combustion	10,200
No. of units or loss of heat by imperfect combustion	4,300

from which it is evident that 29·65 per cent. of the latent heat in coal is wasted by not being converted into sensible heat.

Messrs. Favre and Silbermann made some interesting experiments in the total heat of hydrogen with the following results. The total heat in one pound of hydrogen is 62·032 units, but M. Dulong, another French professor, termed it to be 62·536 units, which gave a mean thus:—

TOTAL HEAT IN ONE POUND OF HYDROGEN.

	British units.
According to MM. Favre and Silbermann	62,032
,, M. Dulong	62,536
	2) 124,568
Mean No. of units	62,284

We may add also, in passing, that there are three French authorities on the carbon question, and that all three have agreed to differ on the amount of heat in one lb. of carbon in the following manner:—

TOTAL HEAT IN ONE POUND OF CARBON.

	British units.
According to MM. Favre and Silbermann	14,500
,, M. Dulong	12,906
,, M. Desprotz	14,040
	3) 41,446
Mean No. of units	13,815

But the 14,500 units are observed as the standard by English engineers generally, and the loss by imperfect combustion about 25 per cent., thus: $14,500 \div 4 = 3825$, then $14,500 - 3825 = 10,675$, as the practical number of units in a pound of coal as it is generally burned in land and marine boilers; inasmuch that it is very plausible to claim the 14,500 units of heat in a pound of carbon, but if by imperfect combustion 25 per cent. is lost, then the sum 10,675 must be used instead of the higher sum, excepting where, by the introduction of more oxygen and ample room and time for combustion, the loss is proportionately reduced; but for general practice, as the boilers are now arranged and set, the 25 per cent. loss is on the charitable side.

AMOUNT OF AIR REQUIRED PER LB. OF FUEL TO CAUSE PERFECT COMBUSTION.—The amount of air required per lb. of fuel to cause perfect combustion and dilution is best known from practical experiment, although formulæ have been introduced; as, for example, a well-known formula is, that 12 lbs. of air is required for combustion, and 12 lbs. of air for dilution per lb. of coal; then the weight of air per lb. of fuel is 24 lbs., while in other cases 6 lbs. of air is required for combustion to 12 lbs. of air for dilution per lb. of coal, thus making only 18 lbs. of air to be required.

We must here explain that the air for combustion is required to "fan" the flame, while the air for dilution is required to release the latent heat or to dilute the gases and admit the air to the ignition of the fuel to commence combustion. Another formula is $W = 12 \times C + 36 (H - \frac{O}{8})$, when weight of air is represented by W, carbon by C, hydrogen by H, and oxygen by O. The numeral 12 is lbs. of air to 1 lb. of carbon; the numeral 36 is lbs. of air to 1 lb. of hydrogen; and the numeral 8 is lbs. of oxygen to 1 lb. of hydrogen.

Taking this question in its broad form, the following table is computed for practical purposes:—

TABLE OF DIFFERENT MATERIALS REQUIRING A CERTAIN AMOUNT OF AIR PER LB. OF FUEL TO CAUSE COMPLETE COMBUSTION.

Material.	Carbon.	Oxygen and Hydrogen.	Weight of Air in lbs. per lb. of Fuel.
Caking coal	0·85	0·10	23·0
Cannel coal	0·84	0·11	23·6
Flaming coal	0·77	0·10	20·8
Lignite coal	0·70	0·25	19·0
Bituminous coal	0·87	0·13	24·0
Anthracite coal	0·915	0·06	24·9
Wood charcoal	0·93	..	22·3
Peat charcoal	0·80	..	19·0
Wood	0·50	..	12·0
Mineral oil	0·85	0·15	32·0

We must now quote two authorities on combustion well known in scientific regions, T. S. Prideaux and Charles Wye-Williams. Mr. Prideaux, in his work on "The Economy of Fuel,"* states :—

"Combustion is, strictly speaking, the development of heat by *chemical combination*, but though this may take place from the union of a variety of bodies, the omnipresent agent, oxygen, plays so vastly more important a *rôle* than all others in the disengagement of light and heat, that the act of its combination with other bodies is preeminently entitled *combustion*, and, except in the mouths of chemists, has quite monopolised the appellation. Since combustion, in the *ordinary* acceptation of the word, is the only means had recourse to in the arts for the development of artificial heat, *perfect combustion* may, for our purpose, be defined to be—the combination of a combustible body with the largest measure of oxygen with which it is capable of uniting. In fact, for all practical purposes, the fuel or combustible body employed may be regarded as composed exclusively of carbon and hydrogen, so that our inquiry becomes narrowed to the combinations of oxygen with these two elementary substances.

"Most of my readers are doubtless aware that chemical combinations take place only in certain definite proportions which are multiples of each other. One atom of A, for instance, combines with one, two, three or more atoms of B; or two of A, with three, five, or more of B; from which fact it follows, as a necessary consequence, that the chemical equivalents (or combining weights) of all bodies may be considered as expressing the relative weight of their atoms. Bodies may be *mingled* together in any proportions, but it is only in certain definite ones that they unite and form one homogeneous whole. Thus six parts (by weight) of carbon combine with eight (by weight) of oxygen to form carbonic oxide, and with twice this quantity, or sixteen parts, to form carbonic acid; and there exists no intermediate combination of these two bodies in which six of carbon is united with more than eight and less than sixteen of oxygen. Carbonic oxide and carbonic acid gases may, it is true, be *mingled* in any proportion, and thus a gas obtained in which six parts of carbon are present with more than eight and less than sixteen parts of oxygen, but this is a *mixture* and not a chemical *unity*, as may be shown by the addition of potash, which will separate the carbonic acid and leave the carbonic oxide behind.

"Different coals vary much in their component parts, and in the proportion of these to each other; carbon and hydrogen, however, are the essential ingredients of all, as far as their heating capabilities are concerned; and throughout this essay I shall

---

* Lockwood and Co.

assume as a convenient standard that 100 parts of coal consist of 80 parts of carbon and 5 of hydrogen, leaving out of view the other elementary substances which enter into their composition (consisting of oxygen, nitrogen, sulphur, and incombustible ashes in various proportions), as only likely to complicate the details without being essential to the argument.

"Assuming 100 lbs. of coal to consist of 80 lbs. of carbon and 5 lbs. of hydrogen, then, since the oxygen is to the carbon, in carbonic acid, as 16 to 6, to effect perfect combustion 80 lbs. of carbon will require $313\frac{1}{3}$ lbs. = 2527 cubic feet of oxygen, to furnish which 967·26 lbs. = 12,635 cubic feet of atmospheric air will be required, air consisting of 1 volume of oxygen to 4 of nitrogen, or 8 parts by weight of the former to 28 parts of the latter; and since oxygen is to hydrogen, in water, as 8 to 1, 5 lbs. of hydrogen will require 40 lbs. = 473 cubic feet of oxygen or 181·5 lbs. = 2,365 cubic feet of atmospheric air.

"967·29 lbs. × 181·5 = 1148·76 lbs. = 15,000 cubic feet of atmospheric air required for the perfect combustion of 100 lbs. of coal.

"And the product resulting will be: 2527 cubic feet of carbonic acid, 946 cubic feet of steam, and 12,000 cubic feet of uncombined nitrogen.

"We thus perceive that each lb. of coal requires 150 cubic feet of air for its perfect combustion, or, in other words, for the conversion of all its carbon into carbonic acid, and all its hydrogen into water; and it must be remembered, that just in proportion as this proper quantity is *deficient* is combustion imperfect and fuel wasted, whilst the supply of a *surplus* quantity is but a change of evils, and equally injurious in an economic point of view, since all the air which passes through a furnace without giving up its oxygen to the fuel serves only to abstract heat, without yielding any in return. However difficult may be the regulation of the admission of just the proper quantity of air to the fuel, it is not the less certain that exactly in proportion as we deviate from the correct standard will be the loss we incur, for the laws of chemical affinity are unerring and inexorable.

"It is commonly but erroneously supposed that when no smoke appears at the chimney-top combustion is perfect. Smoke, however, may be absent, and yet the carbon may only have united with 1 atom of oxygen, forming carbonic oxide (a colourless gas), instead of with 2 atoms, forming carbonic acid, and consequently have only performed half the duty, as a fuel, of which it was capable, whilst the loss of duty on the coal taken as a whole (supposing all its hydrogen to have become oxydised) will be upwards of 40 per cent.

"Hydrogen, having a stronger affinity than carbon in the gaseous state for oxygen, when the supply is short, still seizes on its equivalent, and leaves the carbon minus. Thus, when coal gas, (carburetted hydrogen) is inflamed with an insufficient supply of air to effect the perfect combustion of both its constituents, the hydrogen is still converted into water, whilst the carbon, in different proportions according to the oxygen present, becomes—deposited in the form of soot—converted into carbonic oxide, or partly into carbonic oxide and partly into carbonic acid.

"This great cardinal point in economy of furnace management, viz., the exact apportionment of the supply of air to the wants of the fuel, so as to convert all its carbon into carbonic acid, and all its hydrogen into water, could be achieved with comparative ease were the same conditions always present in the interior of the furnace, so as to cause the quantity of air required by the fuel to be *uniform*. In this case, the average rarefaction in the stack being once attained, a steady supply would enter the furnace according to the area of the grate-bar openings, the size of which once adjusted, the equable and economic working of the furnace would be secured. Unfortunately, however, the reverse is the case, and the great practical difficulty to be overcome in apportioning the supply of air to the demands of the fuel arises from the fact, that in furnaces of the ordinary construction this demand is not only *variable*, but fluctuates within very wide limits.

"Where a fresh supply of coal is put on a briskly-burning fire, the first thing which takes place is, that the coal softens and swells, attended with the evolution of a large quantity of carburetted hydrogen gas, requiring for its combustion a correspondingly large supply of atmospheric air—the coal undergoing, in fact, in the first stage of combustion, just the same process as it does in the retort in the manufacture of gas, but with this difference in the result: that the gas, which in the latter case is preserved and found so valuable a commodity, here escapes unconsumed up the chimney, not only furnishing no heat itself, but abstracting from the heat arising from the combustion of the carbonaceous portion of the fuel the heat for its own gasification—a circumstance which readily explains the fact, that more heat is practically obtained in many kinds of furnaces from coke (or, in other words, coal deprived of ¼ part by weight of that portion of its combustible matter which is richest in furnishing material for heat) than from coal in its pure state, with all its hydrogenous portion intact. From the same cause also (viz., the imperfection of our furnaces) the commercial value of coal is often in the inverse ratio to the quantity of its bituminous constituents and its real heat-giving powers, had we the capacity to render them practically available. It could not in fact be otherwise. A furnace immediately after a fresh supply of fuel requires more than double the quantity of air it did the instant before, whilst we have no contrivance for furnishing such a supply, although without it, throughout the space of time during which rapid gasification of the hydrogenous portion is going on, more than half the fuel consumed is wasted, and passes off unburnt, becoming thereby not only totally unproductive in itself, but absolutely an agent of evil, by robbing the furnace of the heat absorbed in its own volatilization."

Mr. Williams, in his treatise on "Combustion of Coal and Prevention of Smoke,"* states:—

"A consideration of the nature of the products into which the combustible constituents of coal are converted in passing through the furnace and flues of a boiler, will enable us to correct many of the practical errors of the day, and ascertain the amount of useful effect produced and waste incurred. These products are:—

"1st. Steam—highly rarefied, invisible, and incombustible.

"2nd. Carbonic acid—invisible and incombustible.

"3rd. Carbonic oxide—invisible, but combustible.

"4th. Smoke—visible, partly combustible, and partly incombustible.

* Lockwood and Co.

"Of these, the two first are the products of perfect combustion, the latter two of imperfect combustion.

"The first—steam—is formed from that portion of the hydrogen (one of the constituents of coal-gas) which has combined chemically with its equivalent of oxygen from the air, in the proportion of 1 volume of hydrogen to half a volume of oxygen; or, in weight, as 1 is to 8.

"The second—carbonic acid—is formed from that portion of the constituent, carbon, which has chemically combined with its equivalent of oxygen in the proportion of 16 of oxygen to 6 of carbon, in weight; or, in bulk, of 1 volume of the latter to 2 of the former.

"The third—carbonic oxide—is formed from that portion of the carbonic acid which, being first formed in the furnace, takes up an additional portion of carbon in its passage through the ignited fuel on the bars, and is then converted from the *acid* into the *oxide* of carbon; thus changing its nature from an incombustible to a combustible. This additional weight of carbon so taken up, being exactly equal to the carbon forming the carbonic acid, necessarily requires for its combustion the same quantity of oxygen as went to the formation of the acid.

"The fourth—smoke—is formed from such portions of the hydrogen and carbon of the coal-gas as have not been supplied or combined with oxygen, and, consequently, have not been converted either into steam or carbonic acid.

"The hydrogen so passing away is transparent and invisible; not so, however, the carbon, which, on being so separated from the hydrogen, loses its gaseous character, and returns to its natural and elementary state of a black, pulverulent, and finely-divided body. As such, it becomes *visible*, and this it is which gives the dark colour to smoke.

"Not sufficiently attending to these details, we are apt to give too much importance to the presence of the carbon, and have hence fallen into the error of estimating the loss sustained by the blackness of the colour which the smoke assumes, without taking any note of the *invisible* combustibles, hydrogen and carbonic oxide, which accompany it. The blackest smoke is, therefore, by no means a source of the greatest loss; indeed, it may be the reverse; the quantity of invisible combustible matter it contains being a more correct measure of the loss sustained than could be indicated by mere colour.

"This will be still more consistent with truth, should any of the gas (carburetted hydrogen) escape undecomposed or unconsumed, as too often is the case.

"In the ordinary acceptation of the term 'smoke,' we understand *all* the products, combustible and incombustible, which pass off by the flue and chimney. When, however, we are considering the subject scientifically, and with a view to a practical remedy against the nuisance or waste it occasions, we must distinguish between the gas as it is generated and that which is the result of its imperfect combustion. In fact, without precise terms and reasoning, we disqualify ourselves from obtaining correct views either of the evil or the remedy.

"Now, let us look at this gas, which we are desirous of converting to the purposes of heat, under the several aspects in which it may be presented under the varying degrees of temperature or supplies of air.

"In the first instance, suppose the equivalent of air to be supplied in the proper manner to the gas, namely, by jets, for in this respect the operation is the same as if we were supplying gas to the air, as in the Argand gas-lamp. In such case one-half of the oxygen absorbed goes to form steam, by its union with the hydrogen; while the other half forms carbonic acid, by its union with the carbon. Both constituents being thus supplied with their equivalent volumes of the supporter, the process would here be complete—perfect combustion would ensue, and no smoke be formed; the quantity of air employed being *ten times the volume of the gas consumed.*

"Again, suppose that but one-half, or any other quantity, *less* than the saturating equivalent of air were supplied. In such case, the hydrogen, whose affinity for oxygen is so superior to that of carbon, would seize on the greater part of this limited supply; while the carbon, losing its connection with the hydrogen, and not being supplied with oxygen, would assume its original black, solid, pulverulent state, and become *true smoke.* The quantity of smoke then would be in proportion to the deficiency of air supplied.

"But smoke may be caused by an *excess* as well as a *deficiency* in the supply of air. This will be understood when we consider that there are *two* conditions requisite to effect this chemical union with oxygen, namely, a certain degree of temperature in the gas, as well as a certain quantity of air; for, unless the due temperature be maintained, the combustible will not be in a state for chemical action.

"Now, let us see how the condition, as to *temperature,* may be affected by the quantity of air being in *excess.* If the gas be injudiciously supplied with air, that is, by larger quantities or larger jets than their respective equivalent number of atoms can *immediately combine with,* as they come into contact a *cooling effect* is necessarily produced instead of a *generation of heat.* The result of this would be, that, although the quantity of air might be correct, the second condition, the required temperature, would be sacrificed or impaired, the union with the oxygen of the air would not take place, and smoke would be formed.

"Thus we perceive that the *mode* in which the air is introduced exercises an important influence on the amount of union and combustion effected, the quantity of heat developed, or of smoke produced; and, in examining the mode of administering the air, we shall discover the true cause of perfect or imperfect combustion in the furnace as we see in the lamp. This circumstance, then, as regards the manner in which air is introduced to the gas (like the introduction of gas to the air), demands especial notice, as the most important, although the most neglected, feature in the furnace, and in which practical engineers are least instructed by those who have undertaken the task of teaching them.

"We see, then, how palpably erroneous is the idea, that smoke, once formed, can be consumed in the furnace in which it is generated, and how irreconcilable is such a result with the operations of nature. The formation of smoke, in fact, arises out of the failure of some of the processes *preparatory* to combustion, or the absence of some one of the conditions which are essential to that consummation from which light and heat are obtained. To expect,

2 x

then, that smoke, which is the very result of a deficient supply of heat, or air, or both, can be consumed in the furnace in which such deficient supply has occurred, is a manifest absurdity, seeing that, if such heat and air had been supplied, this smoke would not have existed.

"Whence, then, it may be asked, does the visible black of the cloud proceed? Solely from the unconsumed portion of black carbon, insignificant though it may be in weight or volume.

"This *carbon* of the gas being the sole black-colouring element of smoke, it is here necessary to examine the several phases and conditions of its existence and progress, *before*, *during*, and *after* it has been in the state of flame. Flame is not the combustion of the gas. Flame itself has to undergo a further process of combustion, being but a mass of carbon atoms, *still unconsumed*, though at the temperature of incandescence and high luminosity. Flame is then but one of the stages of the process of combustion. Its existence marks the moment, as regards each atom, of its separation from and the combustion of its accompanying *hydrogen*, by which so intense a heat is produced as instantaneously to raise the solid *carbon atom*, then in contact, to that high temperature: thus preparing it the more rapidly to combine with oxygen *so soon as it shall have obtained contact with the air, but not a moment sooner.*

"Instead, however, of administering the air while the carbon is at this high temperature of 3000° (as we see in our gas-burners), our custom is first to allow it, or even *force it*, to cool down, by its contact with metallic tubes, to the state of soot; and then to expect, by some mechanical apparatus, to restore it to the necessary temperature from which it had been so gratuitously reduced.

"But, it may be asked, why allow it to lose its already acquired high temperature? Why create a necessity for the sake of overcoming it? It seems an act of mere stupidity to waste the high temperature the carbon had thus naturally acquired, by allowing the opportunity to pass before we administer the only thing needful, namely, *the air*.

"We have seen how the carbon of the gas, in the absence of air and its oxygen, returns to its normal state of black solid atoms in the form of soot. It will here, then, be useful to illustrate the well-defined stages through which this carbon passes from its invisible state, as a constituent of the gas, to its *visible* state in smoke.

"First stage—*Invisible* and *intangible*, the carbon being then in chemical union, and surrounded by the two atoms of hydrogen, forming carburetted hydrogen gas.

"Second stage—*Visible*, *tangible*, and raised by the heat produced on the combustion of its accompanying hydrogen to the temperature of incandescence, which, by their number, give the white luminous character to flame.

"Third stage—*Invisible* and *intangible*, after its combustion, having then entered into union with two atoms of oxygen, and forming invisible carbonic acid.

"Fourth stage—*Visible* and *tangible*, in the state of lamp-black, or soot, having escaped combustion by not having had access to the air before it was cooled below the temperature required for chemical action.

"*Composition of Smoke.*

"Eight atoms of invisible nitrogen from the four of air that supplied the oxygen both to the hydrogen and carbon of the gas.

"Two atoms of invisible steam from the combustion of the hydrogen of the gas.

"One atom of visible carbon unconsumed, and becoming the colouring matter of smoke.

"Two atoms of invisible carbonic acid from the carbon of solid coke on the bars of a furnace.

"Eight atoms of invisible nitrogen from the four of air that supplied the oxygen for the combustion of the coke of the coal.

"Thus we see that out of the 21 atoms which are the constituents of any given weight of smoke, the only combustible one, the carbon, weighs but 6; the *incombustible and invisible* portion weighing 286. As to volume, we see, as above, the comparatively insignificant space it occupies, although it possesses the power of giving the black tint to the cloudy mass. These volumes are here supposed to be at atmospheric temperature. When, however, we consider that, with the exception of the *carbon*, which alone (being *a solid*) retains its original diminutive bulk, while all the others, being *gaseous*, will be enlarged to *double*, possibly to *treble*, their previous bulk, in proportion to their increased temperature, we are amazed, not only at the comparative insignificance of the carbon, but at our own credulity in believing that this merely blackened cloud could be made available as *a fuel* and a source of heat.

"Generally speaking, this black cloud is supposed to be an aggregate or *mass of carbon*, in the form of a sooty powder. This is, manifestly, an error, since that would assume that the three other products, nitrogen, carbonic acid, and steam, in their great volumes, had been neutralised, or otherwise disposed of. As, however, that is impossible, smoke must be taken as it is, namely, a *compound cloud of all these three* gaseous bodies, together with the portion, more or less, of the solid, uncombined, visible free carbon, then in the *fourth stage*. Here, then, is a definition of smoke, which is susceptible of the most rigorous proof.

"We see the black cloud from a chimney extending for miles along the horizon, and hence conclude that the quantity of carbon must be considerable to produce such an effect. Nothing but strict chemical inquiry could have enabled us to correct this error. By it we ascertain that this black cloud is *tinted, literally but tinted*, by the atoms of carbon, and which, though issuing in countless myriads, are comparatively insignificant in weight or volume, or in commercial value as a combustible. In truth, the eye is deceived as to the mass by the extraordinary colouring effect produced by the minuteness, but great number, of its atoms of carbon.

"And now as to the relative *quantities* of the several constituents of smoke: 1st, of the *invisible nitrogen*. As atmospheric air contains but 20 per cent. of oxygen, the remaining 80 per cent. being the *nitrogen*, passes away, invisible and uncombined. If, then, a ton of coal requires absolutely for its combustion the oxygen of 300,000 cubic feet of air, the 80 per cent., or 240,000 cubic feet of invisible and incombustible nitrogen, forms the first ingredient of this black cloud. 2nd, of the *invisible carbonic acid*. This portion of the cloud may be estimated as equal in volume to the 20 per cent. of oxygen which had effected the combustion of the carbon *both of the gas and the coke* of the coal. 3rd, of the *invisible steam* formed by the combustion of the hydrogen of the gas. In this will be found the great source of the prevailing misapprehension; yet no facts in chemistry are more accurately defined than those which belong to the formation, weight, and volume of the constituents of steam.

"The following extract from a paper read before the Institution of Civil Engineers, being from the report, is much to the point of this inquiry, particularly as regards the great volume of water resulting from the combustion of the coal gas:—

"'All substances used for the purposes of illumination may be represented by oil and coal gas. Both contain carbon and hydrogen, and it is by the combustion of these elements with the oxygen of the air that light is evolved. The *carbon* produces *carbonic acid*, which is deleterious in its nature and oppressive in its action in closed apartments. The *hydrogen* produces *water*. A pound of oil contains about 0·12 of a pound of hydrogen, 0·78 of carbon, and 0·1 of oxygen. When burnt, it produces 1·06 of water, and 2·86 of carbonic acid; and the oxygen it takes from the atmosphere is equal to that contained in 13·27 cubic feet of air. A pound of London gas contains on an average 0·3 of hydrogen, and 0·7 of carbon. It produces, when burnt, 2·07 of water, 2·56 of carbonic acid gas, consumes 4·26 cubic feet of oxygen, equal to the quantity contained in 19·3 cubical feet of air. A pint of oil, when burnt, produces a pint and a quarter of water; and *a pound of gas, more than two and a half pounds of water*, the increase of weight being due to the absorption of oxygen from the atmosphere—one part of hydrogen taking eight parts, by weight, of oxygen to form water. A London *Argand gas-lamp* in a closed shop window will produce, *in four hours, two pints and a half of water*, to condense, or not, upon the glass or the goods, according to circumstances.'

"To say, then, that above 900 lbs. weight of water (nearly half the weight of the ton of coal consumed) passes from the furnace, and by the chimney, in the form of *steam*, though produced by the 5¼ per cent. of hydrogen alone, which the coal contained, may appear exaggeration; nevertheless, the fact is unquestionable, the details of which it is here unnecessary to repeat. Now, when we consider the enormous mass of steam that would be produced by the vapour of this nearly half a ton weight of water (independently of the nitrogen and carbonic acid), we can readily account for the magnitude of the cloudy vaporous column of the smoke.

"The next consideration is, as to the *value of the carbon* which produces the darkened colour of the smoke cloud. Now, the weight of this carbon, in a cubic foot of black smoke, is not equal to that of *a single grain*. Of the extraordinary light-absorbing property and colouring effect produced by the inappreciable myriads of atoms of this finely-divided carbon, forming part of the cloud of the steam alone, some idea may be formed by *artificially* mixing some of it when in the deposited state of soot with water. For this purpose, collect it on a metallic plate held over a candle or gas-jet, and touching the flame. Let a *single grain weight* of this soot be gradually and intimately mixed on a pallet, as a painter would, with a pallet-knife, first, with a few drops of gumwater, enlarging the quantity until it amounts to a spoonful. On this mixture being poured into a glass globe containing a gallon of water, the whole mass, on being stirred, will become opaque and of the colour of *ink*. Here we have physical demonstration of the extraordinary colouring effect of the minutely divided carbon—a *single grain* weight being sufficient to give the dark colour to a gallon of water."

In the beginning of the year 1872, Mr. Prideaux lectured at the United Service Institution, "On Economy of Fuel in Ships of War," and said he:—

"To evaporate the greatest weight of water from the fuel employed involves two considerations: First, the generating from that fuel the utmost amount of heat it is capable of rendering, to be accomplished by effecting perfect combustion in the furnace. Secondly, the transference of the largest practicable proportion of this heat to the water in the boiler, to be attained by the judicious arrangement of its heating surface. The first, a question beset with complication; the second, one of comparative simplicity.

"Coal is so much more extensively used than any other combustible for the production of steam power, that in so condensed an exposition of the subject as the present must necessarily be, it will be most profitable to restrict our remarks to its use.

"At the outset of our inquiry we are confronted with the question, What proportion of the whole heat coal is capable of yielding do we succeed in utilising in raising steam, or, in other words, transferring to the water in the boiler?

"After carefully investigating the subject, the conclusion at which I have arrived is, that we cannot assign to carbon a smaller heating power than 9000° centigrade, which is equivalent to the capacity for evaporating fifteen times its own weight of water from 100° Fahrenheit. The average duty obtained in marine boilers at present must, I apprehend, be set down at not more than 7¼ lbs. of water evaporated by 1 lb. of coal, or, in other words, we only obtain half the duty which theory assigns to the fuel. Let us console ourselves for so humiliating a result by the reflection, that this very large margin for waste is as encouraging to our future prospects as it is discreditable to our present practice.

"The heat obtained from coal is evolved by the chemical union of the hydrogen and carbon of the coal with the oxygen of the atmosphere, forming with the former substance, water, and, with the latter, carbonic acid; and just in proportion to the exactitude with which we transform all the hydrogen of the combustible into water, and all its carbon into carbonic acid, without *admitting any superfluous air*, will be the temperature produced, and (other conditions being equal) the amount of water evaporated.

"Different coals vary in their component parts, and in the proportion of these to each other. Carbon and hydrogen, however, are the essential constituents of all as far as their heating capacities are concerned, and I shall assume as a convenient standard that 100 parts of coal consist of 80 parts of carbon and 5 of hydrogen, leaving out of view the other elementary substances which enter into their composition (consisting of oxygen, nitrogen, sulphur, and incombustible ashes, composed principally of sand and clay), as non-essential to the subject under discussion. There is a convenience in assuming these proportions of 80 carbon and 5 hydrogen for the composition of coal, for since hydrogen furnishes, weight for weight, four times as much heat as carbon, the 5 parts of hydrogen will furnish 20 per cent. of the whole heat, and the 80 parts of carbon 80 per cent., being the same proportionate part of the heat as it forms by weight of the fuel, and thus the heating power of coal, as a whole, may be treated as equivalent to an equal weight of carbon.

"The hydrogen in coal exists in chemical combination with carbon, and when heated without the access of air, passes off with the carbon, with which it is combined in the gaseous form as carburetted hydrogen, a gas consisting of 1 part by weight of hydrogen and 3 of carbon, being in fact the coal-gas we use for illuminating purposes. 100 lbs., therefore, of coal containing 5 parts of hydrogen and 80 parts of carbon in the 100 lbs. would yield 20 lbs., or 525 cubic feet of this gas, and 65 lbs. of solid carbon, or coke.

"As the 5 parts of hydrogen in coal-gas furnish 20 parts of the total heat of the coal, and the 15 parts of carbon 15 parts, the two combined contain 35 per cent. of the total heating power of the combustible. But here I must not omit to draw your attention to the important fact that if, through the defective regulation of the air-supply of the furnace, you cause half this gas to pass off up the chimney unburnt, the heat wasted is more than 17½ per cent., because the gas has robbed the furnace of the heat employed in its own volatilisation or transformation from the solid into the gaseous form. 7½ per cent. is by no means an over-estimate of the heat subtracted from the duty of the carbon by this process, and hence we arrive at the result that, in the case of a furnace so managed as to dissipate half the carburetted hydrogen unburnt, the waste of fuel through this channel alone reaches 25 per cent. That this amount of waste does in practice often take place from this cause is unquestionable, since I have in numerous instances witnessed this percentage of saving affected by simply altering the conditions governing the air-supply of the furnace during the first five minutes after coaling, the period at which, under the ordinary system of furnace management, or rather mismanagement, this great waste of unconsumed gas takes place. In short, I have rarely seen an instance where experiment did not prove that, with bituminous or north-country coal, a saving of not less than two-thirds of this amount, or from 17 to 18 per cent., was obtainable by these means, so that I feel justified in saying that in furnaces as constructed and fired at present, i.e., with no provision for adapting the air-supply to the changing conditions and consequent varying requirements of the fuel, this waste of a large proportion of the gaseous constituents of the coal must be regarded as the normal state of things, and inseparable from the rude, defective, and unintelligent system of furnace management pursued.

"Although, as a general rule, the combustion of the carbon of the fuel is effected without the large proportion of waste attendant upon that of its gaseous constituents, yet occasionally—in such a case, for instance, as one of which we have recently heard, where, in the Southern Pacific, the flames proceeding from the top of the funnel of one of Her Majesty's ships at night were actually mistaken for an eruption from a volcano in one of the peaks of the Andes—a great waste of carbon occurs. Such a state of things as this, however, can only exist with most defective furnace arrangements, involving a really frightful sacrifice of fuel.

"I will endeavour to describe the conditions under which alone I consider such an occurrence as described to be possible. The flame emitted does not extend from the furnace through the flues to the funnel-head, as many suppose, but is produced by carbonic oxide gas (generated in the furnace through deficiency of the air-supply) passing off at such a high temperature as to preserve the temperature necessary for ignition, after mixing with the 2½ volumes of atmospheric air required to furnish the oxygen needful for its transformation into carbonic acid. Now, since an atom of carbon, in combining with one atom of oxygen, and becoming carbonic oxide, only yields one-third of its heating power, the remaining two-thirds being set free upon its combining with the second equivalent of oxygen, and becoming carbonic acid, it follows that two-thirds of the heating power of all the carbon passing off as carbonic oxide is wasted.

"Even this statement does not adequately represent the frightful sacrifice of heating power, of which flames passing off at the funnel-head are the sign; for small as may be the proportion of the heating power of the fuel developed in the furnace, of this small proportion an unusually large part is wastefully dissipated into the atmosphere, as is shown by the high temperature of the gases issuing from the funnel, such high temperature being the result of the diminished subtraction of their heat by the surface of the tubes, consequent upon the latter being coated with soot. Unless there be something more than ordinarily faulty, either in the construction of the furnace, or the mode of firing pursued, the coating of the tubes with soot may be regarded as the necessary precedent to the exhibition of flame at the funnel-head, as in fact begetting the conditions under which it occurs. A deposition of soot on the interior of the tubes simultaneously diminishes the draught and, in a still greater ratio, the steam generated. Heavier firing is resorted to to keep up steam, with the effect of aggravating the previously existing evils. Flame makes its appearance at the top of the funnel, and a vicious circle of evil consequences is produced, which mutually react upon and aggravate each other.

"The waste of a large proportion of the coal-gas, or carburetted hydrogen contained in the fuel, during say the first third, or first five minutes of the coaling interval, which ordinarily takes place under the present system of furnace management, by no means presupposes such conditions in the arrangements and working of the furnace as would subsequently lead to the dissipation of carbon as carbonic oxide. The fatal sign of flame at the funnel-head, however,—announcing as it does a thick layer of ignited carbon on the bars and an insufficient air-supply,—may be accepted as indicative of conditions being present in the furnace which must have been productive of a more than average waste of carburetted hydrogen at an earlier period.

"Great as are the evils which we have been reviewing, proceeding from a deficient air-supply, they are rivalled, and, perhaps, having regard to its greater frequency (during the latter portion of the coaling interval), exceeded in amount as a whole, by the opposite fault of too much air entering the furnace. The excess of air acts prejudicially, not by interfering with the perfect combustion of the carbon, none of which under these altered conditions can escape transformation into carbonic acid, but by diluting the temperature of the furnace by the admixture of cold air, which, after receiving a large accession of temperature, passes off through the funnel, robbing the furnace of heat that should have been employed in raising steam. All good firemen are awake to the evil consequences of allowing their fire to burn through at the back, admitting a free ingress of air, or the still more fatal results of perforations in the door or bridge. Let us analyse the conditions under which these evil consequences ensue, and the causes on which they depend:—

"One hundred pounds of coal require, in order to obtain the equivalent of oxygen necessary for its perfect combustion (or the transformation of all its hydrogen into water and all its carbon into carbonic acid), 15,000 cubic feet of air, and the temperature resulting may, to speak in round numbers, be set down at $4200°$. No such exact adjustment of the air-supply is possible in working a furnace, and there are reasons why, in practice, a better result is obtained by admitting an increased quantity, say one-third more than the amount required to furnish the exact combining equivalent of oxygen. Assuming, therefore, the admission of one-third more, or 20,000 cubic feet, we get a temperature of $3150°$, of which, taking the temperature of the gases when they leave the heating surface of the boiler and enter the funnel at $600°$, 81 per cent. is absorbed by the heating surface of the boiler, and profitably employed. If instead of 20,000 feet twice the quantity, or 40,000 feet, enter the furnace, we only get a temperature of $1575°$; and supposing the temperature of the gases entering the stack to be the same, viz. $600°$, we only get 62 per cent. absorbed by the heating surface and performing the duty of raising steam. In reality, however, since twice the volume of gas traverses the flues in the same time, it must pass over the heating surface at twice the velocity. The area of the heating surface therefore becomes halved in proportion to the quantity of gas passing, and, as a consequence, the products of combustion will pass off at a higher temperature, so that in reality less than 62 per cent. will be absorbed.

"The supply of air which enters a furnace depends upon the ratio existing between the *pull* exerted by its chimney or stack and the *obstruction* caused by the layer of fuel on the grate-bars, and a more or less confined ashpit. The force of the pull, or suction, exerted by the stack is the difference in weight between the column of heated gas it contains and a similar column of air at the temperature of the atmosphere. The higher the temperature and the taller the stack, the greater the force of the draught, and *vice versâ*. The great practical difficulty which stands in the way of obtaining the regular flow of the right quantity of air into the furnace is the fact that the demand is not a *constant*, but a *varying*, quantity.

"Assume a coaling interval of 15 minutes, and that towards its end the carburetted hydrogen having passed off, a layer of ignited carbon alone is left on the bars, through which the current of air has gradually worn for itself a series of channels or perforations, till their united area is more than sufficient to admit the right quantity of air under the existing amount of draught. Let us then examine the condition of the same furnace a few minutes later, after the act of coaling has taken place, and we shall find a thicker layer of fuel on the bars, the channels for the passage of air which previously existed in the thinner layer obliterated, and the temperature of the furnace greatly lowered by the abstraction of heat consequent upon the introduction of the charge of cold fuel, and the volatilisation of its gas going forward. Thus the diminished temperature in the stack diminishes the draught at the period when the resistance offered by the fuel to the passage of air is greatest, and the demand of the furnace for air is at its highest point, in consequence of the large quantity of oxygen required for the ignition of the carburetted hydrogen gas now in course of rapid evolution. As the necessary sequel to these conditions, the gas generated passes off unburnt, worse than simply wasted, because it has abstracted a portion of heat from the ignited carbon in undergoing transformation into the gaseous form. We here get an explanation of the fact that more heat is practically obtained in some kinds of furnaces from coke, or, in other words, coal deprived of one-third part by weight of that portion of its combustible matter which is richest in heating material, than from coal in its pure state. From the same cause also, viz., defective furnace management, the commercial value of coal is often in an inverse ratio to the quantity of its bituminous constituents and heat-giving power, were proper means employed to make them practically available."

Thus far the subject of combustion may be said to be exhausted sufficiently, and we next treat of the temperature of heat in a furnace during the proper process of burning the fuel in it, the specific heat products of which are tabulated as follows by Rankine:—

	Specific Value.
Carbonic acid gas	·217
Nitrogen	·245
Air	·238
Ashes	·200
Steam	·475
	5) 1375
	·275

And if it is required to know the maximum amount of heat in a furnace, the following formula is used:— Divide the total heat in 1 lb. of coal by the amount of air in lbs., plus 1 lb. of fuel multiplied by the mean specific heat, and the quotient equals the number of degrees Fahr. Suppose, for example, the following figures:—

Total heat, say	14,000	H
Amount of air requisite for combustion	22 lbs.	A
Coal in lbs.	1 lb.	C
Specific heat in the fuel	·275	S
Temperature of furnace		T

Then $\dfrac{H}{A + C \times S} = T$, or 2245° Fahr.

We next refer to reliable results in a practical form, as carried out in the years 1857 and 1858, and to others which are compiled from the Admiralty returns as requested in the House of Commons, during the years 1871 and 1872.

344   THE CAUSE AND EFFECT OF COMBUSTION.

Table of the Duty of "Hartley's" Coal, by Messrs. Armstrong, Longridge, and Richardson, by Practical Experiments at Newcastle-on-Tyne, 1857.

	First Series		Second Series		
Work Done.	Hard Firing. Much Smoke.	Hard Firing. No Smoke.	Work Done.	Hard Firing. Much Smoke.	Hard Firing. No Smoke.
Coal burned per square foot of fire grate per hour	Lbs. 18·50	Lbs. 21	Coal burned per square foot of fire grate per hour	Lbs. 21	Lbs. 17·34
Water evaporated from 60° Fah. per square foot of fire grate per hour	Cub. Feet. 2·197	Cub. Feet. 2·932	Water evaporated from 60° Fah. per square foot of fire grate per hour	Cub. Feet. 2·900	Cub. Feet. 2·937
Total evaporation per hour from 60° Fah.	Cub. Feet. 60·5	Cub. Feet. 83·5	Total evaporation per hour from 60° Fah.	Cub. Feet. 56	Cub. Feet. 56¼
Water evaporated from 212° Fah. by 1 lb. of coal	Lbs. 8·61	Lbs. 10·10	Water evaporated from 212° Fah. by 1 lb. of coal	Lbs. 10·06	Lbs. 12·27

Showing an increase of work done of 38 per cent., and a superior economy of fuel of 17 per cent., whilst making no smoke.

In the above series of experiments we had—
Area of fire grate . . . . . . 28¼ square feet.
Heating surface (total) . . . . . 749 ,,
Ratio of fire grate to heating surface . 1 to 26¼.

Table of Average Results from the Experiments with North Country and Welsh Coals, with Williams' Apparatus for Consuming Smoke, to compare the Evaporative Values of the Two Descriptions of Coal.

Date of Experiment.	Description of Coal experimented upon.	Area of Fire Grate.	Results calculated from the latent Heat of Steam, being 966°, as taken by Messrs. Armstrong, Longridge, and Richardson, in their Reports to the Steam Collieries Association of the North of England.				Results calculated from the latent Heat of Steam, being 988°, as adopted by the Admiralty in their Reports of Trials of Coals.		Remarks.
			Economic Value on the Evaporation from 212° by 1 lb. of Coal in lbs. of Water.	Rate of Combustion, or Number of lbs. of Coal burnt per Square Foot of Fire Grate per Hour.	Rate of Evaporation from 60° per Square Feet of Fire Grate per Hour.	Total Evaporation from 60° in Cubic Feet per Hour.	Lbs. of Water evaporated by 1 lb. of Coal, calculated from constant Temperature of 100°.	Cubic Feet of Water evaporated per Hour, calculated from constant Temperature of 100°.	
		Feet.	Lbs.	Lbs.	Cubic Feet.	Cubic Feet.	Lbs.	Cubic Feet.	
1858: 20 June 3 July 26 ,, 29 ,, 30 ,,	North Country Coal, West Hartley District.	22	11·05	26·21	3·82	94·08	10·39	87·16	North Country. Welsh. 10·39 87·16 11·14 86·80 9·84 77·06 10·30 75·85 10·44 70·37 9·93 69·11 40·32 292·60 41·43 253·39 10·08 73·00 10·36 86·84 The whole of the experiments from which these averages have been taken were made with the air-passages of Williams' apparatus open, and no smoke issuing from the top of the boiler chimney, excepting when Welsh coal was burnt, and then the air-passages were closed.
19 ,, 20 ,, 22 ,,	North Country Coal and from Woolwich Dockyard	22	10·97	23·61	3·38	74·35	9·84	77·06	
28 June 2 July	Welsh Coal, Powell's Duffryn	22	12·44	21·53	3·81	83·86	11·14	86·80	
16 ,, 17 ,, 21 ,,	Welsh Coal, Blaengwarn Merthyr, sent from Woolwich Dockyard	22	11·98	20·36	3·33	73·15	10·30	75·85	
23 ,, 24 ,,	North Country small Welsh small	22 22	10·78 11·13	17·31 9·16	2·51 1·36	55·20 30·43	9·65 9·97	57·21 31·54	
2 August 6 July	North Country Coal Welsh Coal . . .	18 18	11·66 11·08	24·89 24·12	3·78 3·71	68·09 66·72	10·44 9·93	70·57 60·11	
5 ,,	North Country Coal	22	9·75	29·53	3·94	86·63	8·73	89·97	This experiment was made to observe the quantity of smoke issuing from the chimney-top, and also to ascertain the evaporative value of the coal when making smoke and when not making smoke.

THE CAUSE AND EFFECT OF COMBUSTION.   345

STATEMENT SHOWING THE QUALITY OF LAND ENGINE AND SMITHERY COALS CONSUMED IN THE DOCKYARD AND FACTORY DURING THE QUARTER ENDED 30TH SEPTEMBER, 1870.

Keyham Yard, 22nd October, 1870.

Description of Coals.	No. of lbs. of Water evaporated by 1 lb. of Coal.	Percentage of		Total.	Smoke.	Remarks.
		Clinker.	Ash.			
Townhill	6·78	3·7	2·85	6·55	Black.	The land engine coals are fit for the service, observing that the evaporative power is low, and the refuse generally high. The smithery coals consumed during this quarter are of the description known as Tanfield Moor, and were found to be uniformly clear and free from sulphur, but small and light, and will not stand a powerful blast. Screened Tanfield Moor issued for use at the steam-hammer furnaces have been of larger size, and well adapted for that service.
Lassodie	6·48	3·12	3·22	6·34	Dark.	
Lassodie and Elgin	6·64	2·27	3·03	5·3	Dark.	
Oakpit	6·93	4·34	2·38	6·72	Dark.	
Mixed Coal	7·37	3·57	2·7	6·27	Black.	
Bates's West Hartley	6·88	2·22	1·75	3·97	Dark.	
Premier Steam Coal	5·69	5·	2·63	7·63	Light.	
Scotch	6·43	4·34	1·63	5·97	Dark.	
Aston Hall Steam Coal	7·27	5·	2·32	7·32	Dark.	
North Wales Coal Pit	7·22	2·5	2·	4·5	Dark.	
Hartley	7·35	4·34	2·27	6·61	Dark.	

STATEMENT SHOWING THE QUALITY OF LAND ENGINE COALS AND SMITHERY COALS CONSUMED IN THE DOCKYARD AND FACTORY DURING THE QUARTER ENDED 31ST DECEMBER, 1870.

Keyham Yard, 24th January, 1871.

Description of Coals.	No. of lbs. of Water evaporated by 1 lb. of Coal.	Percentage of		Total.	Smoke.	Remarks.
		Clinker.	Ash.			
Mixed Coal	7·38	3·84	4·16	8·	Dark.	The land engine coals used during this quarter are fit for the service, observing the evaporative power is very fair, but the percentage of clinker and ash is rather above the average. The smithery coals consumed during the quarter have been Tanfield Moor, and were found to be uniformly clear and free from sulphur, but small and light, and will not stand a powerful blast. Since the 8th September screened Hartley coals have been issued for the use of the steam-hammer furnaces, and they have been found in every respect fit for the service, and superior to the screened Tanfield Moor coals previously issued.
West Hartley	6·	5·26	5·55	10·81	Black.	
Aberdare Black Vein	6·69	5·55	2·43	7·98	Black.	
Hartley	6·38	5·	3·25	8·25	Dark.	
Premier	7·11	3·44	2·38	5·82	Dark.	
Townhill	7·16	3·84	2·43	6·27	Dark.	
Vernon Steam Coal	7·2	5·55	4·76	10·31	Black.	

THE USE OF COAL IN HER MAJESTY'S SHIPS.

Admiralty, 29th July, 1870.

"The Welsh smokeless coal, which, since 1867, has been the only coal supplied to Her Majesty's ships, readily falls to pieces in the bunkers of ships, and especially when stored in tropical climates, and the small coal so produced is almost entirely wasted; a combination, however, of north-country coal with the small Welsh coal enables it to be burnt with great advantage. Her Majesty's ships and the coal depôts abroad have, therefore, been supplied with coal from the South Wales collieries mixed with bituminous coal.

"These coals are to be supplied to Her Majesty's ships in the following proportions:—

"For Royal yachts a special supply of Nixon's navigation steam coal will be obtained.

"For trials at the measured mile and at the six

hours' trial at sea, good hand-picked Welsh coal of any of the following descriptions, viz.:—

"Ynsfaio Merthyr,
"Davis's Merthyr,
"Ferndale,
"Powell's Duffryn,
"Sqwborwen Merthyr,
"Fothergill's Aberdare, or
"Nixon's Navigation.

"No ship, unless ordered to be at a given port by a given date, or unless her safety would be endangered by observing this rule, is ever to steam when she has a *fair* wind capable of sending her between four and five knots, or when she has a *foul* wind sufficiently strong to prevent her carrying royals, unless when going in or out of harbour.

"The maximum supply of coal is to be limited, and the best speed obtainable at that limitation is to be strictly enforced on all ordinary occasions, as follows:—

"For ships which have engines of from 1200 to 1350 horse-power, the maximum consumption is to be limited to 25 cwt. per hour.

"For ships which have engines of 1000 horse-power, but less than 1200, the maximum consumption is to be limited to 22 cwt. per hour.

"For ships of 800 and less than 1000 horse-power, the consumption is to be limited to 20 cwt. per hour.

"For ships of 500 and less than 800 horse-power, the maximum consumption is to be limited to 18 cwt. per hour.

"The consumption of smaller ships is to be limited in the same proportion.

"If a speed of between four and five knots can be obtained by a smaller consumption of fuel than the maximum allowance, it will be the duty of the commander-in-chief to enforce the smaller consumption on all ordinary occasions of steaming.

"On every occasion of a ship making a passage, a summary of the log on the accompanying form is to be transmitted either to the senior officer commanding the station, or to the Secretary of the Admiralty:—

"The number of hours from port to port	220
The number of hours under steam	192
Total distance in knots made by log under steam	883
Total consumption of coal, in tons, under steam	240
Average speed in knots per hour, under steam	4·6
Average consumption of coal, in cwts., per hour, under steam	25
Average force of the wind, in numbers, while under steam	4 to 5
General direction of the wind relative to the ship's course while under steam	ahead.
Maximum force of the wind when steaming	6 to 7
Speed of the ship in knots at that time	3
Consumption of coal per hour, in cwts., at that time	35
Direction of the wind at the time	ahead.
Maximum speed in knots under steam	6·6
Direction of the wind with reference to the ship's course at that time	starboard quarter.
Force of the wind	4 to 5
Consumption of coal, in cwts., per hour, at that time	28 "

TABLE RESPECTING THE COALS CONSUMED ON BOARD SHIP.

Description of Coals.	Date of Supply to Depôt.	Supplied to Ship.		Quantity of each Description, and Total Quantity during the Three Months.		Remarks on the Kind and Quality of Coal.			Percentage of Ash and Clinker.
		From which Depôt.	Date.			Generating of Steam.	Emission of Smoke.	Consumption of Fuel.	
			1871.	Tons.	Cwt.				Per cent.
Davis's Merthyr	20 April	Malta	20 June	50	0	Very good	A little brown	Moderate	5
Hastings' Hartley	21 April	Malta	21 June	50	0	Good	Much dark brown	Moderate	6
Powell's Duffryn	1 June	Gibraltar	1 July	80	0	Very good	Very little	Below average	4
Cowpen Hartley	1 June	Gibraltar	1 July	40	0	Fair	Dark, moderate	Quick burning	6¼
			Total	220	0				

Memo.—Some of Davis's Merthyr rather small; other kinds all in good condition. Coals used mixed, and kept steam well.

THE CAUSE AND EFFECT OF COMBUSTION. 347

Mr. McCulloch and his colleagues, in their report to the Admiralty, dated January, 1872, state:—

"SMOKE NUMBERS.

"No. 0. No smoke visible.
"No. 1. Smoke great, visible.
"No. 2. Light brown, fully visible.
"No. 3. Brown; sky visible through the smoke.
"No. 4. Brown smoke, nearly opaque.
"No. 5. Dark brown, opaque.
"No. 6. Black smoke.
"No. 7. Dense black smoke.

"Point 4 is answered in the answer to point 3.

"As regards point 5, we beg to say, that while it is a matter of the highest importance to have the coal thoroughly mixed, the act of mixing is one very easy of accomplishment, especially at home ports and at places like Malta.

"The plan we recommend, where bags alternately filled with north-country and Welsh cannot be obtained, is for two barges of equal size to be brought alongside, one filled with north-country and one with Welsh, and the contents of the barges started into the bunkers by baskets alternately filled with the two kinds of coal.

"We think it very necessary that the coal should be mixed before or in the process of delivery into ships' bunkers, and that it should not be left to be mixed on board.

"Upon point 6, we beg most emphatically to point out, that however good the conditions under which the furnaces are fitted, however good the coal, and perfect the mixture, these advantages are certain to be, to a large extent, thrown away unless due care be taken to secure proper stoking, and due attention be paid to the air-slides and the other simple elements of smoke consumption. Great as have been the economical results of the experiments lately tried compared with the consumption formerly necessary, we think that even better results might come from more special attention to this point."

CHANNEL SQUADRON TRIAL.

RESULTS OBTAINED DURING A TRIAL OF THREE HOURS' DURATION BY HER MAJESTY'S SHIP "NORTHUMBERLAND," WITH FOUR BOILERS. 2ND OCTOBER, 1871.

Hours.	Mean.				Coal.				Revolutions.			Number of Boilers in Use.	Mean Amount of Smoke generated.	Temperature of Stokehole.	Wind.		Remarks.
	Pressure by Indicator Diagram.	Indicated Horse Power.	Consumption per Indicated Horse Power.	Consumption per Hour.	Burnt on a Square Foot of Fire Grate per Hour.	The Description and Properties.			Per Minute.	Total as shown by Counter.	Distance run by Ship.				Force.	Direction.	
P.M.			Lbs.	Tns. Cwts.	Cwts.	Lbs.					Knots.						
1	12·02	2261·0	4·11	4 4	9·48	27·34	Welsh and N.C. in equal proportions		40·	2,357	8·5	4·	4·83	112	2–3	E.N.E.	Used the four forward boilers fitted with smoke-consuming apparatus per Admiralty plan. During the whole of the trial the slides were kept wide open.
2	11·075	2013·7	3·45	3 2	7·3	20·18	Welsh Yniadw Merthyr, Mixed Powell's Insthyr, N.C. Hastings, Hartley		38·5	2,346	8·5	4·	4·66	110	2–3	E.N.E.	
3	12·06	2220·3	3·5	3 10	8·38	22·75			39·25	2,308	8·35	4·	5·4	118	2–3	E.N.E.	

348   THE CAUSE AND EFFECT OF COMBUSTION.

CHANNEL SQUADRON TRIAL—*continued.*

Time of commencing observations.	Amount of Smoke as per Number at the undermentioned Periods of Five Minutes each, after commencing Observations.												Coal.	
													Description.	Proportion of North Country to Welsh used.
H. M.	5 m.	10 m.	15 m.	20 m.	25 m.	30 m.	35 m.	40 m.	45 m.	50 m.	55 m.	60 m.		
1 0	6	5	5	6	5	4	4	5	5	4	4	5	Welsh; Ynsfaio Merthyr, Mixed; Powell's Duffryn, N. C. Hastings' Hartley.	Welsh and north country in equal proportions.
2 0	5	3	5	5	4	4	4	5	6	3	5	6		
3 0	6	6	4	5	5	4	6	6	6	5	7	6		

Smoke emitted from the funnel to be represented by Numbers:
    0. No smoke visible.   4. Dark brown, nearly opaque.
    1. Smoke just visible.   5. Dark brown, opaque.
    2. Light brown fully visible.   6. Black smoke.
    3. Brown, sky visible through the smoke.   7. Dense black smoke.

HER MAJESTY'S SHIP "TOPAZE" AT VIGO, 23rd October, 1871.

RESULTS OBTAINED WHILE PERFORMING EVOLUTIONS UNDER STEAM, ON THE 18TH AND 21ST OCTOBER, 1871.

Date.	Hour.	Mean.		Coal.				Revolutions.		Number of Boilers in Use.	Temperature of Safetyhole.	Wind.		Remarks.
		Pressure by Indicator Diagram.	Indicated Horse Power.	Steam.				Minimum per Minute.	Maximum per Minute.			Force.	Direction.	
				Consumed per Indicated Horse Power.	Consumed per Hour.	Quantity burnt on a Square Foot of Fire Grate.	Description.	Proportion used.						
1871:														
18 October	11 a.m.	12·5	818	4·1	3,360	12·07		30	34	3	98	2	N.N.W.	Coals small and of inferior quality, burn very fast.
,,	12 ,,	12·5	818	4·1	3,360	12·07	Powell's Duffryn, Hastings' Hartley.	30	34	3	98	2		
,,	3 p.m.	14·1	1,058	4·65	4,928	13·27		30	30	4	100	2 to 3		
,,	4 ,,	14·1	1,058	4·65	4,928	13·27		30	30	4	100	Calm.		
21 October	11 a.m.	13·25	815	4·94	4,032	14·49	Equal Quantities.	26	36	3	90	2		
,,	12 ,,	13·4	851	4·73	4,032	14·40		30	36	3	90	2	N.E.	
,,	3 p.m.	14·5	1,004	4·35	4,368	15·68		34	38	3	95	2		

Date.	Time of commencing Observation.	Amount of Smoke as per Number at the undermentioned periods of Five Minutes each, after commencing Observations.												Coal.	
														Description.	Proportion.
	H. M.	5 m.	10 m.	15 m.	20 m.	25 m.	30 m.	35 m.	40 m.	45 m.	50 m.	55 m.	60 m.		
1871:															
18 October	10 0	2	4	2	3	1	2	0	6	4	4	2	2	Powell's Duffryn, Hastings' Hartley.	Equal Quantities.
,,	11 0	0	2	2	0	1	0	4	4	3	1	1	2		
,,	2 0	0	1	1	1	3	2	3	2	2	2	1	1		
21 October	3 0	1	2	2	2	2	0	2	4	3	2	2	2		
,,	10 0	0	3	0	1	1	3	1	2	1	0	2	1		
,,	11 0	0	4	2	1	1	2	1	2	0	0	2	4	3		
,,	2 0	1	3	4	4	3	4	1	2	0	1	3	2		

THE CAUSE AND EFFECT OF COMBUSTION. 349

HER MAJESTY'S SHIP "MONARCH," AT VIGO, 23rd October, 1871.
RESULTS OBTAINED WHILE PERFORMING EVOLUTIONS UNDER STEAM, ON THE 18TH AND 21ST OCTOBER, 1871.

Days.	Hour.	Mean.				Coal.		Revolutions.		Number of Boilers in Use.	Temperature of Stokehole.	Wind.	
		Pressure by Indicator Diagrams.	Indicated Horse Power.	Consumption per Indicated Horse Power.	Consumption per Hour.	Quantity burnt on a Square Foot of Grate Surface.	Description.	Proportion used.	Minimum per Minute.	Maximum per Minute.			Force.	Direction.
		Lbs.		Lbs.	Cwt.	Lbs.			No.	No.	No.			
1871: 18 October	11 a.m.	6·34	1262·65	4·25	48	12·3	Powell's Duffryn and Yniddu Merthyr, Hastings' Hartley.	Half of each.	16	41	4	121°	2	N.W.
,,	12 a.m.	7·23	1589·63	3·64	51·75	13·3			14	44	4	120°	2	N.W.
,,	3 p.m.	7·23	1589·63	3·64	51·75	13·3			24	40	4	130°	2	S.W.
,,	4 p.m.	6·24	1080·28	3·94	44·50	11·5			24	43	4	128°	2	S.W.
21 October	11 a.m.	6·22	1070·06	3·63	34·75	8·9			16	39	4	118°	2	E.N.E.
,,	12 a.m.	6·22	1070·06	3·03	34·75	8·9			16	40	4	118°	2	N.E.
,,	3 p.m.	6·22	1070·06	3·98	38	9·8			16	40	4	116°	3	N.E. by N.

Days.	Time of commencing Observation.	Amount of Smoke as per Number at the undermentioned periods of Five Minutes each, after commencing Observations.											Coal.	
		5 m.	10 m.	15 m.	20 m.	25 m.	30 m.	35 m.	40 m.	45 m.	50 m.	55 m.	60 m.	Description.	Proportion.
1871:	H. M.														
18 October	10 0	2	4	4	1	3	3	4	6	3	2	3	4	Powell's Duffryn, Yniddu Merthyr, Hastings' Hartley.	Half of each.
,,	11 0	4	5	3	4	3	3	2	3	2	3	3	2		
,,	2 0	4	3	2	2	3	2	2	1	0	2	0	2		
,,	3 0	0	5	3	4	3	4	5	1	5	0	4	4		
21 October	10 0	2	3	6	0	5	2	6	2	6	3	4	2		
,,	11 0	0	2	5	4	3	3	2	4	3	3	4	4		
,,	2 0	2	3	3	2	2	1	4	3	2	1	4	2		

TRIALS OF MIXED COAL (NORTH COUNTRY AND WELSH) MADE DURING THE TRIALS OF THE ENGINES OF HER MAJESTY'S SHIP "ACTIVE."

Three hours' trial of "Active" at Sea, 22nd August, 1870.

The coal used (Powell's Duffryn and Cowpen Hartley, mixed) generated steam well. The amount of smoke during the six hours, as per scale from 0 to 7, was 3·069; it never fell below 2. There was no period clear of smoke during the trial.

Six hours' trial of Her Majesty's Steam Vessel "Active," 1870.
Three Hours' Running.

Half Hours.	Pressure of Steam.		Temperature of Steam.			Vacuum in Condensers.		Revolutions of Engines.		Mean Pressure per Square Inch on Piston.	Indicated Power.	Temperature of Engine-room.		Temperature of forward Stokehole.				Total Consumption of Coals.	
Number.	Boilers.	Engine-room.	At Superheaters.	At Cylinders.	Pyrometer not fitted.	Forward.	Aft.	Counter.	Per Minute.		Horses.	Stb'd, forward.	Aft.	Port, aft.	Starb'd, aft.	Mid.	Aft.	Tons.	Cwt.
1	31·75	29·00	278	26·50	26·5	2217·	73·90	20·450	3899·49	65	..	84	96	113	77	1	18
2	31·00	28·25	283	26·50	26·5	2219·	73·96	20·750	3950·01	86	..	85	85	113	75	2	14
3	32·00	28·25	290	26·25	26·25	2230·	74·33	21·050	4037·25	91	..	87	84	111	81	3	10
4	30·25	27·50	286	26·5	26·5	2197·	73·23	20·425	3859·41	91	..	87	85	111	82	3	8
5	29·75	27·00	289	26·5	26·5	2178·	72·60	10·775	3704·44	90	..	85	90	113	82	3	8
6	29·25	26·50	291	26·5	26·75	2197·	73·23	20·375	3649·96	80	..	65	114	120	70	0	16
Mean.	30·66	27·91	..	26·45	26·50	2206·3	73·54	20·470	3884·20	..	..	..	..	..	..		
															Total.	15	14

350   THE CAUSE AND EFFECT OF COMBUSTION.

Average Results of Experiments.—Evaporative Power of different Descriptions of Coal and Coke in Locomotive Engines.
Proceedings of Institution Mechanical Engineers, 1860.

Numbers of Experiments.	District.	Description of Fuel.	Distance run.	Fuel consumed.		Water evaporated.		Percentage of Ashes and slag.	Pressure of Steam per Square Inch.
				Total.	Per Mile.	Total.	Per lb. of Fuel.		
Numbers.		COAL.	Miles.	Lbs.	Lbs.	Lbs.	Lbs.	Per cent.	Lbs.
1–11		Staveley Coal	1408	68,824	48·9	473,481	6·9	4·7	99
12–14		Staveley Coal	438	11,312	25·8	73,593	6·5	6·1	85
15–22		Beggarlee Coal	1024	54,544	53·3	360,849	6·6	4·5	106
23–25		Pentrich Coal	384	10,136	26·4	67,329	6·6	5·6	106
26–29		Grassmore Coal	477	24,864	52·1	160,639	6·5	8·4	97
30–32	Derbyshire	Babbington Coal	384	20,384	53·1	131,694	6·5	9·4	92
33–34		Molyneux Coal	256	14,336	56·0	92,341	6·4	11·3	75
35–38		Portland Coal	538	13,160	24·5	85,031	6·5	5·5	81
39–42		Shipley Coal	484	12,320	25·5	80,635	6·5	9·9	77
43–48		Wingerworth Coal	816	23,380	28·7	154,379	6·6	5·8	85
49–50		Riddings Coal	318	9,744	30·6	62,673	6·4	4·8	78
51–56		Shire Oak Coal	888	18,981	21·4	133,185	7·0	6·8	105
57–58	Durham	St. Helen's Coal	256	12,656	49·4	96,992	7·7	4·9	96
59–61		Kepsier Coal	425	16,016	37·7	121,005	7·6	4·3	107
62–63	South Yorkshire	Edmund's Main Coal	220	5,264	23·9	41,110	7·8	4·7	79
64–67		Kilnhurst Coal	440	11,200	25·5	85,096	7·6	5·3	81
68–69	Leicestershire	Ibstock Coal	220	5,628	25·6	36,796	6·5	7·0	82
70–73		Swadlincote Coal	472	14,266	30·2	90,281	6·3	4·8	81
74–77	South Staffordshire	Lord Ward's Coal	484	11,746	24·3	69,707	5·9	8·7	86
78–80		Lord Ward's Coal	384	18,480	48·1	123.274	6·7	5·5	105
81–85	Forest of Dean	Forest of Dean Coal	640	32,816	51·3	228,040	6·9	7·4	100
86–88	Bristol	Yate Coal	384	19,712	51·3	150,722	7·6	6·7	105
89		Kingswood Coal	128	6,720	52·5	48,078	7·2	8·3	72
90–93	South Wales	Aberdare Coal	521	21,504	41·3	180,866	8·4	7·4	95
94–99		Rhondda Valley Coal	888	16,658	18·8	129,755	7·8	8·2	103
		COKE.							
100–102		Tapton Coke	384	24,052	62·6	170,501	7·1	..	97
103–104	Derbyshire	Tapton Coke	296	6,772	22·9	52,580	7·8	..	61
105–107		Staveley Coke	444	11,144	25·1	75,414	6·8	..	92
108–109		Staveley Coke	200	9,426	47·1	63,562	6·7	9·7	94
110–116		Brancepeth Coke	896	49,723	55·5	399,646	8·0	..	94
117–118	Durham	Brancepeth Coke	296	7,244	24·5	54,492	7·5	..	79
119–127		Pease's West Coke	1152	64,268	55·8	507,711	7·9	..	96
128–129		Pease's West Coke	296	6,692	22·6	53,350	8·0	..	73
		STATIONARY ENGINE.							
130–141	Derbyshire	Staveley Coal		62,440	..	420,275	6·7	4·9	50
142–147	Durham	Brancepeth Coke		30,128	..	219,681	7·3	4·6	50
148–153		Pease's West Coke		26,824	..	203,299	7·6	6·4	50
154–159		Mixed Coal and Coke		27,160	..	190,187	7·0	4·4	50

CHAPTER XIV.

ACTION OF FLAME AND RAISING OF STEAM.

THE action of flame relates to the opportunity of the heat to pass through the boiler plate or tube to enter the water to raise steam, which to accomplish in the best manner is, that the flame must "beat" or cause an impact on the plate or tube. With land boilers of the Cornish and Lancashire types, set in brickwork, the flame acts on the flue tubes in an undulating manner, and therefore only one-half of the total surface can be said to be really heating surface; and when the flame reaches the back end, it in most cases divides to act on a portion of that part, and then returns through the side flues to the forward end. Now, the action of the flame, when passing through those flues, is entirely dependent on their width, because, if they are too wide, the flame has space to undulate, and glides lightly on the plate; and also, if they are too narrow, the flame is choked in its passage. But with a proper width—which we have given in the rules—the flame impinges, at an angle, of course on the plate, and therefore the heat has time, which is "opportunity," to enter through to the water. The flame, on descending to the bottom flue, glides fiercely on the plate, because it turns at right angles in its line of progression to the bottom flue, where the undulation is again to a great extent in action and the heat the least.

The best means of reducing the undulation of flame in the flues is to bridge them with slabs projecting towards the boiler plate; the flame will then at the points of bridging impinge with more effect, because the line of progression is broken, and the area of passage is reduced at intervals.

In locomotive boilers, the action of the flame is mostly in the fire box, which is the reason for making that portion so comparatively large, with bridges and other means for dividing the flame, as shown in Chapter VII. Combustion in that instance being in relation to space combined with time, while the action of the flames through the tubes is undulating also, and the longer they are, the more time is permitted for the "opportunity" of heat passage; while, if too long, a choking occurs, as with a too-narrow flue.

Taking marine boilers next into consideration, we must remark, that but very little improvement has been made in them practically during the last ten years. We have certainly grown from the square shell to the round, where the pressure demanded it, but beyond that the differences are but small.

Supposing the common arrangement, with the fire boxes under the tubes and the smoke box at the front end, the action of the flame will be that only a portion of the crown of the fire box will be acted on—say one-third as the least, and one-half as the most—then the flame, after passing the bridge, will, of course, incline towards the tube openings, but not towards the bottom tubes first; in fact, they receive the residue of the worst part of the flame, and therefore become choked earliest, and, consequently, are of but little use for evaporation throughout an interval of a voyage. The cause for this is, that the heavier portion of the flame, by its weight, will fall, and it therefore impedes the action of the lower portion of the flame in the combustion chamber, and as that falling is at right angles to the line of progression, the flame carries into the tubes more solid matter below than above, while the speed, by being reduced, allows time for congregation.

The action of the flame in tubes is much the same as in a flue tube, on account of the proportionate quantity of flame being the same in relation to the contents of the tube in nearly all cases. It becomes then a question of the least amount of tube-openings leading from the combustion chamber a boiler should have, and not the most; because it is evident, that if the openings permit a rapid passage of flame there is no time for the absorption of heat by the water. Looking, then, at this fact as imperative, we must conclude that the capacity of the combustion chamber regulates the evaporative power of a boiler to a great extent, because the flame has to *rest* therein before entering the tube openings; but the word "rest" in this case does not mean "still," but rather time for amalgamation without passage; and by that circumstance the process of combustion is better performed.

With a modern tube boiler, with the structure composed of tubes and the water inside them, there is a much better chance for the action of the flame impinging on the surfaces, adapted to receive the blow from impact; because, by the alternate position of the tubes, the flame is compelled to encircle them to a great extent, although not entirely by impact.

Our next subject is the "raising of steam," which is another term for the description of the "circulation of heat" in the water from which the steam is raised.

We may well here correct the erroneous notion that the "circulation of water" is the main feature to be noticed, for the reason that the water being naturally passive, it is only set in motion by the heat; therefore the heat must have attention first, and the water will follow of its accord, i.e., give the heat permission to act, and the water will be "taken up."

Fig. 1137.

An illustration of the action of flame sliding along the plate, thus raising steam very slowly, because there is no impact or time allowed for the flame to drive its heat through the plate.

Leading, then, from that conclusion, we direct attention to Fig. 1137, which shows the result of the flame acting on a plate parallel with the line of progression. In that case the particles of heat, it will be noticed, are very shallow above the flame plate; while in the next example, shown by Fig. 1138, the plate being at an angle, the contrast is

Fig. 1138.

An illustration of the action of flame against an angular surface, showing that by the impact of the flame the water immediately on the plate is converted into steam rapidly.

Fig. 1139.

An illustration of the action of flame against a plate at right angles to the line of progression, thus raising steam most rapidly, because the blows from the impact drives the heat through the plate without cessation.

Fig. 1140.

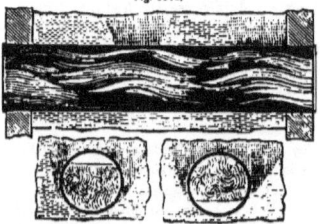

An illustration of the action of flame in a tube, showing the unequal conversion of steam, and the loss therefrom by the flame acting on nearly one-half of the tube surface only.

obvious; and the difference is still more expressed by the illustration, Fig. 1139, where the plate is

ACTION OF FLAME AND RAISING OF STEAM.

shown at right angles to the line of the flame's progression.

In a nearly similar manner tubes are operated on inside, as illustrated by Fig. 1140, from which it is evident that if the tube is not properly *filled* with flames a proportionate amount of evaporative loss is incurred, that illustrates the value of conical tubes, which we have advocated for some years.

Those tubes, it must be explained, are open at each end and connected to plate boxes thereat, forming what is now known as the "tube boiler," which, if not designed and carried out from scientific principles, will not permit the circulation of heat properly.

A contrary method of connecting the tubes to one plate only is illustrated by Fig. 1143, which

Fig. 1141.

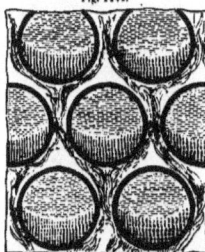

An illustration of the action of flame amongst tubes, and showing also the circulation of the heat under the water in the upper part of the tube from which the steam is produced.

Fig. 1142.

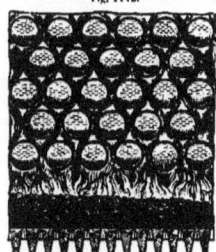

An illustration of the section of fire bars, coals, and flames, above which are a series of water tubes, illustrating the formation of steam in water tube boilers.

Fig. 1143.

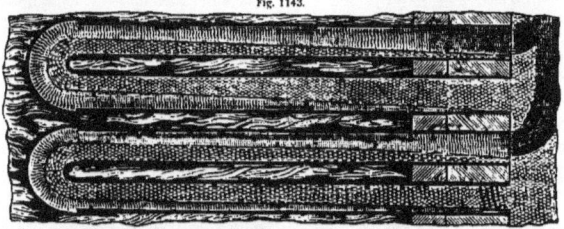

An illustration of the action of flame around the syphon tubes of Moy and Shill's patent "non-radiating engine," showing that the evaporation commences in the lower tube and increases in the upper tube of each syphon, by which the circulation of *heat* in the water is permitted without cessation, and thereby the steam rises through the water in the casing. Patented in the year 1871.

When the water is contained in the tube and the flame acts on the exterior, the result is best expressed by the illustration, Fig. 1141, which clearly shows the "licking" or impact of the flame around the lower portion of the tube, and also the circulation of the heat in the water at that part; while Fig. 1142 illustrates a series of tubes with the coals and the fire grate under them.

consists of a syphon-tube arrangement connected to a casing or reservoir surrounding the engine cylinders.

The generation of the steam is attained most rapidly and efficiently. The flame attacks the lowest surfaces of the syphon tubes first, and it is natural that the water contained therein, as it is rendered lighter by the heat, should ascend to the

2 z

upper return part of the tubes, the water in the reservoir keeping up the supply for the lower tubes, and the steam bubbles in the upper tubes ascending through the upper body of water in the reservoir into the space above, from which the steam is used in the engine cylinder; thus no radiation of heat occurs during its working, because the cylinder is surrounded by heat as hot as the steam.

The result of this close connection of the heat in the water and fire also permits of more rapid conduction of heat, and thus the evaporation of the water is proportionately increased; as, for example, according to the various authorities on the subject of evaporation, it was agreed that 14,500 thermal units equal 1 lb. of coal, and 1241° Fahr. equal 300 lbs. on the square inch—including latent heat —of steam; and the formula for evaporation is—

$\frac{T}{H-F}$ = about 14 lbs. of water per lb. of coal, when T = thermal heat, H = temperature of the steam, and F = feed water at a temperature of 210° Fahr. Now, it will be observed, in this instance no notice is taken of the actual temperature of the fire relative to the temperature of the steam. Next, as the temperature of a proper fire equals 3000° Fahr., and as the temperature of steam at 300 lbs. on the square inch is 25 per cent. above the temperature of 60 lbs. on the square inch, the formula is extended thus—14 ÷ 4 = 3·5; then 14 + 3·5

= 17·5 lbs. of water evaporated per lb. of coal consumed when steam at 300 lbs. on the square inch is kept up in a proper boiler, composed entirely of tubes, and arranged that the flame encircles them, and the heat is continuous in the fire or combustion

Fig. 1144.

An illustration of the action of flame in Burgh's treble combustion chamber tubular boiler, causing the flame to have "time" for prolonged impact, and thereby increased evaporation and raising of steam.

chamber, while also the quantity of water in the tubes is not more than ten times the quantity of water forming the steam used per minute.

Fig. 1145.

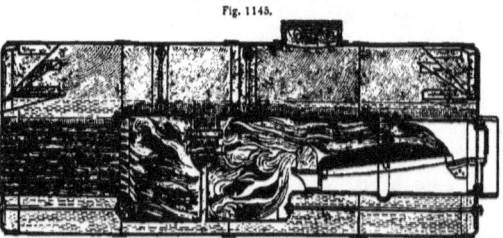

An illustration of the action of flame in Burgh's horizontal cylindrical tubular boiler, having a properly proportioned combustion chamber to the area of the fire grate, by which the evaporation is increased 30 per cent.

The next method of raising steam is illustrated by Fig. 1144, which is a vertical boiler, designed by ourselves in the year 1871, and for quick steaming

ACTION OF FLAME AND RAISING OF STEAM.

in a small compass is not excelled. Another of our designs is illustrated by Fig. 1145, which refers to our horizontal arrangement, with a very large combustion chamber between the fire boxes and the tube plate—the water bridge and vertical tube in the combustion chamber permitting the circulation of heat. In this boiler no smoke is possible, because the combustion chamber prevents that occurrence.

The action of flame and raising of steam in an ordinary cylindrical marine boiler may not be uninteresting. We therefore illustrate that by Fig. 1146; and also as in a pair of marine boilers of general construction by Fig. 1147.

As the temperature of the burning of the fuel in the fire box has been the main question in the formula expressed in this chapter, we have thought it best to finish it by illustrations of pyrometers.

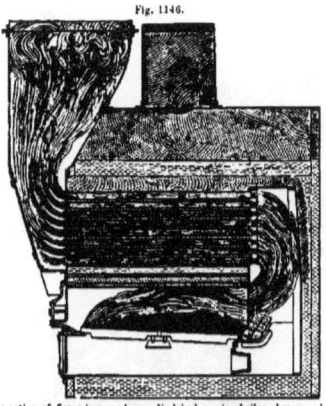

Fig. 1146.

An illustration of the action of flame in a modern cylindrical marine boiler, showing also the surfaces from which the evaporation is most effective, and also the proportion of the steam room to the water space required according to the arrangement. In the uptake is illustrated the "boil" of the smoke, subject to a small combustion chamber.

Fig. 1147.

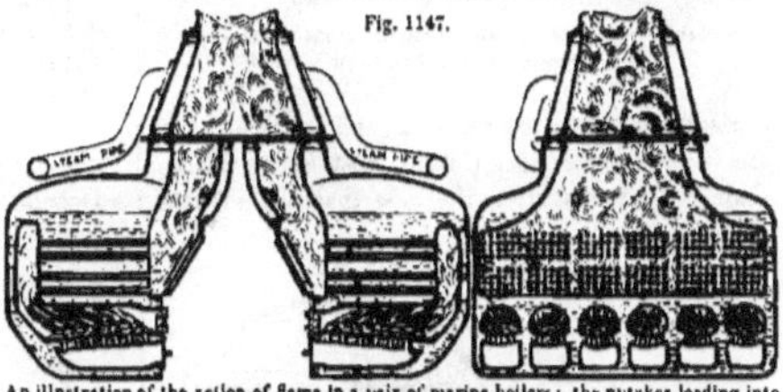

An illustration of the action of flame in a pair of marine boilers; the uptakes leading into one chimney, with a steam space around it for superheating; the fire bars being dispensed with, and perforated water spaces in their place. See also Fig. 186, on page 65.

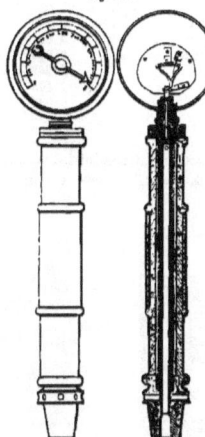

Fig. 1148.

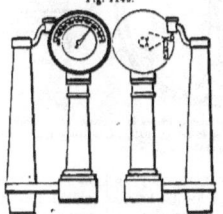

Fig. 1149. Fig. 1150.

Casartelli's Pyrometer, consisting of a central tube connected to the lever of the quadrant rack that actuates the hand of the dial plate; the expansion of the tube indicates the increase of temperature. Patented in the year 1872.

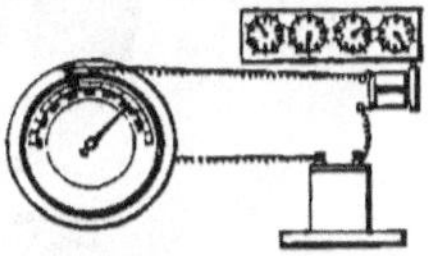

Bailey's Pyrometer, in connection with a galvanic battery and extra dial plates, which assists the indication of the temperature and also rings a bell. Patented in the year 1868.

Wood's Pyrometer, consisting of a tube in connection with a quadrant rack that actuates the hand of the dial plate, to indicate the temperature by the expansion of the tube. Patented in the year 1866.

2 z 2

CHAPTER XV.

CAUSES OF BOILER EXPLOSIONS.

The causes of boiler explosions are more venturesome than accidental, inasmuch that it is often very wonderful that some boilers have not exploded from neglect before they are worn out, and on examination the illustration is merciful in most cases; as, for example, it is too often the case that the internal deposit is so extensive on the plate acted on by the flame, that the heat could not reach the water before burning the plate, as shown by Fig. 1151, and the illustration, Fig. 1152, shows where the sediment settles mostly in a boiler.

Fig. 1151.

An illustration of a burnt plate, on account of the sediment on it being a non-conductor of heat to the water.

In a lecture we gave at the Society of Arts, London, in the end of the year 1867, "On the Principles that govern the future Development of the Marine Boiler Engine and Screw Propeller," we stated that the properties of sea water are—

" Water	964·745
Chloride of sodium	27·059
Chloride of potassium	0·766
Chloride of magnesium	3·666
Bromide of magnesium	0·029
Sulphate of magnesia	2·296
Sulphate of lime	1·406
Carbonate of lime	0·033
	100·000

"Generally there are only traces of iodine and ammoniacal salt, and the average specific gravity of the water is about 1·0274 at 60° Fahr. Now it is evident from the above figures that about $\frac{3}{8}$th of the total bulk is soluble matter.

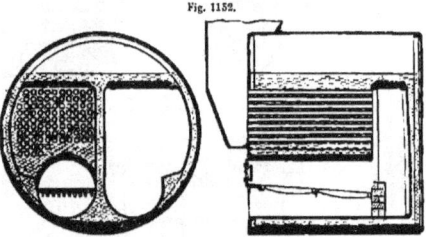

Fig. 1152.

Illustration of the main surfaces in a marine boiler where sediment is certain to settle if allowed to collect at all, also showing the necessity of surface "blow off" and bottom "blow out."

"The effect of incrustation on the heating surfaces is, that in proportion to the amount of solid matter accumulated, combined with its non-conductive property, so will the evaporation be retarded, and the relative plate exposed to the action of the flame be burnt, so that two evils are produced actually from one cause. Now there is no difficulty in the present day in preventing incrustation entirely, simply by using surface condensed steam water as feed water; but when this is used alone, the boiler suffers from the galvanic action of the mineral property and grease in the water, which the latter has robbed from the surface condensers and cylinder. It will be remembered we have stated that the incrustation formed by the sea water is almost a non-conductor of heat; it is also a non-conductor of galvanic action to some extent; therefore, by allowing a certain amount of sea water to mingle with the surface condenser feed water, the internal surface of the boiler below the water level is coated to a thickness of about one-sixteenth of an inch, and thus the pitting or wearing of the plate is prevented."

Another cause of explosion arises from the water becoming below the level of the roof of the combustion chamber, or fire box, or flue tube; and a third cause is from the tubes—in marine boilers—becoming covered with salt and saline substances.

The internal wear of a boiler is due mostly to galvanic action, which is caused by the water from the surface condenser being surcharged with acids from the tallow in the cylinders of the engines, that rob the tubes of the condenser of some of the metal which is conveyed into the boiler with the feed-water, and impinging on the plate, an electric current is set up that corrodes or eats the metal, known as "pitting"; and as an illustration of this we quote from our work, " Modern Marine Engineering " :—

" In port, before starting, a number of new boilers were filled with fresh water, while another number were filled with salt water. An examination after the first voyage, during which only distilled water had been used for feeding the boilers, showed the following effects, which were increased in every subsequent voyage, until the practice was adopted of feeding with, say, from one-sixth to one-tenth of salt water. First: Both above and below the water line, the surfaces of the plates, tubes, and rivets were covered with a deposit resembling hydrated oxide of iron, which when the water was evaporated was in the state of a fine impalpable brownish coloured powder. This deposit was thickest above the water line, sometimes averaging $\frac{3}{4}$ inch thick. When the boilers were emptied a thick slimy deposit adhered all over the inside, an analysis of which showed that it consisted of—

"Oxide of iron	77·50
Moisture	19·75
Grease	0·85
Sulphate of lime	0·80
Oxide of copper	0·60
Traces of alumina and chloride of sodium and magnesium	..
Loss	0·50
	100·00

"Secondly: Underneath this deposit the plates and tubes were found to be eaten into, indented, or ' pitted.' The indentations varied in diameter from the smallest speck to $\frac{3}{8}$ inch, and in depth from the merest impression to the entire thickness of the plates or tubes. And although they were formed all over the boilers, they were most frequently found and were most numerous just over the fireplaces, and in those parts immediately in connection with the greatest heat. In some of these parts the surface was entirely covered with the indentations; while in other parts as much as a square foot of plate, although subjected to the greatest heat, was free from them. The plates and tubes in all cases have been of the best iron and by good makers, and the 'pittings' occur in what looks like iron of good quality, with a good fibre, no slag or cinder being perceptible. So destructive was this ' pitting ' in boilers, using the same water over and over again, that in one instance the tubes of new boilers were actually eaten through at the end of two or three voyages, extending over only a few months altogether, and it became necessary to put in new tubes, and to use a portion of salt water for feed, to keep up an incrustation, so that the boilers should not be acted upon. If the iron of the boilers had been all of one make, it would naturally have been concluded that the 'pitting' was due to the quality of the iron; but as the iron of different boilers had been obtained from different makers from time to time, the quality of the iron could not be blamed.

"The presence in the boiler of a soft metal, such as copper from the condenser tubes, it was considered, would induce a galvanic action such as might affect the iron in some way. But the analysis which was made of the deposit scraped from the boiler shows that there was scarcely a trace of any foreign metal there. Indeed it might have been concluded that a soft metal could not be present, for the tubes of the condenser and the copper pipes were all in a perfect condition. Even at the joints, made tight by india-rubber, hardened by vulcanising, there was scarcely a speck of corrosion.

"A search was then made to ascertain whether the gluey deposit was present that arises from the decomposition of the tallow and oil used for lubrication. For the purpose of ascertaining this, the mud cocks of a vessel were not opened for some time before arriving in port; and the fires were then put out on arrival and the mud discharged, when the only substance found was the watery brownish deposit before referred to. The deposit remaining in the bottom of the boiler was carefully examined, but here again there was only the same deposit. As it was believed that the lubricating material carried into the boilers with the feed might, by continued subjection to heat, form an acid capable of producing the effects observed, the kind of lubricating material employed was noticed, in order to ascertain whether animal or vegetable oils acted most injuriously; but it was found that the action went on as much with the one oil as with the other. In case, however, a fat acid, formed as already mentioned, might be the cause, pieces of chalk were put into the boilers, and from time to time fresh pieces were added; carbonate of soda was also mixed with the feed water in regular doses; but all to no purpose, the action went on getting worse and worse.

"No other alternative was therefore left but to feed the boilers with a portion of salt water sufficient to keep a thin incrustation over the surface of the iron. It was suggested that the deposit was nothing else than rust or oxide of iron, and that it was formed by the chlorine present in the small proportion of salt water, which would combine with the iron to form chloride of iron; and this being readily decomposed by oxygen, oxide of iron would result. The difficulty here, however, was to know whence the oxygen was obtained; for the quantity of air entering with the feed water must have been very small indeed. It was also suggested that hydrochloric acid might be present from the small quantity of sea water that may have found its way into the boilers; but then the difficulty was to know where a quantity of the acid was to come from, sufficient to act over such an extended surface, and as rapidly as the results showed," which was the experience of a marine engineering firm. We may add, in passing, that by Fig. 1153 is illustrated the

Fig. 1153.

An illustration of the corrosion of a boiler plate by galvanic action, commonly termed "pitting;" showing also the proportionate extent of the excavation to the thickness of the plate.

appearance of a "pitted" plate, showing a series of indents of unequal depths and shapes, not as occurring by burning where the metal is shrunk up and cracked, but the indents appear as if they had been "cut" out with a dull tool, and the recessed surface highly polished afterwards with a coating.

There have been explosions occur after a boiler has been "standing," or at rest, when on starting the engine, or rather the boiler, an immediate explosion occurred. The cause of that was from a sudden release of latent heat in the boiler, because the sensible heat, being locked up, when unlocked converted the latent heat into its own nature, and the sudden expansion overcame the strength of the boiler.

But the worst of all causes of boiler explosions is bad workmanship and worse material; there being no excuse in those cases. The main fault lies not so much with the manufacturer as with the buyer, because it is often the case that tenders are sent in for the delivery of a boiler, or set of boilers, to a purchaser, and he selects the cheapest, or rather the lowest in cost, irrespective of pressure of steam, in relation to the risk which the manufacturer must entail from cheap material and cheap workmanship, the both meaning the use and result of unskilled labour and the weaker material. What boiler-makers do, who understand their profession, is to take all the risk of explosion, and charge fairly for the responsibility; and thus the loss of life and destruction of property are reduced, while the boiler is properly worked and duly inspected. In fact, the inspection of boilers must be imperative, and the Association formed for that purpose deserves great credit for their exertions. Mr. E. H. Marten, at Stourbridge, is the Engineer, and has contributed much valuable information on the subject to the Institution of Mechanical Engineers; and in his paper, read in 1870, he summarised the causes of explosions as follows:—

CAUSES OF EXPLOSIONS SUMMARISED BY MR. MARTEN IN A PAPER READ AT THE INSTITUTION OF MECHANICAL ENGINEERS, 1870.

Faults in construction or repair.
Faults which should be detected by periodical examination.
Faults which should be prevented by careful attendants.
Causes extraneous or uncertain.

Practically those faults are—

Weak tubes.	Internal corrosion.
Weak combustion chambers.	Shortness of water.
	Burning of plates.
Weak ends.	Scale or mud.
Weak domes.	Undue pressure.
Weak manholes.	Too much flame on one part.
Bad repair.	
External corrosion.	Want of stays.

Boilers.

Cornish.	Crane locomotive.
Lancashire.	Marine domestic feed-water heaters, and
Plain cylindrical elephant.	
Upright agricultural.	Rag-steamers.

The causes of explosions from bad workmanship refer principally to the "drifting," the punched holes, and riveting, after which the connection is as shown by Fig. 1154. Bad caulking, also, is a cause

Fig. 1154.

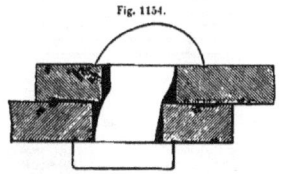

An illustration of the result of defective punching, fitting, drifting, riveting, and caulking, therefrom causing liability to explosion, and permitting leakage at the least.

for leakage, which leads to explosion, should the leakage not be choked by sediment or rust. But should the holes be drilled, and the plates properly fitted so that the rivet holes are "in a line," then the riveting and caulking will remain as shown by Fig. 1155.

Fig. 1155.

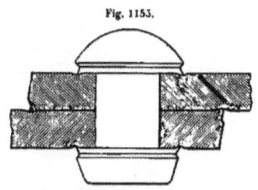

An illustration of the result of proper drilling, fitting, riveting, and caulking, therefrom preventing leakage and liability to explosion.

Boilers that have exploded from the collapsing of the fire boxes are very numerous, and the cause of the disaster lies in two paths of error: the one is the burning of the plate seamed longitudinally; and the other is the burning of the seams when across the crown.

CAUSES OF BOILER EXPLOSIONS.

The illustration Fig. 1156 shows the collapse of a fire box, and the illustration Fig. 1157 shows the collapse of two fire boxes that occurred in the same steamer and at the same moment. In those three cases the seam was on the crown longitudinally.

Fig. 1156.

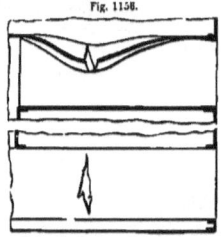

Illustrations of the collapse of one of the fire boxes in one of the marine boilers in H.M.S. "Thistle," in the year 1869.

Fig. 1157.

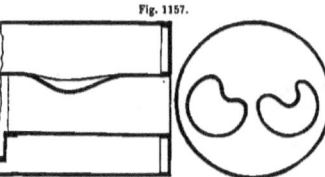

Illustrations of the collapse of the fire boxes of one of the marine boilers in H.M.S. "Thistle," in the year 1869.

The illustration of the cross seam is shown by Fig. 1158, which collapsed more violently than the previous examples.

Besides those matters explained, there are others of equal importance, which refers to the "setting"

Fig. 1158.

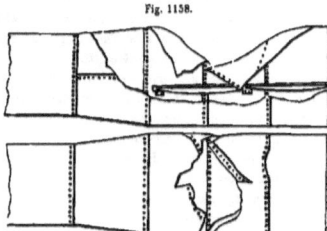

Illustrations of the collapse of the fire box of a Cornish boiler, in the year 1867.

being non-examinable, as the cause of explosion; as, for example, a boiler may be "set" so that any leakage runs into the flues and poses on the side of the boiler, from which a corrosion occurs, where the brickwork and the plate are in contact, and as that circumstance is hidden, the fault accrues until the disaster stops it.

The remedy for all those liabilities is good workmanship, attention in working, and scientific inspection; indeed, no owner of boilers, in work, should be permitted to neglect that.

We look, therefore, on the causes of boiler explosions, not as errors now, but rather the apathy on the part of our legislation in empowering qualified professional men to instruct those who are too wise in their own wisdom.

CHAPTER XVI.

BOILER-MAKING.

BOILER-MAKING requires more primary consideration, talent, scientific education, and practical knowledge in construction than any other branch of engineering yet known. To begin with, the nature of the strains that have to be resisted demand investigation, which of course depends mainly on the shape of the shell. Commencing with the Cornish or Lancashire arrangement, we must notify that it is the strongest shape yet known, but made in the weakest manner in proportion to its dimensions; those being in all cases too large for high pressures, but for steam pressure below 100 lbs. on the square inch the sizes are admirable. The weak points in those boilers are the weight of the material in proportion to the normal strength of the structure, and the unequal method of heating the boiler for raising the steam by the flame acting inside and outside the shell at intervals rather than equally throughout: as, for example, when the side flues are too large, the flame acts only here and there on the shell of the boiler, and when the boiler front is open to the atmosphere, the flame acts on but a portion of the back end only, and therefore the strain on the material by unequal expansion must be considerable, while also with long boilers the strain is increased by the upper weight of the metal pressing downwards, and causing unequal strains.

The result of this unequal action of the flame does not confine itself to the loss of evaporation, but it is extended to the general deterioration of the boiler, and so much so, that the portions the least acted on are often the first to give way, which consist of the ends connected to the flue tubes. Those connections often puzzle the best of boiler-makers; and many are the devices to allow for expansion and contraction by heating and cooling during the intervals between working and rest of the boiler. The original connection of the flue tube with the boiler front was by angle iron, and it is much adopted now, while the later method is shown by Fig. 1159, which consists of a bold curved flange formed with the end plates of the flue tube. In cases where the angle-iron connection is still used, the flue tube is formed

Fig. 1159.

Expansion Joint for the flue tube of Cornish and Lancashire boilers. Introduced in the year 1851.

Fig. 1160.

Adamson's Expansion Joint, fitted in flue tubes of boilers so as to preserve the end plate connections from disturbance during the relative actions from heat and cold. Patented in the year 1850. See Fig. 678, on page 258.

Fig. 1161.

Hill's Expansion Ring, fitted in flue tubes of boilers so as to preserve the end plate connections from disturbance during the relative actions from heat and cold. Patented in the year 1860. See Fig. 355, page 133.

in short lengths, with curved flanges and a ring between them, as shown by Fig. 1160. This was invented by Mr. Adamson as far back as the year 1850, when boiler-making might be termed "sledge-hammer engineering." In the year 1860, Mr. Hill came forward with his idea of an expansion connection, as shown by Fig. 1161.

The connecting of plates by riveting is the general method at present, and in some cases it is well done, while in more plentiful cases "piece-work" is resorted to for economy, and therefore bad workmanship results. This cheap method consists of "roughing off" the edges of the plates in a shearing machine, punching the holes, fitting the plates by bolts and nuts, and then "drifting" the holes to bring the two or three plates—as the case is—in position, next riveting with rivets often of bad iron and at quick speed, and lastly, caulking the edges of the plates only, and supposing that the rivet is sufficiently jointed by snapping without caulking. This then is the category of bad workmanship without mechanical appliances, whereas with proper means and good workmanship the difference is at least 40 per cent. on the safe side against explosions. We will therefore explain how boiler-making is properly done.

(1.) Level or flatten the plates by passing them between the rollers of the "bending machine," to make the plates straight and take away any indent or uneven surfaces, as also to test the quality of the material, whether it is liable to crack or become hollow from imperfect manufacture.

(2.) Lay the plates on the rack table and screw-clamp them for fixing; next punch, but better drill, the rivet holes on one side only: each plate is then pegged in position to make the holes at the other side and ends.

(3.) "Clamp" each plate on the planing machine table, and plane the edges perfectly clear from any imperfections.

(4.) Pass the plates between the rollers of the bending machine for curving, the rolls being available to be fixed for that purpose, as for levelling or flattening.

(5.) Connect the curved plates or straight plates that form the sides, top, and bottom of the shell together by bolts and nuts at about a foot pitch—commonly termed "fitting."

(6.) All the rivet holes being "fair" or "true" with each other, commence to rivet, which can be done by machine or by hand, the machine being the strongest and the hand the "handiest."

(7.) Take away shell from machine—if need—and rivet on back end, piece by piece or in full, as most available, the former means being mostly adopted.

(8.) Construct fire boxes and combustion chamber similarly as the shell, and "caulk" or "set back" the edges of the plates and rivets to make the joints perfect; but this can be done after in the shell.

(9.) Connect fire boxes and combustion chamber by constructing the front end of the shell.

(10.) Caulk all joints—edges of plates and rivets of shell; this may be done before "staying," in some cases, but it is always desirable to permit the "set" to occur as much as possible before the final caulking.

(11.) Fit in all the stays and tubes after the front tube plate is fitted.

(12.) Test the boiler to at least double the working steam pressure.

(13.) Attach fire box door frames, smoke box doors, fire bar bearers, general doors for clearing deposit and inspection, gauge fittings and cocks, blow off and blow out cocks, scum trough, safety valve, man-hole door, and any other appliance required.

(14.) Fit on uptake and connect lower part of chimney to it, and arrange the connections for air casings; and should the chimney be telescopic, as in Plate 31, arrange for the working gear and chain motion to be within the casing.

The practical difficulties in boiler-making are illustrated by Plates A, B, and C, and are so complete and named in detail, that they need no description here; but, in passing, we add that one of the main features in boiler-making is the uniform strength of the entire structure, because if one part is stronger than the other, an unequal contraction and expansion is always straining the weakest portion of the boiler, and the fracture is certain to occur sooner or later.

In punching the holes, care should be taken that the punch *cuts* the iron or steel without tearing it; but the safest method is by drilling, which is 10 per cent. strength added to a punched hole. In riveting, care should be taken not to burn the rivets; and in closing, to strike fairly.

LAPS.—The laps of plates demand much attention, and it will be noticed that, in the Plates A, B, and C

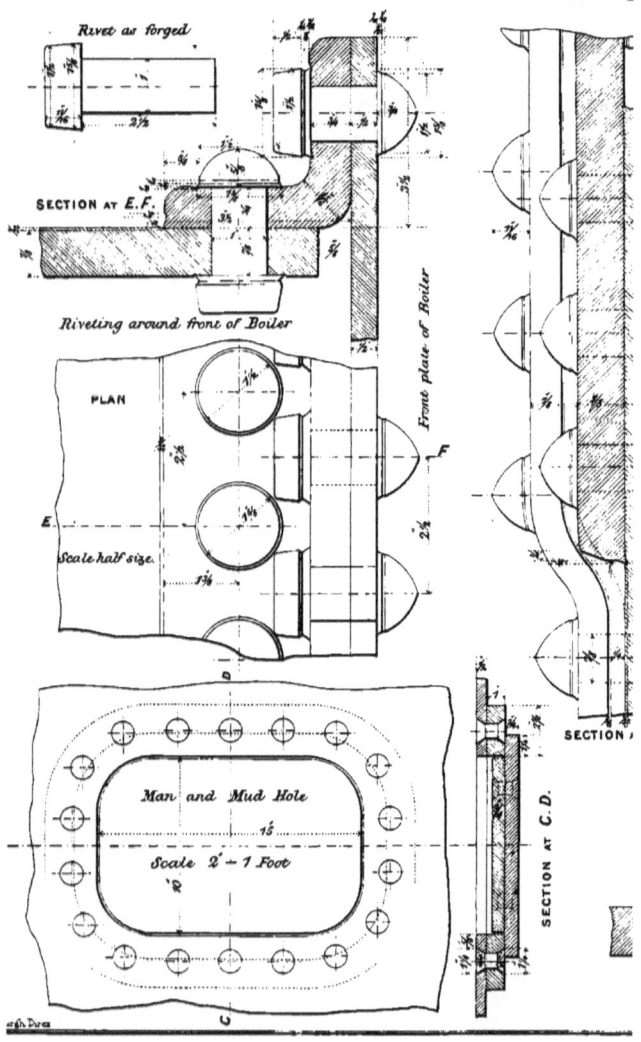

MESSRS MAUDSLAY SONS & FIELD. 1873.

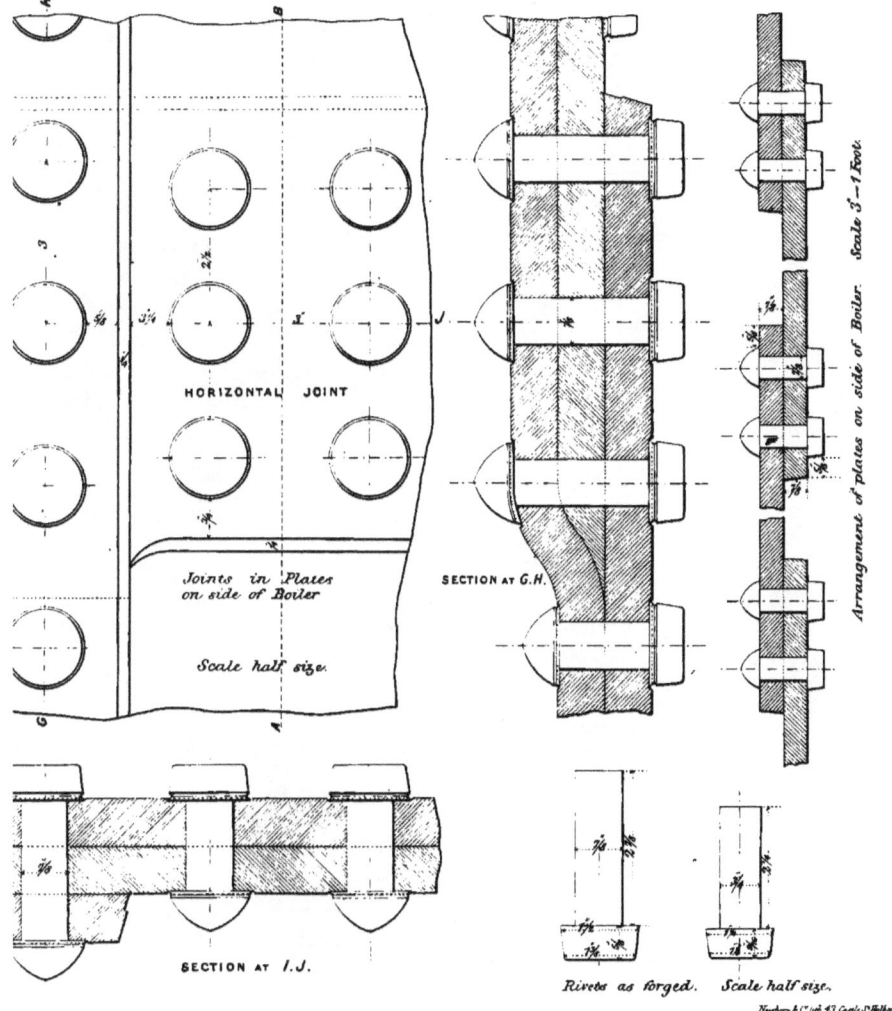

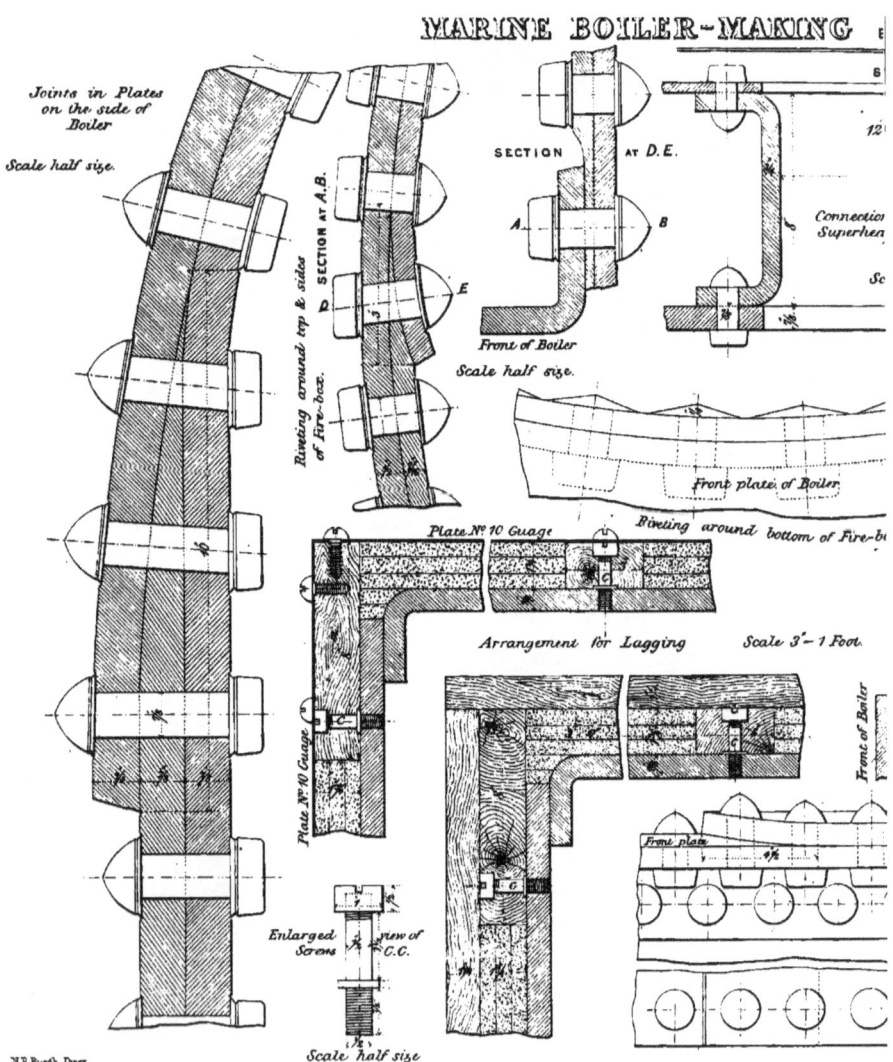

Plate B.

MESSRS MAUDSLAY SONS & FIELD, 1873.

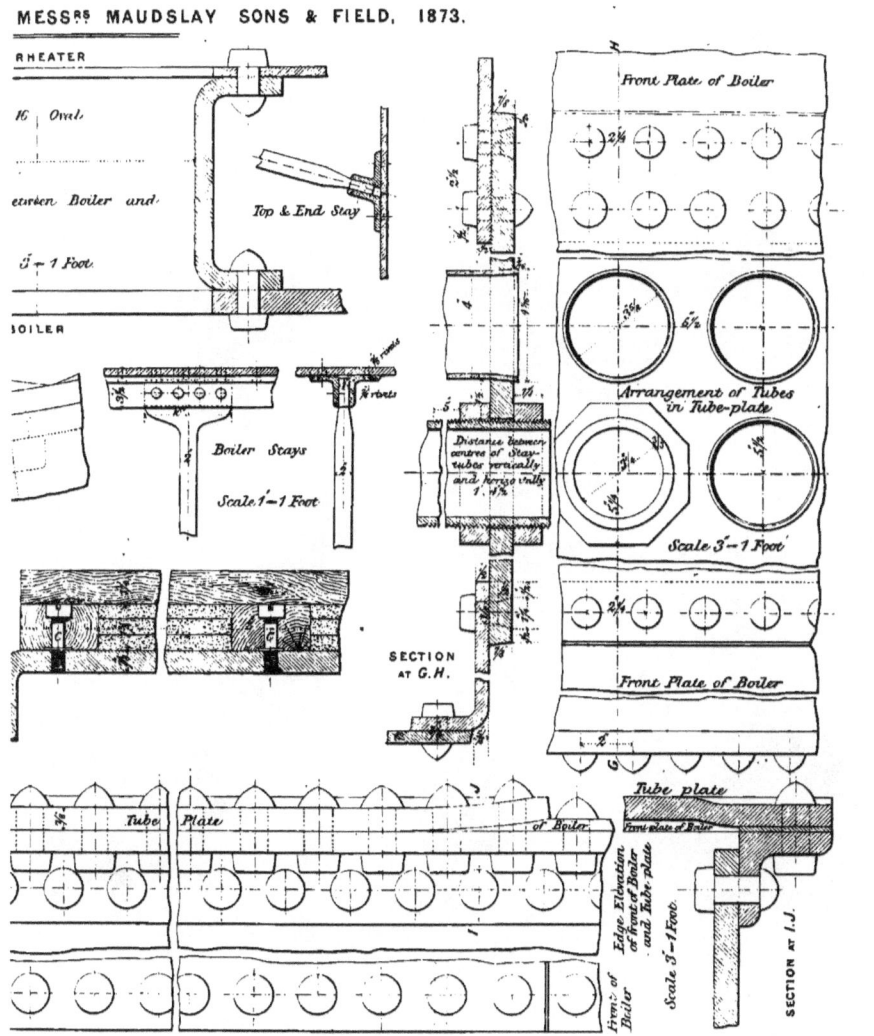

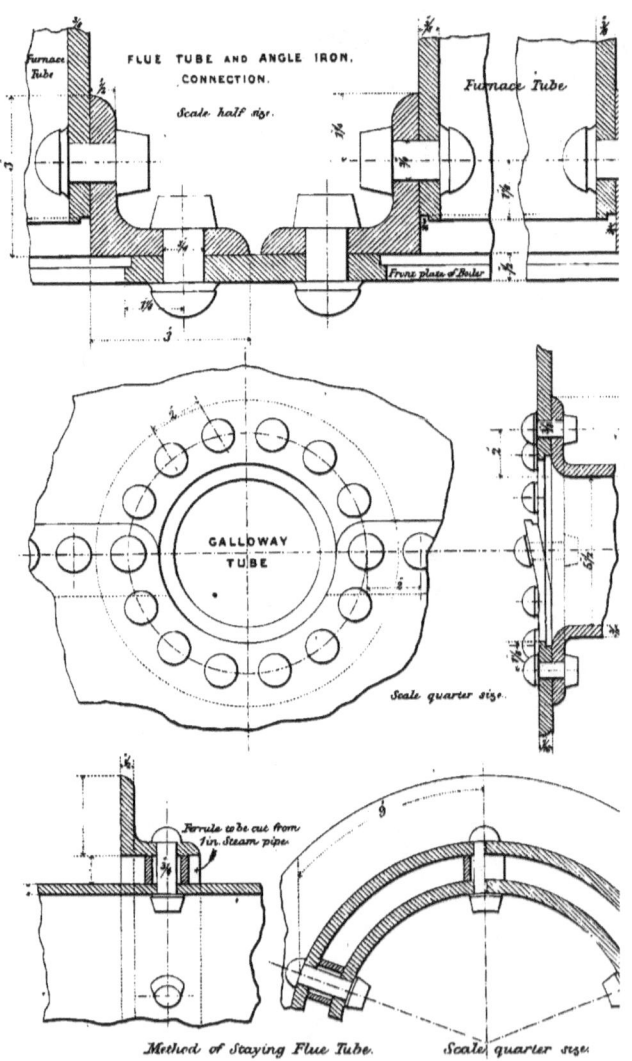

R^S HODGE & SONS, 1873.

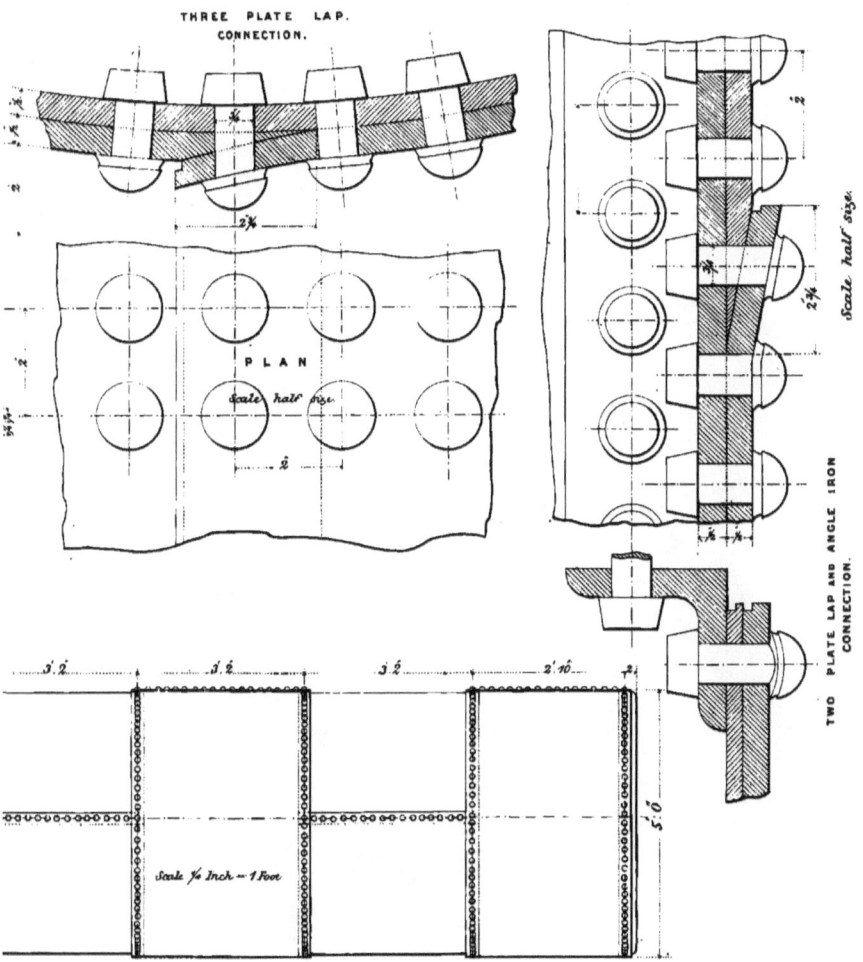

that branch has been particularly illustrated, even to the extent of from single, double, treble, and quadruple laps.

The single lap is two plates joined, the double lap is two plates "butting" and a third plate connecting the joint, the treble lap is two plates "butting" and two plates on each side connecting the joint, and the quadruple lap is three plates that are "bound" by a fourth to make good the connection.

Fig. 1162.

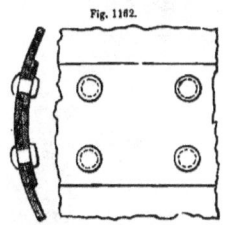

Beattie's method of assisting the strength of two welded plates by a third plate riveted on each side of the weld. Patented in the year 1858.

Fig. 1163.

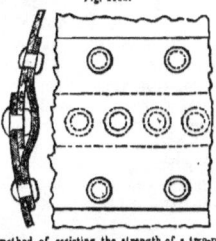

Beattie's method of assisting the strength of a two-plate single riveted connection by a third plate single riveted on each side of the connection. Patented in the year 1858.

Mr. Beattie paid some attention to this subject, as shown by Figs. 1162 and 1163.

LAGGING.—The term lagging consists of covering a boiler with the best non-radiating material possible to keep the heat in; and the most general means at present in use is by felt or coarse flannel, with wooden strips over it, to prevent any wear from damp or ripping off, and then by a sheet of lead protect the wood, where wet is liable or foot pressure is requisite. We perhaps had better explain that the wooden strips are secured to other strips attached by studs to the plates, as shown in Plate B.

Lagging, however, is but imperfectly carried out at present, inasmuch that the radiation from the front of the boiler is entirely neglected—which, by the way, is where the radiation is most—and permits from 20 to 30 per cent. of the total heat in the fuel to be wasted in some boilers from that cause only, to say nothing of other neglects.

REPAIRING.—This subject is entirely apart from boiler-making, and as such we must treat it, although included in the same chapter. In the event of a boiler requiring repair, the first step should be to examine the strongest portions nearest the weakest, for the purpose of knowing where to best make the jointing of the new plate with the old; because if the new plate is connected to a weak portion of the old plate, a leakage will sooner or later occur from the unequal tenacity of the two materials. Indeed, the great art in repairing a boiler is to make it of as much equal strength as possible.

We have seen an instance where a portion of the top of a marine boiler shell was worn by decay very thin, or an eighth of an inch in thickness, and yet withheld a pressure of twenty pounds on the square inch, but on fitting a three-eighth inch plate to cover the weak part, it split that part in less than a week, proving the truth of our prior remarks.

In no case whatever should a boiler be what is termed "patched," which is really bolting a plate on instead of riveting, because by bolting, the joint cannot be made perfect, on account of the caulking, which if not done will permit an early leakage.

The correct way then to repair a boiler, is to cut away all the weak parts and make the boiler as nearly of equal strength as possible, as we said before.

CHAPTER XVII.

TABLES, RULES, AND MEMORANDA FOR BOILER-MAKING.

TABLE OF THE STRENGTH OF MATERIALS USED IN BOILER-MAKING IN THE YEAR 1873.

Name of Material.	Tension. Breaking strain per square inch in lbs.	Tension. Yielding strain per square inch in lbs.	Compression. Breaking strain per square inch in lbs. 30 diameters.	Compression. Yielding strain per square inch in lbs. 30 diameters.	Torsion. Breaking strain per square inch in lbs.	Torsion. Yielding strain per square inch in lbs.	Shearing. Breaking strain per square inch.	Shearing. Yielding strain per square inch.	Practical working strains one-tenth of breaking strains.			
									Tensile.*	Compression.*	Torsion.*	Shearing.*
Steel bars	89,000	67,200	49,280	47,040	25,497	20,397	64,960	51,968	8,960	4,928	2,549	6,496
Wrought-iron bars	64,960	51,968	37,000	36,000	17,000	11,000	50,000	40,800	6,496	3,700	1,700	5,000
Wrought-iron plates	50,000	40,000	36,000	34,000	..	..	..	..	5,000	3,000	..	..
Cast iron	17,000	13,000	90,000	85,000	7,000	6,300	20,000	19,000	1,700	9,000	700	2,000
Gun metal	35,000	28,000	12,000	11,000	9,000	8,000	25,000	20,000	3,500	1,200	900	2,500
Copper sheets	28,000	24,000	15,000	12,200	..	..	..	..	2,800	1,500	..	..
Copper bars	33,600	26,680	18,000	14,400	11,200	8,800	21,000	19,000	3,300	1,800	1,120	2,100

* Lbs. on the square inch.

Sir William Fairbairn has contributed much information on the strength of boilers; and although many recent experiments have been made on the strength of riveted plates, no more reliable conclusion has been arrived at than Sir William's.

GENERAL SUMMARY OF RESULTS AS OBTAINED FROM EXPERIMENTS.

No. of experiments.	Cohesive strength of the plates. Breaking weight in lbs. per square inch.	Strength of double-riveted joints of equal section to the plates, taken through the line of rivets. Breaking weight in lbs. per square inch.	Strength of single-riveted joints of equal section to the plates, taken through the line of rivets. Breaking weight in lbs. per square inch.
1	57,724	52,352	45,743
2	61,579	48,421	36,606
3	58,322	58,286	43,141
4	50,983	54,594	43,515
5	51,130	53,879	40,249
6	49,281	53,879	44,715
7	43,805	..	37,161
8	47,062	..	..
Mean.	52,486	53,635	41,590

The relative strengths will therefore be—

For the plate 1,000
Double-riveted joint . . . 1,021
Single-riveted joint . . . 791

From the above it will be seen that the single-riveted joints have lost one-fifth of the actual strength of the plates, whilst the double-riveted have retained their resisting powers unimpaired. These are important and convincing proofs of the superior value of the double joint; and in all cases when strength is required this description of joint should never be omitted. It appears when plates are riveted in this manner, that the strength of the joints is to the strength of the plates of equal sections of metal as the numbers—

Plate.	Double-riveted joint.	Single-riveted joint.
1,000 :	1,021 and 791
In a former analysis it was 1,000 :	933 and 781
Which gives us a mean of 1,000 :	977 and 761

which in practice we may safely assume as the correct value of each. Exclusive of this difference, we must however deduct 30 per cent. for the loss of metal punched out for the reception of the rivets; and the absolute strength of the plates will then be to that of the riveted joints as the numbers 100, 68, and 46. In some cases, where the rivets are wider apart, the loss sustained is not so great; but in boilers and similar vessels, where the rivets require to be close to each other, the edges of the plates are weakened to that extent. In this estimate we must take into consideration the circumstances under which the results were obtained, as only two or three rivets came within the reach of experiment; and, again, looking at the increase of strength which might be gained by having a greater number of rivets in combination, and the adhesion of the two surfaces in contact, which is considerable, in the compressed rivets by machine, we may fairly assume the following relative strengths as the values of plates with their riveted joints:—

Taking the strength of the plate at	100
The strength of the double-riveted joint will be	70
And the strength of the single-riveted . . .	56

These proportions may therefore in practice be safely taken as the standard value of joints such as are used in vessels where they are required to be steam or water-tight, and subjected to pressure varying from 10 to 100 lbs. on the square inch.

RIVETS.—On this subject we have to consider the diameter, pitch, and length necessary to be observed in forming sound and tight joints without injury to the plates beyond the amount of metal punched out for the reception of the rivets. I have investigated the subject with great care, and, from my own personal knowledge and that of others, have collected a number of practical facts, such as long experience alone could furnish. From these data I have been enabled to compute the following table, which for practical use I have found highly valuable in proportioning the distances and strengths of rivets in joints requiring to be steam or water-tight:—

TABLE EXHIBITING THE STRONGEST FORMS AND BEST PROPERTIES FOR RIVETED JOINTS, AS DEDUCED FROM THE EXPERIMENTS AND PRACTICAL APPLICATIONS.

Thickness of plates in inches.	Diameter of rivets in inches.	Length of rivets in inches.	Distance of rivets from centre to centre in inches.	Quantity of lap in single joints in inches.	Quantity of lap in double joints in inches.
·19 = 3/16	·38 } 2	·88 } 4·5	1·25 } 6	1·25 } 6	For the double-riveted joint add two-thirds of the depth of the single lap.
·25 = 4/16	·50 } 2	1·13 } 4·5	1·50 } 6	1·50 } 6	
·31 = 5/16	·63 } 2	1·38 } 4·5	1·63 } 5	1·88 } 6	
·38 = 6/16	·75 } 2	1·63 } 4·5	1·75 } 5	2·00 — 5·5	
·50 = 8/16	·81 } 1·5	2·25 } 4·5	2·00 } 4	2·25 } 4·5	
·63 = 10/16	·94 } 1·5	2·75 } 4·5	2·50 } 4	2·75 } 4·5	
·75 = 12/16	1·13 } 1·5	3·25 } 4·5	3·00 } 4	3·25 } 4·5	

The figures 2, 1·5, 4·5, 6, 5, &c., in the preceding table are multipliers for the diameter, length, and distance of rivets, also for the quantity of lap allowed for the single and double joints. These multipliers may be considered as proportionals for the thicknesses of the plates to the diameter, length, distance of rivets, &c. For example, suppose we take three-eighths plates, and required the proportionate parts of the strongest form of joint, it will be—

·375 × 2 = ·750 diameter of rivet, ¾ inch.
·275 × 4½ = 1·688 length of rivet, 1½ inches.
·375 × 5 = 1·875 distance between rivets, 1⅞ inches.
·375 × 5½ = 2·063 quantity of lap, single-riveted joint, 2 inches.
·375 × 5½ + ⅜ = 3·438 quantity of lap, double-riveted joints, 3½ inches.

·75, 1·68, 1·87, 2·06, and 3·43 are therefore the proportionate quantities necessary to form the strongest steam or water-tight joints on plates three-eighths of an inch thick.

TABLE SHOWING THE BURSTING AND SAFE WORKING PRESSURE OF BOILERS, AS DEDUCED FROM EXPERIMENT WITH A STRAIN OF 34,000 LBS. ON THE SQUARE INCH AS THE ULTIMATE STRENGTH OF RIVETED JOINTS.

Diameter of boilers.		Working pressure for ⅜-inch plates.	Bursting pressure for ⅜-inch plates.	Working pressure per ½-inch plates.	Bursting pressure for ½-inch plates.
Ft.	In.	Lbs.	Lbs.	Lbs.	Lbs.
3	0	118	708¼	157¼	944¼
3	3	109	653¾	145¼	871¾
3	6	101	607	134¾	809½
3	9	94½	566½	125¾	755½
4	0	88½	531	118	708¼
4	3	83¼	500	111	666½
4	6	78¾	472	104¾	629½
4	9	74½	447¼	99¼	596¼
5	0	70¾	425	94¼	566½
5	3	67¼	404¾	89¾	539½
5	6	64¾	386¼	85¾	515
5	9	61½	369½	82	492¾
6	0	59	354	78¾	472
6	3	56½	340	75½	453¼
6	6	54¼	326¾	72½	435¾
6	9	52¼	314¾	69¾	419½
7	0	50½	303½	67¼	404½
7	3	48¾	293	65	396¾
7	6	47	283¼	62¾	377½
7	9	45½	274	60¾	365½
8	0	44	265¾	59	354
8	3	42¾	257½	57	343¼
8	6	41½	250	55½	333¼

TABLES, RULES, AND MEMORANDA FOR BOILER-MAKING.

Rule for ⅜th-inch plates.—Divide 4250 by the diameter of the boiler in inches; the quotient is the working pressure, being one-sixth of the strength of the joints.

Rule for ½-inch plates.—Divide 5666·6 by the diameter of the boiler in inches, and the quotient will be the greatest pressure that the boiler should work at when new; that is, at one-sixth the actual strength of the punched iron.

The strength of the flue tubes of boilers varies in the inverse ratio of their diameters, inversely as the lengths and directly as a power of the thickness; or it may be stated that the strengths decrease in the ratio of the increase of the diameters and the lengths, and increase nearly as the square of the thickness of the plates.

Rule for the collapsing pressure in lbs. per square inch to crush in flue tubes = $\dfrac{T^2 \times 67166}{D \times L}$

when T = thickness in inches,
  D = diameter in feet,
  L = length in feet.

Rule for the bursting pressure in lbs. per square inch of cylindrical boilers along the sides =

$$B = \dfrac{T \times t}{R}$$

when T = tensile breaking strain of the material in construction, such as riveted joints,
  t = thickness of the plate in inches,
  R = radius of boiler's diameter in inches,
  B = bursting pressure.

Divide the tabular No. opposite the thickness of shell plating and under the heading of the respective class of riveting by the extreme diameter of the boiler in inches; the quotient will be the pressure in lbs. per square inch at which the boiler may be worked while in good order.

Pressure per square inch on solid stays to be not more than . . . . } Lbs. 6000

Pressure per square inch on screw stays, taking the diameter over the threads . . . . } 5000

Rule for the area in square inches of gussets and stays for the flat ends of cylindrical boilers when the thickness of the plates = thickness of side plates × ·6, then $T = \dfrac{A \times P \times F}{B}$, and

A = area of end of boiler in square inches,
P = pressure of steam in lbs. per square inch,
F = factor of safety,
B = breaking tensile strain in lbs. per square inch.
T = total area of stays and gussets in square inches.

TABLE OF THE STRENGTHS OF FLUE TUBES (FAIRBAIRN).

Diameter of flue tube.	Collapsing pressure in lbs. on the square inch.	Thickness of plates in inches.		
		Flue tube 10 feet long.	Flue tube 20 feet long.	Flue tube 30 feet long.
Ft. Ins.				
1 0		·291	·399	·480
1 6		·350	·480	·578
2 0		·399	·548	·659
2 6	450	·442	·607	·730
3 0		·480	·659	·794
3 6		·516	·707	·851
4 0		·548	·752	·905

TABLE TO FIND THE WORKING PRESSURE FOR CYLINDRICAL BOILERS WITH A GIVEN THICKNESS OF SHELL PLATING.

Thickness of plate.	Single riveting	Double riveting.	Thickness of plate.	Single riveting.	Double riveting.
1/16	274	346	11/32	4,715	5,894
3/32	554	693	3/8	4,992	6,241
⅛	831	1,039	13/32	5,269	6,587
5/32	1,109	1,387	⅞	5,547	6,935
3/16	1,306	1,733	2¼	5,824	7,281
7/32	1,664	2,080	15/32	6,102	7,620
¼	1,941	2,426	15/16	6,379	7,974
9/32	2,219	2,774	½	6,657	8,322
5/16	2,496	3,120	17/32	6,934	8,668
11/32	2,773	3,467	9/16	7,211	9,015
⅜	3,050	3,813	19/32	7,488	9,361
13/32	3,328	4,161	⅝	7,766	9,709
7/16	3,605	4,507	21/32	8,043	10,055
15/32	3,882	4,854	11/16	8,321	10,402
½	4,159	5,200	23/32	8,598	10,748
17/32	4,438	5,548	¾	8,876	11,096

CORNISH AND LANCASHIRE BOILERS FOR LAND PURPOSES.
Proportions in practice.

Nom. H.P.	No. of square feet of heating surface per nom. H.P.	Divisor.	Length of shell.	Diameter of shell.	Diameter of single tube.
			Ft. Ins.	Ft. Ins.	Ft. Ins.
10	12·375	7·22	17 0	4 3	2 1¼
12	12·33	7·97	18 6	4 6	2 3
15	11·93	8·97	20 0	5 0	2 6
20	11·33	9·42	23 6	5 6	2 9
25	10·96	10·32	26 6	6 0	3 0
30	10·27	10·83	28 6	6 3	3 1¼
40	10·00	10·52	30 0	7 0	3 6

Nom. H.P.	Diameter of two tubes.	Working pressure of steam—lbs. per square inch.	Thickness of shell, double riveted.	Thickness of single flue tube, T-iron rings.	Thickness of two flue tubes, T-iron rings.
	Ft. Ins.		Ins.	Ins.	Ins.
10	1 6¼	100			
12	1 7¼	100			
15	1 10	100			
20	2 0½	100			
25	2 3	100			
30	2 4	100			
40	2 8½	100	1¼		

TABLES, RULES, AND MEMORANDA FOR BOILER-MAKING.

Length of boiler $= \dfrac{\text{Total heating surface of boiler}}{\text{divisor as per table}}$

Diameter of boiler $= \dfrac{\text{length}}{4 \text{ to } 4 \cdot 5}$

Diameter of single tube $= \dfrac{\text{diameter of boiler}}{2}$

Depth of water line from top of boiler $= \dfrac{\text{diameter of boiler}}{3}$

Height of water line from top of tube $= \dfrac{\text{diameter of boiler}}{12}$

When two tubes are used in the boiler, their position from top of boiler must be as for a single tube.

Vertical length of flues = diameter of tube.

Heating surface of tube = total surface of tube × ·5.

Heating surface of bottom flue = total surface of boiler exposed in flue.

Heating surface of side flues = surface of boiler exposed in both side flues × ·5.

Grate or fire bar surface = 1 to $\tfrac{3}{4}$ of a square foot per H.P.

Area of side flues $= \dfrac{\text{area of grate surface}}{4}$

Area of bottom flue = area of side flues.

Width of bottom flue = diameter of tube.

Width of side flue $= \dfrac{\text{area}}{\text{length}}$

Length of fire bar of grate $= \dfrac{\text{area of grate surface}}{\text{diameter of flue}}$

Height of bridge = $\tfrac{2}{3}$ of diameter of flue.

SAFETY VALVES.—Area of safety valve = 1 to ·75 square inch per H.P. of boiler.

Length of lever as to design $= \dfrac{\text{diameter of boiler}}{3}$

Weight in lbs. on end of lever =
$\dfrac{\text{pressure against the valve in lbs.} \times \text{by distance from centre of valve to centre of suspension}}{\text{distance from centre of suspension to centre of weight}}$

Pressure on valve in lbs. =
$\dfrac{\text{weight in lbs. on end of lever} \times \text{length of lever from centre of suspension to centre of weight}}{\text{distance from centre of valve to centre of suspension}}$

Distance from centre of suspension to centre of valve $= \dfrac{\text{length of lever}}{9 \text{ to } 10}$

Pressure in lbs. per square inch $= \dfrac{\text{pressure in lbs. on or against the valve}}{\text{area of the valve in square inches}}$

Pressure on or against the valve in lbs. = pressure per square inch × area of valve in square inches.

SQUARE AND CYLINDRICAL SHELLS, MARINE BOILERS.

Total heating surface of the tubes = H.P. × 12 to 10 as a minimum, for boilers above 200 H.P.; H.P. × 14 to 16 as a maximum below 150 H.P.

Diameter of tubes externally = 2 to 3 inches.

Length of tubes = 5 to 7 feet.

Number of tubes $= \dfrac{\text{total surface}}{\text{surface of one tube}}$

Rake or inclination of tubes = $\tfrac{5}{8}$ to $\tfrac{3}{4}$ of an inch per foot.

Water space = 4 to 6 inches.

Diameter of stays = 1 to $1\tfrac{1}{4}$ inch.

Position of stays at right angles above fire boxes = 14 to 16 inches.

Position of stays at sides and bottom of fire boxes = 12 to 14 inches.

These rules for stays are for a pressure of 40 lbs. per square inch on the total surface of boiler.

Number of tubes to one fire box should never exceed 125.

Width of fire box at tube = pitch of tubes × number of tubes transversely.

Fire bar or grate surface = H.P. × ·75 to ·5, using the latter for boilers above 150 H.P.

Length of fire bar grate surface = 7 feet as a maximum, 5 to 6 feet being generally adopted.

Width of fire box at grate $= \dfrac{\text{surface of grate}}{\text{length of grate surface}}$

Radius for top and bottom curves of fire box = width of fire box.

Radii of small curves $= \dfrac{\text{width of fire box}}{4 \text{ to } 5}$

Width of fire door opening 18 inches as a minimum; above this width the fire door opening = width of fire box × ·875.

Area of fire box at grate = grate surface × ·5.

Area of space above bridge $= \dfrac{\text{area of grate surface}}{4}$

Cubic contents (in feet) of steam capacity = H.P. × 2 as a minimum, and H.P. × 4 as a maximum.
Height of water line above fire box at tube end = 6 to 8 inches.
Width of fire box at back end = 18 inches; this will allow room for closing or riveting the end of the tubes when renewed.
Width of smoke box at bottom = 14 inches as a minimum.
Area of opening in uptake = total area of tubes as a minimum; total area of tubes × 1·25 as a maximum.

$$\text{Area of chimney} = \frac{\text{total area of grate surface}}{8 \text{ to } 11}$$

In war ships the following should be observed:—
Top of boiler should be one foot below water line as a minimum; funnel to be telescopic, raised and lowered by two chains on a barrel, keyed on a shaft, to which motion is given by a worm and wheel on each side of the funnel.

Diameter of shaft = 2 to $3\frac{1}{2}$ inches.
Diameter of wheel = 18 to 24 inches.
Pitch of teeth = $1\frac{1}{2}$ to 2 inches.

$$\text{Diameter of worm} = \frac{\text{diameter of wheel}}{4}$$

Radius of handle = 14 inches.

In order to reduce the temperature between deck and the stokehole, the funnel is surrounded by two casings, 4 to 6 inches of space between each, commencing on the main or weather deck, and terminating on the orlop or lower deck; by this means a continuous current of air passes through. The stokehole is further ventilated, and draught increased in some cases by tubes, the tops of which are termed cowls, from being enclosed semicircularly, having the opening at the side, the top being rotative, and its position subservient to the wind.

MARINE SAFETY VALVES.—These valves are mostly weighted, as shown on Plate 4A, and the following are the rules:—

Area of valve in square inches =
$$\frac{\text{total area of grate's surface in feet}}{3}$$

$$\text{Diameter of valve spindle} = \frac{\text{diameter of valve}}{4}$$

Diameter of weight = diameter of valve × 2.
Pressure in lbs. against the valve = pressure per square inch × area of the valve.
Cubical contents of weight, including weight of valve and spindle =
$$\frac{\text{pressure in lbs. against the valve}}{\cdot 4108 \text{ if lead and } \cdot 2631 \text{ if cast iron}}$$

$$\text{Length of weight} = \frac{\text{cubic contents of weight}}{\text{area of weight}}$$

Thickness of casing = $\frac{1}{2}$ to $\frac{3}{4}$ of an inch.
Depth of guide ribs of valve = diameter of valve × ·5.
Diameter of lifting lever weight shaft = diameter of valve spindle.
Length of lifting lever = diameter of weight + $\frac{1}{4}$ inch for clearance between weight.

$$\text{Lift of valves} = \frac{\text{diameter of valve}}{4}$$

When spring safety valves are used, their design is shown by Plates 13A and 45, from which also the proportions can be obtained.

FIRE BARS.—Length should never exceed 3 foot 6 inches.
Inclination for marine boilers = 2 inches per foot.
Inclination for land boilers = 1 inch per foot.
Depth of bar at centre = $1\frac{1}{2}$ to $1\frac{3}{4}$ of an inch per foot of length.
Depth of bar at ends = $\frac{3}{4}$ of an inch per foot of length.
Width of bar at ends = $\frac{3}{4}$ to 1 inch.
Taper of sides of bar = $\frac{1}{8}$ of an inch per inch.
Clearance for ashes = $\frac{1}{4}$ to $\frac{3}{8}$ of an inch.
Depth of centre bearing bar = depth of fire bar at centre.
Width of centre bearing bar = depth of fire bar at end × 2.
Width of end bearing bar = depth of fire bar at end.

MARINE COAL BUNKERS, &c.—Thickness of plates, &c.:—
Top plates, $\frac{1}{4}$ of an inch.
Bottom plates, $\frac{7}{16}$ of an inch.
Radii of curves, 6 to 12 inches.
Corner angle iron, $1\frac{1}{2} \times 1\frac{1}{2} \times \frac{1}{4}$.
Stay angle iron, $2 \times 2 \times \frac{3}{16}$.
Stays, 3 foot pitch.

Temperature tubes, number = 1 per 30 tons of coals, in bunkers containing above 200 tons.

Number of cubic feet per ton of coals = 46.

Space between boilers, or width of stokehole = 9 to 10 feet.

Minimum space allowed for passing behind cylinders or thrust block in screw alley = 12 inches; maximum space 18 inches.

TABLE FOR CALCULATING THE WEIGHT IN LBS. PER SQUARE FOOT OF DIFFERENT MATERIALS IN PLATES.

Thickness in inches.	Wrought Iron.	General Steel.	Wrought Copper.	Zinc.	Lead.	Brass.
$\frac{1}{16}$	2·5	2.59	2·903	2·301	3·701	2·705

When the plate exceeds $\frac{1}{16}$th of an inch in thickness, multiply the proportional excess by the weight of the normal thickness in lbs. = the total weight in lbs.: as, for example, suppose a wrought-iron plate = $\frac{3}{4}$ inch in thickness, then as $\frac{3}{4}$ inch = $\frac{12}{16}$ of an inch, 12 × 2·5 = 30 lbs. per square foot, which will give the actual result as if from a table, without the liability of confusion.

TABLE OF THE WEIGHTS OF ANGLE IRON OF EQUAL SIDES.

Width of each side in inches.	Thickness in inches at root.	Thickness in inches near edge.	Weight in lbs. per lineal foot.
$1\frac{1}{2}$	$\frac{5}{16}$ base	$\frac{1}{4}$ base	2·653
$1\frac{3}{4}$	$\frac{5}{16}$	$\frac{1}{4}$	3·251
2	$\frac{5}{16}$ full	$\frac{1}{4}$ full	3·874
$2\frac{1}{4}$	$\frac{7}{16}$	$\frac{5}{16}$ full	5·011
$2\frac{1}{2}$	$\frac{1}{2}$	$\frac{3}{8}$	6·512
$2\frac{3}{4}$	$\frac{9}{16}$	$\frac{7}{16}$	8·251
3	$\frac{5}{8}$	$\frac{1}{2}$	10·381
$3\frac{1}{2}$	$\frac{5}{8}$ full	$\frac{1}{2}$ full	12·101
4	$\frac{11}{16}$	$\frac{9}{16}$	14·561

TABLE OF THE WEIGHT OF A LINEAL FOOT OF ROUND AND SQUARE BAR IRON IN LBS.

Diam. or side.	Square Bars.	Round Bars.	Diam. or side.	Square Bars.	Round Bars.	Diam. or side.	Square Bars.	Round Bars.
$\frac{1}{4}$	·209	·164	$1\frac{1}{4}$	5·25	4·09	3	30·07	23·60
$\frac{5}{16}$	·326	·256	$1\frac{3}{8}$	6·35	4·96	$3\frac{1}{4}$	35·28	27·70
$\frac{3}{8}$	·470	·369	$1\frac{1}{2}$	7·51	5·90	$3\frac{1}{2}$	40·91	32·13
$\frac{7}{16}$	·640	·502	$1\frac{5}{8}$	8·82	6·92	$3\frac{3}{4}$	46·97	36·89
$\frac{1}{2}$	·835	·656	$1\frac{3}{4}$	10·29	8·03	4	53·44	41·97
$\frac{9}{16}$	1·057	·831	$1\frac{7}{8}$	11·74	9·22	$4\frac{1}{4}$	60·32	47·38
$\frac{5}{8}$	1·305	1·025	2	13·86	10·49	$4\frac{1}{2}$	67·63	53·12
$\frac{11}{16}$	1·579	1·241	$2\frac{1}{8}$	15·08	11·84	$4\frac{3}{4}$	75·35	59·18
$\frac{3}{4}$	1·879	1·476	$2\frac{1}{4}$	16·91	13·27	5	83·51	65·58
$\frac{13}{16}$	2·205	1·732	$2\frac{3}{8}$	18·84	14·79	$5\frac{1}{4}$	92·46	72·30
$\frac{7}{8}$	2·556	2·011	$2\frac{1}{2}$	20·87	16·39	$5\frac{1}{2}$	101·03	79·35
$\frac{15}{16}$	2·936	2·306	$2\frac{5}{8}$	23·11	18·07	$5\frac{3}{4}$	110·43	86·73
1	3·34	2·62	$2\frac{3}{4}$	25·26	19·84	6	120·24	94·43
$1\frac{1}{8}$	4·22	3·32	$2\frac{7}{8}$	27·61	21·68	..	..	..

To convert into weight of other metals, multiply tabular No. for cast iron by ·93, for steel × 1·02, for copper × 1·15, for brass × 1·09, for lead × 1·47, for zinc × ·92.

TABLE OF SPECIFIC GRAVITIES.

	Weight of a Cubic Inch in Lbs.
Copper, Cast	·3178
Iron, Cast	·2631
Iron, Wrought	·2756
Lead	·4103
Steel	·2827
Gun Metal	·3177

TABLE OF GRAVITY OF WATER.

1 cubic foot	=	6·25 imperial gallons.
11·2 imperial gallons	=	1 cwt.
224 ,,	=	1 ton.
1 cubic foot of sea water	=	64·2 lbs.
34·9 ,, ,,	=	1 ton.
277·274 cubic inches	=	1 imperial gallon.
1 gallon of fresh water	=	10 lbs.
1 gallon of sea water	=	10·25 lbs.

TABLE OF HEAT-CONDUCTING POWER OF METALS.

Copper	1,000
Brass	468
Wrought Iron	336
Cast Iron	311
Lead	161
Brick	10

TABLE OF THE TEMPERATURES IN DEGREES FAHR. WHEN CERTAIN MATERIALS MELT.

Wrought Iron	3,800°
Cast Iron	3,350°
Copper	2,600°
Brass	2,000°
Zinc	700°
Lead	599°

ALGEBRAIC SIGNS AS APPLIED IN MECHANICAL CALCULATIONS.

= Sign of equality, and signifies equal to, as 2 added to 5 = 7.

+ Sign of addition, and signifies plus or more, as 4 + 2 = 6.

− Sign of subtraction, and signifies minus or less, as 7 − 5 = 2.

× Sign of multiplication, and signifies multiplied by, as 7 × 6 = 42.

÷ Sign of division, and signifies divided by, as 20 ÷ 5 = 4.

√ Sign of square root } evolution, or the extraction of
∛ Sign of cube root } roots, thus √ 81 = 9 ∛ 729 = 9.

3 D

TABLE OF THE PROPERTIES OF SATURATED STEAM.

Pressure above the Atmosphere.	Sensible Temperature in Fahrenheit Degrees.	Total Heat in Degrees from Zero of Fahrenheit.	Weight of One Cubic Foot of Steam.	Relative Volume of the Steam compared with the Water from which it was raised.	Pressure above the Atmosphere.	Sensible Temperature in Fahrenheit Degrees.	Total Heat in Degrees from Zero of Fahrenheit.	Weight of One Cubic Foot of Steam.	Relative Volume of the Steam compared with the Water from which it was raised.
Lb.	Deg.	Deg.	Lb.		Lb.	Deg.	Deg.	Lb.	
1	216·3	1179·4	·0411	1515	55	302·9	1205·8	·1648	378
2	219·6	1180·3	·0435	1431	56	303·9	1206·1	·1670	373
3	222·4	1181·2	·0459	1357	57	304·8	1206·3	·1692	368
4	225·3	1182·1	·0483	1290	58	305·7	1206·6	·1714	363
5	228·0	1182·9	·0507	1229	59	306·6	1206·9	·1736	359
6	230·6	1183·7	·0531	1174	60	307·5	1207·2	·1750	353
7	233·1	1184·5	·0555	1123	61	308·4	1207·4	·1782	349
8	235·5	1185·2	·0580	1075	62	309·3	1207·7	·1804	345
9	237·8	1185·9	·0601	1036	63	310·2	1208·0	·1826	341
10	240·1	1186·6	·0625	996	64	311·1	1208·3	·1848	337
11	242·3	1187·3	·0650	958	65	312·0	1208·5	·1869	333
12	244·4	1187·8	·0673	926	66	312·8	1208·8	·1891	329
13	246·4	1188·4	·0696	895	67	313·6	1209·1	·1913	325
14	248·4	1189·1	·0719	866	68	314·5	1209·4	·1935	321
15	250·4	1189·8	·0743	838	69	015·3	1209·6	·1957	318
16	252·2	1190·4	·0766	812	70	316·1	1209·9	·1980	314
17	254·1	1190·9	·0789	789	71	316·9	1210·1	·2002	311
18	255·9	1191·5	·0812	767	72	317·8	1210·4	·2024	308
19	257·6	1192·0	·0835	746	73	378·6	1210·6	·2044	305
20	259·3	1192·5	·0858	726	74	319·4	1210·9	·2067	301
21	260·9	1193·0	·0881	707	75	320·2	1211·1	·2089	298
22	262·6	1193·5	·0905	688	76	321·0	1211·3	·2111	295
23	264·2	1194·0	·0929	671	77	321·7	1211·5	·2133	292
24	265·8	1194·5	·0952	655	78	322·5	1211·8	·2155	289
25	267·3	1194·9	·0974	640	79	323·3	1212·0	·2176	286
26	268·7	1195·4	·0996	625	80	324·1	1212·3	·2198	283
27	270·2	1195·8	·1020	611	81	324·8	1212·5	·2219	281
28	271·6	1196·2	·1042	598	82	325·6	1212·8	·2241	278
29	273·0	1196·6	·1065	585	83	326·3	1213·0	·2263	275
30	274·4	1197·1	·1089	572	84	327·1	1213·2	·2285	272
31	275·8	1197·5	·1111	561	85	327·9	1213·4	·2307	270
32	277·1	1197·9	·1133	550	86	328·5	1213·6	·2329	267
33	278·4	1198·3	·1156	539	87	329·1	1213·8	·2351	265
34	279·7	1198·7	·1179	529	88	329·9	1214·0	·2373	262
35	281·0	1199·1	·1202	518	89	330·6	1214·2	·2393	260
36	282·3	1199·5	·1224	509	90	331·3	1214·4	·2414	257
37	283·5	1199·9	·1246	500	91	331·9	1214·6	·2435	255
38	284·7	1200·3	·1269	491	92	332·6	1214·8	·2456	253
39	285·9	1200·6	·1291	482	93	333·3	1215·0	·2477	251
40	287·1	1201·0	·1314	474	94	334·0	1215·3	·2499	249
41	288·2	1201·3	·1336	466	95	334·6	1215·5	·2521	247
42	289·3	1201·7	·1364	458	96	335·3	1215·7	·2543	245
43	290·4	1202·0	·1380	451	97	336·0	1215·9	·2564	243
44	291·6	1202·4	·1403	444	98	336·7	1216·1	·2586	241
45	292·7	1202·7	·1425	437	99	337·4	1216·3	·2607	239
46	293·8	1203·1	·1447	430	100	338·0	1216·5	·2628	237
47	294·8	1203·4	·1469	424	101	338·6	1216·7	·2649	235
48	295·9	1203·7	·1493	417	102	339·3	1216·9	·2674	233
49	296·9	1204·0	·1516	411	103	339·9	1217·1	·2696	231
50	298·0	1204·3	·1538	405	104	340·5	1217·3	·2738	229
51	299·0	1204·6	·1560	399	105	341·1	1217·4	·2759	227
52	300·0	1204·9	·1583	393	106	341·8	1217·6	·2780	225
53	300·9	1205·2	·1605	388	107	342·4	1217·8	·2801	224
54	301·9	1205·5	·1627	383	108	343·0	1218·0	·2822	222

TABLES, RULES, AND MEMORANDA FOR BOILER-MAKING.

TABLE OF THE PROPERTIES OF SATURATED STEAM—*continued*.

Pressure above the Atmosphere.	Sensible Temperature in Fahrenheit Degrees.	Total Heat in Degrees from Zero of Fahrenheit.	Weight of One Cubic Foot of Steam.	Relative Volume of the Steam compared with the Water from which it was raised.	Pressure above the Atmosphere.	Sensible Temperature in Fahrenheit Degrees.	Total Heat in Degrees from Zero of Fahrenheit.	Weight of One Cubic Foot of Steam.	Relative Volume of the Steam compared with the Water from which it was raised.
Lb.	Deg.	Deg.	Lb.		Lb.	Deg.	Deg.	Lb.	
109	343·6	1218·2	·2845	221	150	366·0	1224·9	·3714	169
110	344·2	1218·4	·2867	219	155	368·2	1225·7	·3821	164
111	344·8	1218·6	·2889	217	160	370·8	1226·4	·3928	159
112	345·4	1218·8	·2911	215	165	372·9	1227·1	·4035	155
113	346·0	1218·9	·2933	214	170	375·3	1227·8	·4142	151
114	346·6	1219·1	·2955	212	175	377·5	1228·5	·4250	148
115	347·2	1219·3	·2977	211	180	379·7	1229·2	·4357	144
116	347·8	1219·5	·2999	209	185	381·7	1229·8	·4464	141
117	348·3	1219·6	·3020	208	195	386·0	1231·1	·4668	135
118	348·9	1219·8	·3040	206	205	389·9	1232·3	·4872	129
119	349·5	1220·0	·3060	205	215	393·8	1233·5	·5072	123
120	350·1	1220·2	·3080	203	225	397·5	1234·6	·5270	119
121	350·6	1220·3	·3101	202	235	401·1	1235·7	·5471	114
122	351·2	1220·5	·3121	200	245	404·5	1236·8	·5670	110
123	351·8	1220·7	·3142	199	255	407·9	1237·8	·5871	106
124	352·4	1220·9	·3162	198	265	411·2	1238·8	·6070	102
125	352·9	1221·0	·3184	197	275	414·4	1239·8	·6268	99
126	353·5	1221·2	·3206	195	285	417·5	1240·7	·6469	96
127	354·0	1221·4	·3228	194	335	430·1	1252·3	·6643	88
128	354·5	1221·6	·3250	193	385	444·9	1256·8	·6921	73
129	355·0	1221·7	·3273	192	435	456·7	1277·6	·7200	66
130	355·6	1221·9	·3294	190	485	467·5	1286·5	·7456	59
131	356·1	1222·0	·3315	189	585	487·0	1305·7	·7681	50
132	356·7	1222·2	·3336	188	686	504·1	1321·3	·7842	43
133	357·2	1222·3	·3357	187	785	519·5	1357·7	·9010	38
134	357·8	1222·5	·3377	186	885	533·6	1349·5	·9231	34
135	358·3	1222·7	·3397	184	985	546·5	1361·5	·9400	31
140	361·0	1223·5	·3500	179	1000	600·6	1414·8	·9682	26
145	363·4	1224·2	·3607	174	1500	750·8	1550·8	1·0928	19

TABLE OF THE PROPORTIONS AND WEIGHTS OF MARINE BOILERS.

No. of Plate.	Nominal Power of Boiler.	Area of Fire Grate.	Area of Tube Surface.	Area of Fire Box and Combustion Chamber Surface.	Diameter of Shell.	Length of Shell.	Weight of Water in Boiler.	Weight of Boiler.	Maker's Name.
		Square ft.	Square ft.	Square ft.	′ ″	′ ″	Tons. cwts. qrs. lbs.	Tons. cwts. qrs. lbs.	
1	50	37·76	977·76	142·02	9 2¾	9 8	7 19 3 20	15 19 2 24	N. P. Burgh.
10	30	25·6	674·55	143·9	7 0	15 0	7 4 3 6	10 14 0 0	Messrs. Laird & Son.
11	160	38·0	1615	216·74	11 0	12 8	16 14 0 7	22 6 3 12	Messrs. R. Napier & Sons.
12	60	24·74	697·06	206·02	8 6	14 0	10 18 1 6	12 10 1 4	Messrs. Hodge & Sons.
13	30	22·5	548·8	92·12	6 7½	15 1½	10 0 0 0	18 2 1 12	Messrs. James Watt & Co.
14	117	70·72	1306	367	9 10½	20 1½	17 8 1 5	23 2 0 0	Messrs. James Watt & Co.
15	55	17·2	1050·6	161·36	10 4½	9 0	9 16 0 0	14 13 3 22	Messrs. Maudslay, Sons, & Field.

These examples of boilers are selected as the best in modern practice; the general fittings are not included in the weight of each boiler.

Weight of Boiler on Plate 11.

Details.	Tons.	Cwts.	Qrs.	Lbs.
Shell	7	6	3	12
Ends of Boiler	2	10	0	10
Fire Boxes	2	7	1	15
Doors and Fittings	0	10	0	0
Bearing Bars	0	1	0	0
Fire Bars	1	7	2	8
Brick Bridges	1	7	1	11
Combustion Chamber	1	2	1	8
Tubes	4	0	5	0
Smoke Boxes	0	4	0	2
Angle Iron	0	8	1	15
Stay Bolts and Rods	0	13	3	11
Dogs	0	5	1	12
Steam Pipe	0	1	1	20
Total weight	22	6	3	12

We have selected this example as a contrast to the opposite Table, because the arrangements are entirely different, which can be understood on comparing the plates.

Weight of Boiler on Plate 1.

Details.	Tons.	Cwts.	Qrs.	Lbs.
Shell	5	13	1	8
Back End	1	4	2	16
Front End	0	15	0	0
Laps of Ends	0	10	0	10
Fire Boxes	1	6	0	10
Doors and Fittings	0	5	0	0
Bearing Bars	0	5	0	0
Fire Bars	1	18	2	24
Upright Tube in Combustion Chamber	0	3	1	14
Horizontal Tube in Combustion Chamber	0	6	0	19
Back Plate of Combustion Chamber	0	10	1	22
Tubes	2	5	0	21
Back Tube Plate	0	6	3	5
Smoke Box	0	2	0	0
Angle Iron	0	5	3	2
Stay Rods	0	1	3	2
Stay Bolts	0	0	1	11
Total Weight	15	19	2	24

Table of the Weights of Wrought Iron Pipes (Hurst).

Inside diam. in Inches.	Thickness in Inches and Length One Foot.							
	$\frac{1}{16}$	$\frac{1}{8}$	$\frac{3}{16}$	$\frac{1}{4}$	$\frac{5}{16}$	$\frac{3}{8}$	$\frac{7}{16}$	$\frac{1}{2}$
	Lbs.	Lbs.	Lbs.	Lbs.	Lbs.	Lbs.	Lbs.	Lbs.
$\frac{1}{4}$	·208	·497	·869	1·324	1·861	2·481	3·184	3·969
$\frac{3}{8}$	·289	·661	1·116	1·653	2·273	2·976	3·761	4·629
$\frac{1}{2}$	·372	·827	1·364	1·984	2·687	3·472	4·340	5·291
$\frac{5}{8}$	·455	1·092	1·612	2·315	3·100	3·968	4·919	5·952
$\frac{3}{4}$	·537	1·157	1·860	2·645	3·513	4·464	5·497	6·613
$\frac{7}{8}$	·620	1·323	2·108	2·976	3·927	4·960	6·076	7·274
1	·703	1·488	2·356	3·307	4·340	5·456	6·654	7·936
$1\frac{1}{4}$	·868	1·819	2·852	3·968	5·167	6·448	7·812	9·258
$1\frac{1}{2}$	1·033	2·149	3·348	4·629	5·993	7·440	8·969	10·581
$1\frac{3}{4}$	1·199	2·480	3·844	5·291	6·820	8·432	10·126	11·904
2	1·364	2·811	4·340	5·952	7·646	9·424	11·284	13·226
$2\frac{1}{4}$	1·529	3·131	4·836	6·613	8·473	10·416	12·441	14·549
$2\frac{1}{2}$	1·695	3·472	5·332	7·274	9·300	11·408	13·598	15·872
$2\frac{3}{4}$	1·860	3·803	5·828	7·936	10·126	12·400	14·756	17·194
3	2·025	4·133	6·324	8·607	10·953	13·392	15·913	18·517

INDEX OF PLATES.

	PLATE		PLATE
Adamson's Land Stationary Boiler and Setting	19	Elder and Co.'s Boilers, fitted in H.M.S. "Hydra" and "Cyclops"	23
Allibon and Noyes' Vertical Marine Boilers, as fitted in S.S. "Kirkstall"	16		
Allison's Patent Vertical High Pressure Boiler	36A	Fairbairn's Land Sationary Boiler, fitted with twin water space fire tubes	42
Ashton's Vertical Water Tube Marine Boiler	36		
Beattie's Locomotive Boiler and Fittings	22	Galloway's Patent Forge Furnace Boilers	37
Boiler-making (Marine), by Messrs. Maudslay, Sons, and Field	A & B	Galloway and Son's Patent Galloway Boiler	38
		Gifford's Feed Injector	35
Boiler-making (Land), by Messrs. Hodge and Sons	C	Hart and Co.'s Improved Jukes' Patent Smoke Preventing Furnace	43
Brotherhood and Hardingham's "Paragon" Donkey Pump	26	Hawkesley, Wild, and Co.'s Safety Steam Boiler	18
Burgh's Boiler	1	Hawkesley, Wild, and Co.'s Flanged-flued Boiler	18
Burgh's Arrangement of Boilers, Fittings, and Superheater	2	Hayward, Tylor, and Co.'s Patent "Universal" Steam Pump	27
Burgh's Arrangement of Superheater and Safety Valve Fittings of Boilers	3	Hodge and Son's Return Tubular Boiler	12
		Howard's Patent Safety Marine Boiler	20
Burgh's Arrangement of Combustion Casing and Funnel of Boilers	4	Howard's Patent Stationary Safety Boiler	21
Burgh's Details of Boilers	4A	Laird Bro.'s Marine Boiler and Superheater	5 & 5A
Burgh's Patent "Warranted" Steam Engine Pump ½ Gallon	32	Laird and Son's Tubular Marine Boiler	10
		Laird and Son's Uptake, Casing, Lifting Gear, and Chimney, as fitted in American S.S. "Alabama"	31
Burgh's Improved Safety Tube Boiler	44		
Burgh's Vertical and Horizontal Boilers	39	Laird Bro.'s Cylindrical Valve Donkey Engine	40
Burgh's Spring Safety Valve	45	Laird Bro.'s Steam Launch Boiler and Fittings	40
Day, Summers, and Co.'s Boilers for the Royal Mail Co. S.S. "Liffey"	24	Maudslay, Sons, and Field's Superheating Apparatus, as fitted in H.M.S. "Sirius"	29
Despatch Boats' Boilers	6 & 7	Maudslay, Sons, and Field's Tubular Boilers	15
Dudgeon's Marine Boilers of S.S. "Ruahino"	8 & 9		
Dudgeon's Launch Boiler	41	Napier and Son's Boilers, fitted in S.S. "Africa"	11

INDEX OF PLATES.

	PLATE
Perkins and Son's Marine Tubular Boiler	17
Ravenhill, Hodgson, and Co.'s Superheater and Fittings, as fitted in S.S. "Nubia"	30
Root's Tube Boiler	25
Smeaton and Co.'s Vertical Tubular Boiler Plate	36B
Tweddell's Patent Hydraulic Rivetter and Accumulator	28
Walker's Direct Acting Steam Pump	27
Watt and Co.'s Boilers, as fitted in H.M.S. "Foam"	13
Watt and Co.'s Safety Valve, as fitted in H.M.S. "Foam"	13A
Watt and Co.'s Three Boilers, as fitted in S.S. "Atrato"	14
Watt and Co.'s Donkey Engine	33
Watt and Co.'s 2½ in. Donkey Engine, for Launch Engines	34

INDEX OF ILLUSTRATIONS.

CHAPTER I.

LAND STATIONARY VERTICAL BOILERS.

Fig.	Date.	NAME.	Page	Fig.	Date.	NAME.	Page
43	1855	Atkinson's Vertical Cylindrical Water Tube Boiler	22	157 to 160	1865	Barclay's Vertical Cylindrical Boiler	62, 63
161, 162, 163 and 164	1866	Adamson's Vertical Cylindrical Boiler	63, 64	202 and 203	1868	Bèzy's Vertical Cylindrical Boiler	77
201	1867	Allibon's Vertical Cylindrical Boiler	76	222 and 223	1869	Barclay's Vertical Cylindrical Boiler	84
212	1868	Arnold's Vertical Cylindrical Boiler	81				
236	1869	Allibon's Vertical Boiler	89	226 and 227	1859	Barran's Vertical Cylindrical Boiler	85
259 and 260	1871	Ashton's Vertical Cylindrical Boiler	98				
267	1871	Adams's Vertical Cylindrical Boiler	100	228 and 229	1855	Barran's Vertical Cylindrical Boiler	86
24	1852	Balmorth's Vertical Cylindrical Tubular Boiler	14	232 and 233	1869	Barker's Vertical Cylindrical Boiler	87
25 and 26	1852	Bellford's Vertical Cylindrical Tubular Boiler	14	244 and 245	1870	Barker's Vertical Cylindrical Boiler	92
38 and 39	1853	Bellford's Vertical Cylindrical Shell Water-spaced Boiler	19	10	1780	Cylindrical Boiler with Spiral Flue	4
				11	1822	Clark's Vertical Tubular Boiler	4
64 to 67	1856	Bougleux's Vertical Cylindrical Boiler	28	22 and 23	1852	Craddock's Vertical Tubular Boiler	12, 13
70 and 71	1858	Bowman's Vertical Cylindrical Boiler	31	27	1852	Cameron's Vertical Cone-central and Cone-annular Water-spaced Cylindrical Boiler	15
92 and 93	1860	Burch's Vertical Boiler	37	31	1853	Cowper's Vertical Cellular Boiler	16
94 and 95	1860	Burch's Vertical Boiler	37	40, 41 and 42	1855	Chaplin's Vertical Cylindrical Boiler	21
96 and 97	1860	Burch's Vertical Cylindrical Boiler	38	75, 76 and 77	1859	Chaplin's Vertical Cylindrical Boiler	32
102 and 103	1861	Bremner's Vertical Square Boiler	40	78	1859	Chaplin's Vertical Boiler	33
104, 105 and 106	1861	Bremner's Vertical Square Boiler	41	107 and 108	1861	Cater's Vertical Cylindrical Boiler	43
	1861	Burch's Vertical Boiler	42	147	1865	Chaplin's Vertical Cylindrical Boiler	59

INDEX OF ILLUSTRATIONS.

Fig.	Date.	NAME.	Page	Fig.	Date.	NAME.	Page
241	1869	Chaplin's Vertical Cylindrical Boiler.	91	207	1868	Galloway's Vertical Cylindrical Boiler	79
269 and 270	1871	Clark's Vertical Cylindrical Boiler	101	219	1869	Green's Vertical Cylindrical Boiler	83
				242	1870	Green's Vertical Tube Boilers	91
				17 and 18	1827	Hancock's Vertical Flat Cellular Boiler	9
46 47 and 48	1856	Dunn's Vertical Cylindrical Semi-globular End Boiler	24	20	1836	Holmes's Vertical Annular Flue Boiler	11
				28	1852	Huddart's Vertical Cylindrical Boiler	15
49	1856	Dunn's Angular Conical Semi-gobular One End Boiler	24	56 to 59	1856	Holt's Vertical Cylindrical Boiler	26
50 to 55	1856	Dunn's Vertical Boiler	25	60 to 63	1856	Holt's Vertical Cylindrical Boiler	27
144	1865	Durand's Vertical Cylindrical Boiler.	57				
150 and 151	1865	Davis's Vertical Cylindrical Tubular Boiler	60	109 to 111	1861	Hughes's Vertical Water Tube Coil Boiler	44
179	1866	Dickins's Vertical Cylindrical Boiler.	68	122 and 123	1861	Hewett's Vertical Cylindrical Boiler	49
187 to 190	1867	Dunn's Vertical Cylindrical Boiler	71	165	1866	Howard's Vertical Tube Boiler	64
220	1869	Desvigne's Vertical Cylindrical Boiler	83	169 to 172	1866	Holt's Cylindrical Boiler	66
271 to 274	1853	Dunn's Vertical Cylindrical Boiler	102	194 and 195	1867	Holt's Vertical Cylindrical Boiler	74
45	1856	Ferinhough's Vertical Water-log Boiler	23	201	1868	Howard's Vertical Tube Boiler	76
68	1857	Fowler's Vertical Cylindrical Boiler .	29				
177 and 178	1866	Field's Vertical Cylindrical Boiler .	68	148 and 149	1865	Jordan's Vertical Tube Boiler	59
182 to 184	1867	Fiskon's Vertical Boiler	70	253 and 254	1871	Jeffery's Vertical Cylindrical Boiler.	96
192	1867	Field's Angular Tube Boiler .	73				
196 and 197	1867	Fiskon's Vertical Wedge-combined Boiler	74	32 and 33	1853	Kendrick's Vertical Radial Flat Water-spaced Boiler	17 18
205	1868	Fawcett's Vertical Tube Boiler	78	34 and 35	1853	Kendrick's Vertical Radial Flat Water-spaced Boiler	18
224 and 225	1869	Fletcher's Vertical Cylindrical Boiler	85	36	1853	Kendrick's Vertical Cylindrical Boiler	18
				37	1853	Kendrick's Vertical Cylindrical Boiler	19
239 and 240	1869	Fraser's Vertical Cylindrical Boiler .	90	116 and 117	1861	Kinsey's Vertical Cylindrical Boiler .	47
				235	1869	Kinsey's Vertical Cylindrical Boiler .	88
29 and 30	1853	Galloway's Vertical Tubular Boiler .	16	250	1870	Kenyon's Vertical Conical Tube Boiler	95
44	1855	Golding's Screw Flued Vertical Boiler	23	251 and 252	1871	Lee's Vertical Cylindrical Boiler	96
90 and 91	1860	Giles's Vertical Cylindrical Boiler .	36				
112 and 113	1861	Galloway's Vertical Cylindrical Boiler	45	185 and 186	1867	Lochhead's Vertical Cylindrical Boiler	70
				221	1869	Loader's Vertical Cylindrical Boiler .	83
114 and 115	1861	Galloway's Vertical Cylindrical Boiler	46	263 and 264	1871	Laharpe's Vertical Cylindrical Boiler	99
173 to 176	1866	Green's Vertical Cylindrical Boiler .	67	265 and 266	1871	Laharpe's Vertical Cylindrical Boiler, with twin Fire-boxes	100

INDEX OF ILLUSTRATIONS.

Fig.	Date.	NAME.	Page	Fig.	Date.	NAME.	Page
12	1824	Moore's Vertical Tube Boiler	5	132			
19	1834	McDowall's Vertical Tube Boiler	10	and 133	1863	Roberts's Vertical Cylindrical Boiler	53
72 73 and 74	1848	Millward's Vertical Cylindrical Boiler	31	193 247	1867 1870	Regan's Vertical Cylindrical Boiler Riche's Vertical Cylindrical Boiler	73 93
98 and 99	1861	Matheson's Vertical Boiler	39	2 5 9	1698 1711 1769	Savery's Steam Boiler Savery's Cylindrical Vertical Boiler Smeaton's Haycock Cylindrical Boiler	1 2 4
127	1862	Meriton's Vertical Cylindrical Boiler	51	21	1851	Stenson's Pot-Vertical Return Tubular Boiler	12
129 130 and 131	1862	Merryweather's Vertical Cylindrical Boiler	52	69 124	1858	Soame's Vertical Cylindrical Boiler	30
136 and 137	1863	Meyn's Vertical Cylindrical Boiler	54	125 and 126	1861	Selby's Vertical Cylindrical Boiler	50
140 and 141	1864	Marshall's Vertical Square Boiler	56	134 145 167	1863 1865	Shand's Vertical Cylindrical Boiler Smith's Vertical Cylindrical Boiler	54 57
180 and 181	1866	Miller's Vertical Boiler	69	and 168	1866	Schaubel's Vertical Boiler	65
198 to 200	1867 1869	Messenger's Vertical Cylindrical Boiler	75	191 209 and 210	1867 1868	Shand's Vertical Cylindrical Boiler Shand's Vertical Cylindrical Boiler	72 80
204	1868	Moreland's Vertical Cylindrical Boiler	77	211	1868	Smith's Vertical Cylindrical Boiler	80
206	1868	Morris's Vertical Cylindrical Boiler	78	213 to 218	1868	Smart's Vertical Cylindrical Boiler	82
230 and 231	1869	Miller's Vertical Cylindrical Boiler	87	237 and 238	1869	Salisbury's Vertical Cylindrical Boiler	89
243	1870	Montgomery's Vertical Cylindrical Boiler	92				
268	1871	Martin's Vertical Cylindrical Boiler	101	13	1825	Teissier's Circular and Vertical Tube Boiler	6
3 and 4	1710	Newcomen's Globular Vertical Boiler	2	14	1825	Teissier's Vertical Angular and Curved Tube Boiler	7
6	1714	Newcomen's Globular Vertical Boiler	3	15 and 16	1825	Teissier's Vertical and Angular Water-spaced Flue Boiler	8
7	1711	Newcomen's Haycock Cylindrical Boiler	3	128	1862	Tolhausen's Vertical Cylindrical Boiler	51
8	1711	Newcomen's Haycock Cylindrical Boiler	3	142 and 143	1865	Thomson's Vertical Cylindrical Boiler	56
138	1864	Oakley's Vertical Cylindrical Boiler	55	234	1869	Thirion's Vertical Cylindrical Boiler.	88
261 and 262	1871	Oram's Vertical Cylindrical Boiler	98	100 and 101	1861	Vavasseur's Vertical Tubular Boiler	40
85 to 87	1860	Pullan's Vertical Cylindrical Boiler	35	1	1663	Worcester, Marquis of, Steam Water-lift	1
88 and 89	1860	Pullan's Vertical Cylindrical Boiler	36	118 119 and 120	1861	Williamson's Vertical Annular Water Steam Tube Boiler	47 48
248 and 249	1870 and 1871	Paxman's Vertical Cylindrical Boiler	94	121	1861	Williamson's Twin Vertical Annular Water Steam Tube Boiler	49
255 to 257	1871	Pendred's Vertical Cylindrical Boiler	97	135	1863	Winan's Vertical Cylindrical Boiler	54
79 to 84	1860	Rowan's Vertical Tubular Boiler	34	139	1864	Winstanley's Vertical Cylindrical Boiler	55
				146	1865	Wise's Vertical Tubular Boiler	58
				152	1865	Wheeler's Vertical Cylindrical Boiler	60

3 c

378   INDEX OF ILLUSTRATIONS.

Fig.	Date.	NAME.	Page	Fig.	Date.	NAME.	Page
153 to 156	1865	Wilson's Vertical Cylindrical Boiler	61 62	208	1868	Wilkins's Vertical Square Shell Boiler	79
				246	1870	Wilkins's Vertical Cylindrical Boiler	93
166	1866	Woodward's Vertical Cylindrical Boiler	65	258	1871	Winstanley's Vertical Cylindrical Boiler	97

CHAPTER II.

LAND STATIONARY HORIZONTAL BOILERS.

Fig.	Date.	NAME.	Page	Fig.	Date.	NAME.	Page
346	1858	Adshead's Cylindrical Boiler	129	362	1862	Eastwood's Cylindrical Boiler	137
368	1865	Amos's Cylindrical Combined Boiler	139	432	1871	Edge's Cylindrical Boiler	162
402	1868	Arnold's Cylindrical Boiler	152				
422	1870	Arnold's Cylindrical Boiler	158	312	1853	Fearnley's Cylindrical Boiler	116
426	1871	Atkins's Cylindrical Boiler	160	361	1861	Fanshawe's Spiral Boiler	136
				378	1866	Field's Cylindrical Boiler	142
279	1800	Brindley's Granite Boiler	103	406	1869	Foster's Cast Wagon Flue Boiler	153
286	1811	"Butterley" Horizontal Cylindrical Boiler	105	416	1869	Fraser's Cylindrical Boiler	156
				421	1870	Fairbairn's Cylindrical Boilers	158
290	1820	"Breoches" Tube Boiler	106				
291	1822	"Breoches" Twin Tube Boiler	106	303			
313	1853	Bellhouse's Cylindrical Twin Boilers	116	and	1853	Galloway's Horizontal Boiler	111
318	1853	Barran's Cylindrical Boiler	118	304			
384	1867	Beeley's Cylindrical Boiler	145	305	1853	Galloway's Horizontal Boiler	112
425	1871	Boulton's Cylindrical Boiler	159	321	1853	Green's Horizontal Curved-end Boiler	119
				322	1853	Green's Horizontal Twin Boiler	120
284	1810	Cornish Boiler	105	345	1858	Green's Horizontal Boiler	128
285	1811	Cornish Boiler (Water Tube)	105	356			
311	1853	Culpin's Horizontal Boiler	115	and	1860	Galloway's Horizontal Boiler	133 134
330	1855	Cowburn's Cylindrical Boiler	122	357			
341	1857	Cater's Cylindrical Boiler	126	364			
382	1866	Chevalier's Horizontal Boiler	144	and	1863	Galloway's Retort Boiler	138
398	1868	Chamberlain's Cylindrical Boiler	149	365			
411	1869	Crosland's Elephant Boiler	154	396	1867	Guyet's Horizontal and Vertical Tubular Boiler	149
412	1869	Crosland's Elephant Boiler	155
417	1869	Cockey's Cylindrical Boiler	156	413	1869	Gemmell's "Elephant" Boiler	155
423	1870	Crosland's Elephant Boiler	159				
433	1871	Crosland's Elephant Boiler	162	295	1836	Holmes' Wagon-Flued Boiler	108
				296	1836	Holmes' Saddle Boiler	108
293	1825	Double Horizontal Boilers	107	300	1852	Hopkins's Cylindrical Horizontal Boiler	110
309	1853	Dunn's Cylindrical Egg-end Boilers	113
310	1853	Dunn's Cylindrical Flat-end Boilers	115	316	1853	Horton's Cylindrical Boiler	117
331 to 337	1855	Dunn's Retort Boiler	123 to 125	324			
				and	1854	Holt's Cylindrical Boiler	121
				325			
379	1866	Daglish's Cylindrical Boiler	143	326			
386	1867	Dunn's Cylindrical Boiler	145	and	1855	Henley's Cylindrical Boiler	121
387	1867	Dunn's Cylindrical Boiler	146	327			
388	1867	Dunn's Retort Boilers	146	339			
389	1867	Dunn's Horizontal and Vertical Boiler	146	and	1856	Holt's Cylindrical Boilers	126
				340			
390	1867	Dunn's Cylindrical Boiler	147	348	1858	Hopkinson's Cylindrical Boiler	129
391	1867	Dunn's Cornish Boiler	147	350	1859	Hunt's Cylindrical Boiler	130
428	1871	Davidson's Cylindrical Boiler	160	352	1859	Harman's Cylindrical Boiler	130
				354	1859	Horton's Cylindrical Boiler	132
				355	1860	Hill's Cylindrical Boiler	133
282	1804	Evans's Cylindrical Horizontal Boilers	104	359	1861	Harlow's Cylindrical Egg-ends Boiler	135
292	1825	Elephant Boiler	107	360	1861	Harlow's Two-Flue Cylindrical Boiler	135
294	1830	Elephant Horizontal Boiler (Fire Box)	107	377	1866	Holt's Cylindrical Boiler	142
308	1853	Evans's "Wagon" Boiler	113	393	1867	Holt's Cylindrical Boiler	148
317	1853	Erard's Cylindrical Boiler	118	397	1867	Hopkinson's Cylindrical Boiler	149

INDEX OF ILLUSTRATIONS. 379

Fig.	Date.	NAME.	Page	Fig.	Date.	NAME.	Page
400	1868	Hepworth's Cylindrical Boiler	151	358	1861	Roddewig's Cylindrical Boiler	134
401	1868	Hepworth's Egg-ends Cylindrical Boiler	151	281	1800	Setting of a Cylindrical Horizontal Boiler with Curved Ends	104
414	1869	Horton's Cylindrical Boiler	155				
415	1869	Hawkesley's Horizontal Boiler	156	283	1810	Setting of a "Single Tube" Horizontal Cylindrical Boiler	104
418	1869	Hargreave's Cylindrical Boiler	157				
419	1869	Hamilton's Cylindrical Boiler	157	284 and 285	1810 1811	Setting of Horizontal Cornish Boilers	105
420	1870	Hopkins's Cylindrical Boiler	157				
434	1871	Hawsley's Cylindrical Boiler	162				
435	1871	Heywood's Cylindrical Boiler	163	286	1811	Setting of a "Butterly" Boiler	105
363	1863	Inglis' Cylindrical Boiler	138	287 288 and 289	1811 1814 1819	Setting of "Lancashire Boilers"	105 106
302	1852	Johnson's Cylindrical Boiler	111				
328	1855	Jeffrey's Cylindrical Boiler	122	290	1820	Setting of the "Breeches" Tube Boiler	106
383	1867	James's Cylindrical Boiler	144	291	1822	Setting of a "Twin Breeches" Tube Boiler	106
314 and 315	1853	Kendrick's Cylindrical Boilers	117	292	1825	Setting of the "Elephant" Boiler	107
				293	1825	Setting of the "Double" Boilers	107
394 and 395	1867	Kendrick's Cylindrical Boiler	148	294	1830	Setting of the "Tubular Double" Boiler	107
				299	1852	Schofield's Cylindrical Boiler	109
403	1868	Kensey's Corrugated Water and Flame-speed Boiler	152	329	1855	Stevens's Cylindrical Boiler	122
				366 and 367	1863	Stewart's Cylindrical Boiler	139
287	1811	"Lancashire" Horizontal Cylindrical Boiler	106	370	1865	Smith's Cylindrical Boiler	140
288	1814	Lancashire Boiler (Egg Ends)	106	385	1867	Storey's Cylindrical Boiler	145
289	1819	Lancashire Boiler (Return Tube)	106	404 and 405	1868	Smart's Wagon-shaped Shell Boiler	152 153
297	1849	Leigh's Cylindrical Boiler	109				
298	1852	Lawes's Cylindrical Boiler	109				
369	1865	Lake's Cylindrical Boiler	140	342 and 343	1857	Taylor's Water Chamber Elliptical Boiler	127
429 to 431	1871	Laharpe's Cylindrical Boilers	161				
				344	1857	Taylor's Water Tube Elliptical Boiler	128
306	1853	Murgatroyd's Cylindrical Boiler	112	351	1859	Tapp's Cylindrical Boiler	130
353	1859	Musgrave's Cylindrical Boiler	132	371 to 373	1865	Townsend's Cylindrical Boilers	140
381	1866	Miller's Cylindrical Boiler	144				
410	1869	Miller's Horizontal Boiler	154				
424	1871	Mack's Cylindrical Boiler	159	376	1866	Thomson's Cylindrical Boiler	141
				380	1866	Twibill's Cylindrical Boiler	143
407 408 and 409	1869	Ormson's Cast Wagon Flue Boiler	153				
				427	1871	Vansteenkiste's Tubular Boiler	160
	1869	Ormson's Cast Wagon Flue Boiler	154				
307	1853	Pearce's Cylindrical Boilers	113	275	1788	Watt's Wagon Boiler	103
338	1856	Pearce's Horizontal Corrugated Boiler	125	276	1789	Watt's Wagon Boiler	103
347	1858	Price's Cylindrical Boiler (Flue Tube)	129	277 and 278	1793	Watt's Wagon Boiler	103
348	1858	Price's Cylindrical Boiler	129				
392	1867	Pollit's Cylindrical Boiler	147				
				280	1800	Watt's Wooden Boiler	104
301	1852	Ramsell's Twin Fire Tube Boiler	110	323	1854	Weatherley's Cylindrical Boiler	120
319 and 320	1853	Rèmond's Cylindrical Boiler	119	374	1865	Wilson's Cylindrical Boiler	141
				375	1866	Woodward's Cylindrical Boiler	141
				399	1868	Whittle's Cylindrical Boiler	105

CHAPTER III.

LAND STATIONARY TUBE BOILERS.

Fig.	Date.	NAME.	Page	Fig.	Date.	NAME.	Page
435	1840	Alban's Tube Boiler	164	12	1824	Moore's Vertical Tube Boiler	164
467 to 469	1869	Ashbury's Tube Boilers	181	439	1858	Meiklejon's Horizontal Tube Boiler	167
				442	1861	Matheson's Coil Tube Boilers	170
				458	1868	Mackie's Egg-end Boiler	178
479	1871	Allen's Hanging Tube Boiler	185	464	1869	Miller's Angular Circulating Tube Boiler	180
480	1872	Allen's Tube Boiler	188				
437	1856	Brayshay's Tube Boiler	166	481 and 482	1871	Mirchin's Return Angular or Zigzag Tube Boilers	186
451	1865	Belleville's Syphon Tube Boiler	174				
461	1869	Babbitt's Cast Tube Boiler	179	486	1871	Mirchin's Horizontal Tube Boiler	187
454	1867	Carville's Tube Boiler	176	474 to 478	1871	Norton's Angular Tube Boilers	184
484	1871	Dodge's Vertical Tube Boiler	187				
446	1862	Elder's Spiral Coil Tube Boiler	172	440	1859	Perkins's Combined Tube Boiler	168
				456	1868	Perkins's Tube Boiler	177
444	1861	Green's Vertical Tube Boiler	170	485	1871	Pastré's Vertical Tube Boiler	187
443	1861	Green's Horizontal Ring Tube Boiler	170	487	1871	Pursell's Vertical Tube Boiler	188
445	1862	Harrison's Angular Tube Boiler	171	455	1869	Root's Angular Tube Boiler	176
452	1867	Howard's Tube Boiler	175	460	1869	Root's Angular Tube Boiler	178
470 and 471	1869	Howard's Tube Boilers	182	465 and	1865	Rowan's Tubular Boilers	181
480	1871	Howard's Tube Boiler	185	466 472 and 473	1870	Root's Angular Tube Boilers	183
490	1872	Harrison's Six-Ball-Unit Cast-iron Tube Boiler	189				
447 and 448	1863	Inglis's Tube Boiler	173	441	1860	Traye's Saddle Shell Double Tube Boiler	169
457	1868	Inglis's Tube Boiler	177	449	1865	Twibill's Angular Tube Boiler	173
438	1857	Joly's Vertical Tube Boiler	166	450	1865	Twibill's Vertical Tube Boiler	174
453	1867	Lochhead's Angular Tube Boiler	175	462	1869	Wigzell's Tube Boiler	179
459	1869	Loader's Syphon Tube Boiler	178	483	1871	Watt's Horizontal Tube Boiler	186
463	1869	Luder's Angular Tube Boiler	180	488	1871	Westerman's Cross Tube Boiler	188

CHAPTER IV.

INJECTION BOILERS.

Fig.	Date.	NAME.	Page	Fig.	Date.	NAME.	Page
498	1848	Armstrong's Egg-end Cast-iron Injection Boiler	193	494	1825	Gilman's Injection Tube Boiler	191
				506	1860	Grimaldi's Revolving Cylindrical Injection Boiler	199
499	1852	Belleville's Injection Tube Boiler	194	508	1861	Grimaldi's Revolving Injection Boiler	200
504	1858	Benson's Tube Injection Boiler	197	509	1861	Grimaldi's Stationary Injection Boiler	200
510	1865	Brown's Egg-end Shell Revolving Injection Boiler	201	516	1867	Gould's Injection Boiler	203
505	1859	Carr's Coil Tube Injection Boiler	198	500	1852	Hyde's Dish-shaped Injection Boiler	194
				502	1857	Hodiard's Coil Tube Injection Boiler	196
518	1853	Duncan's Revolving Tube Injection Boiler	204	511	1838	McCurdy's Flat-surface Water and Steam Casing Injection Boilers	201

INDEX OF ILLUSTRATIONS. 381

Fig.	Date.	NAME.	Page	Fig.	Date.	NAME.	Page
517	1867	Mitchell's Revolving Injection Boiler	203	495	1825	Raddatz's Molten Boiler	191
				513	1865	Romminger's Injection Boiler	202
491 492	1822	Perkins's Injection Boiler	190	503	1857	Scott's Revolving Coil Tube Injection Boiler	196
and 493	1824	Paul's Injection Tube Boilers	191	507	1860	Sautter's Injection Coil Tube Boiler	199
496 and 497	1827	Perkins's Injection Tube Boiler	192	514 and 515	1865	Sturgeon's Spherical Injection Boilers	203
501	1855	Perkins's Tube Injection Boiler	195	512	1865	Thayer's Injection Boiler	202

CHAPTER V.

MARINE BOILERS.

Fig.	Date.	NAME.	Page	Fig.	Date.	NAME.	Page
541	1852	Adamson's Marine Boiler	212	540	1852	Glasson's Marine Boiler	212
573 and 574	1861	Anthony's Marine Boilers	221	556	1856	Galloway's Marine Boiler	216
				571	1860	Galloway's Marine Boiler	220
582	1866	Adamson's Marine Boiler	224	550	1854	Hyde's Marine Boiler	214
598	1866	Allibon's Vertical Marine Boiler	228	551	1855	Henley's Marine Boiler	214
612	1871	Ashton's Vertical Marine Boiler	233	562	1859	Hunt's Marine Boiler	218
				577	1862	Howden's Marine Boiler	224
549	1853	Bristow's Marine Boiler	214	580	1863	Howden's Marine Tube Boiler	223
600	1868	Bczy's Marine Boiler	229	583	1866	Hall's Marine Boiler	224
606 and 607	1869	Bennett's Marine Boilers	231	585 and 586	1866	Holt's Marine Boilers	225
611	1871	Brough's Marine Boiler	232	594 595 and 596	1867	Holt's Marine Boilers	227
620	1871	Bartlett's Tube Boiler	235
520	1829	Church's Marine Boiler	205
522	1830	Church's Marine Boiler	206	601	1868	Hawthorn's Cylindrical Marine Boiler	229
584	1866	Cochrane's Marine Boiler	225	603	1869	Howard's Marine Tube Boiler	230
604 and 605	1869	Crawford's Marine Boilers	230	619	1871	Howard's Marine Tube Boiler	235
613	1871	Crighton's Marine Boiler	233	537	1852	Knowles' Marine Boiler	211
615	1856	Crosland's Marine Boiler	234	542 to 547	1853	Kondrick's Marine Boilers	212 213
527 and 528	1842	Dickson's Marine Vertical Boilers	208	597	1867	Kendrick's Marine Boiler	228
529 and 530	1848	Dundonald's Marine Boilers	209	552	1855	Lee's Marine Boiler	215
				587 and 588	1866	Lewis's Marine Boilers	225
554 and 555	1856	Dunn's Marine Boilers	215	610	1870	Lee's Marine Boiler	232
591 592 and 593	1867	Dunn's Marine Boilers	226	621 and 622	1871	Laharpe's Marine Boilers	235
				523	1833	Maudslay's Marine Boiler	207
				532	1851	Mill's Marine Boiler	209
548	1853	Erard's Marine Boiler	214	533 and 534	1851	Mill's Marine Boilers	210
614	1853	Erard's Marine Boiler	233
609	1869	Fraser's Marine Boiler	231	557 and 558	1857	Morrison's Marine Boilers	216
539	1852	Glasson's Marine Boiler	211

INDEX OF ILLUSTRATIONS.

Fig.	Date.	NAME.	Page	Fig.	Date.	NAME.	Page
565, 566, and 567	1859	Miller's Marine Boilers	219	559	1858	Parson's Marine Boiler	217
				524	1835	Rickard's Marine Boiler	207
				560	1858	Rowan's Marine Boiler	217
569 and	1860	Macnab's Vertical Marine Boilers	220	561	1858	Robson's Marine Boiler	218
570, 575, and	1861	Macnab's Vertical Marine Boilers	221	563 and 564	1859	Randolph's Marine Boilers	218
576, 578, and	1862	Meriton's Marine Boilers	223	519	1827	Steenstrup's Marine Boiler	205
				521	1830	Summer's Annular Vertical Tube Boiler	206
579				535 and 536	1852	Selby's Marine Boilers	210
590	1867	Mace's Marine Boiler	226				
602	1868	Morduc's Combined Marine Boiler	230	553	1855	Stevens' Marine Boiler	215
608	1869	Miller's Marine Boiler	231	581	1864	Stirner's Marine Boiler	223
617 and 618	1854	MacFarlane's Marine Boilers	234, 235	589	1867	Storey's Marine Boiler	226
616	1855	Montoby's Marine Boiler	234	568	1859	Tapp's Marine Boiler	219
572	1861	O'Hanlon's Marine Boiler	221	526	1841	Whitehouse's Marine Flue Boiler	208
				538	1852	Whytehead's Marine Boiler	211
525	1838	Price's Marine Boiler	207	599	1868	Whittle's Marine Boiler	228
531	1850	Pim's Marine Boiler	209	623	1871	Watt's Marine Boiler	236

CHAPTER VI.
LIQUID FUEL BOILERS.

Fig.	Date.	NAME.	Page	Fig.	Date.	NAME.	Page
625	1862	Biddle's Liquid Fuel Land Boiler	237	630 and 631	1865	Sim's Liquid Fuel Tubular Retort Apparatus	240
640	1867	Barff's Liquid Fuel Boiler	243				
652	1868	Cutler's Gas Fuel Boiler	250	632 and 633	1865	Sim's Liquid Fuel Tubular Boilers	241
641	1868	Dorsett's Liquid Fuel Apparatus	244				
				642	1868	Stevens' Liquid Fuel Boiler	245
639	1866	Lee's Liquid Fuel Boiler	243	646	1868	Smith's Liquid Fuel Boiler	246
643 and 644	1868	Lafone's Liquid Fuel Boilers	246	647	1868	Sauvage's Liquid Fuel Locomotive Boiler	247
				649	1868	Spartali's Liquid Fuel Marine Boiler	249
628 and 629	1864	Richardson's Liquid Fuel Apparatus	238, 239	624	1822	Viney's Liquid Fuel Boiler	237
648	1868	Ravel's Liquid Fuel Coil Tube Boiler	248	650	1869	Taylor's Liquid Fuel Boiler	249
651	1869	Robinson's Liquid Fuel Boiler	250				
626 and 627	1863	Schmidt's Liquid Fuel Land Boilers	238	634 to 638	1865	Wise's Liquid Fuel "Stop" Slab Grate	242
				645	1868	Weir's Liquid Fuel Marine Boiler	246

CHAPTER VII.
LOCOMOTIVE BOILERS.

Fig.	Date.	NAME.	Page	Fig.	Date.	NAME.	Page
678	1852	Adamson's Locomotive Tubular Boiler	258	679	1853	Beattie's non-Chimney Locomotive Boiler	258
688	1853	Allan's Locomotive Fire Box	261				
676	1852	Barran's Locomotive Fire Box	257	680	1853	Beattie's non-Chimney Locomotive Boiler	259
677	1852	Brosson's Locomotive Vertical Boiler	257	689	1854	Beattie's Locomotive Boiler	261

INDEX OF ILLUSTRATIONS. 383

Fig.	Date.	NAME.	Page	Fig.	Date.	NAME.	Page
690, 691, 692, and 693	1854	Beattie's Locomotive Boilers	262, 263	667 to 672	1847, and 1848	Johnson's Locomotive Fire Box	256
694 to 698	1854	Beattie's Locomotive Fire Box	264	685, 686, and 687	1853	Kendrick's Locomotive Boilers	260, 261
699	1854	Blavier's Locomotive Boiler	265				
701 to 709	1855	Beattie's Locomotive Boilers and Fire Boxes	266, 267	664	1842	Lawthwaite's Locomotive Boiler	255
				733	1865	Loubat's Vertical Cylindrical Locomotive Boiler	273
723, and 724	1858	Blinkhorn's Locomotive Agricultural Boilers	270	745 and 746	1871	Laharpe's Angular Tube and Barrel Locomotive Boilers	277
735	1865	Belleville's Locomotive Tube Boiler	274				
739	1868	Bèzy's Locomotive Boiler	275	700	1855	Monteley's Locomotive Boiler	265
				721	1857	Molino's Locomotive Boiler	269
665	1842	Crampton's Locomotive Boiler	255	726	1861	Martin's Locomotive Boiler	271
710 to 720	1856	Crosland's Locomotive Boiler	267, 268, 269	727	1864	Martin's Locomotive Smoke Box	271
				740	1869	Miller's Locomotive Boiler	275
722	1858	Clare's Locomotive Boiler	270	656	1826	Neville's Return Tubular Boiler	252
				658	1831	Napier's Locomotive Boiler	253
684	1853	Dunn's Locomotive Fire Box	259	683	1853	Newton's Locomotive Fire Box	259
				743	1871	Norton's Locomotive Boiler	276
659	1833	Fraser's Locomotive Boiler	253				
660	1833	Field's Water Tube Locomotive Boiler	253	747	1836	Perkins' Single Steam Action Locomotive Engine	277
729 and 730	1864	Fairlie's Locomotive Combined Boilers	272	748	1836	Perkins' Locomotive Boiler	278
731	1864	Fairlie's Locomotive Combined Boilers	273				
734	1865	Fairlie's Locomotive Central Fire Box	274	657	1829	Stephenson's "Rocket" Locomotive	252
742	1869	Fox's Locomotive Boiler	275	661	1833	Stephenson's Tubular Locomotive	253
				675	1851	Stenson's Tubular Locomotive	257
744	1871	Girdwood's Vertical Locomotive Boiler	276	681	1853	Scott's Locomotive Fire Box	259
				682	1853	Shaw's Locomotive Boiler	259
662 and 663	1839	Hawthorn's Return Tubular Boilers	254	728	1852	Selby's Locomotive Boiler	272
				732	1851	Stenson's Locomotive Boiler	273
666	1847	Hackworth's Locomotive Boiler	255	653 to 655	1803, 1804	Trevithick's Locomotive Boilers	251
673	1850	Hodge's Locomotive Boiler	256				
674	1850	Hodge's Return Tubular Boiler	257				
725	1859	Hunt's Locomotive Boiler	271	741	1869	Thirion's Locomotive Boiler	275
737	1866	Holt's Locomotive Boiler	274				
738	1867	Holt's Locomotive Boiler	274	736	1866	Woodward's Locomotive Boiler	274

CHAPTER VIII

BOILER STEAM SAFETY VALVES AND GEAR.

Fig.	Date.	NAME.	Page	Fig.	Date.	NAME.	Page
803 and 804	1868	Ashcroft's Safety Valves	284	780	1859	Clayton's Safety Valve	281
				797	1867	Cooke's Safety Valve	283
				798	1867	Cameron's Safety Valve	283
839	1872	Annular Seat Safety Valve	288	802	1868	Church's Safety Valve	284
				812	1871	Cowburn's Safety Valve	285
768 to 775	1857	Bodmer's Safety Valves and Regulating Feed Apparatus	280, 281	837	1872	Cazier's Safety Valve	287
790 and 791	1866	Baldwin's Safety Valves	283	840	1872	Double Seat Safety Valve	288
				826	1871	Field's Safety Valve	286

INDEX OF ILLUSTRATIONS.

Fig.	Date.	NAME.	Page	Fig.	Date.	NAME.	Page
781 and 782	1860	German Safety Valves	281	756	1850	Twin Safety Valve	279
				759	1853	Tyler's Safety Valve	279
783	1861	Galloway's Safety Valves	281	815 and 816	1871	Taylor's Safety Valves	285
784	1861	Galloway's Safety Valves	282				
789	1864	German Balance Piston Safety Valve	282	827 to 835	1872	Turton's Safety Valve Springs	287
822 to 825	1872	Giles' Safety Valves	286				
				752	1800	Vacuum Safety Valve	279
765, 766 and 767	1856	Holt's Safety Valves	280	751	1800	Watt's Safety Valve	279
				810 and 811	1871	Wilke's Marine Safety Valve	284
776	1858	Haste's Safety Valves	280	818	1871	Watson's Marine Safety Valve	285
779	1859	Harman's Safety Valves	280	819	1871	Watson's Marine Safety Valve	286
803 to 809	1870	Hopkinson's Safety Valves	284	**BOILER ALARM SAFETY VALVES AND GEAR.**			
777	1858	Illingworth's Safety Valve	281	842 and 843	1840	Alarm Whistle, Float, Lever, and Balance	288
764	1855	Johnson's Safety Valve	280	880	1858	Archer's Alarm Valve	293
753	1831	Locomotive Safety Valve	279	881	1858	Archer's Reverse Seats Safety Valves	293
755	1840	Locomotive Safety Valve	279	909	1871	Adamson Alarm Safety Valve Apparatus	295
817	1871	Loe's Safety Valve	285				
836	1872	Lockwood's Safety Valve	287	877 and 878	1858	Bodmer's Alarm Valve and Whistle	292
788	1863	Mash's Safety Valve	282	892 to 895	1866	Bray's Alarm Safety Valve Apparatus	294
813 and 814	1871	Mirrlin's Safety Valves	285				
820 and 821	1872	MacDonald's Spring and Piston Valve and Box	286	902	1868	Benson's Alarm Safety Valve Water-pipe	295
				852 and 853	1855	Cowburn's Water Level, Float, and Lever	289
785 to 787	1863	Naylor's Safety Valves	282	854	1855	Cowburn's Water Level, Float, Wheel, and Chain	290
749	1695	Papin's Safety Valve	279	898 to 900	1867	Cowburn's Alarm Fusible Plugs	294
793 to 796	1867	Parson's Safety Valves	283	910	1871	Cowburn's Alarm Safety Valve and Apparatus	296
841	1867	Pollard's Safety Valves	288				
760	1855	Ramsbottom's Safety Valves	279	885	1860	Davies' Alarm Safety Valve and Curved Pipe	293
761 to 763	1855	Ramsbottom's Safety Valves	280	886	1860	Davies' Double Seat Alarm Safety Valve and Casing	293
799 and 800	1867	Richardson's Safety Valves	283	887	1861	Galloway's Alarm Valve Apparatus	293
				851	1855	Hall's Alarm Safety Valve Apparatus	289
750	1698	Savery's Safety Valve	279	864	1857	Hurton's Safety Valves	291
754	1833	Stephenson's Safety Valve	279	869 and 870	1858	Haste's Safety Valves	291
755	1840	Safety Valve, Locomotive	279				
757	1852	Spencer's Safety Valve	279	871	1858	Haste's Safety Valves	292
758	1853	Scott's Twin Safety Valve	279	891	1866	Hackett's Alarm Valves	294
778	1858	Smith's Safety Valve	281				
792	1866	Swann's Safety Valve	283	903 and 904	1869	Hugh's Alarm Safety Valve Apparatus	295
801	1868	Sander's Safety Valve	283				
838	1872	Safety Valve ("Engineer")	287				

INDEX OF ILLUSTRATIONS. 385

Fig.	Date.	NAME.	Page
908	1870	Hopkinson's Alarm Safety Valve Apparatus	295
882 and 883	1858	Illingworth's Alarm Safety Valve Apparatus	293
844	1852	Johnson's Indicating Alarm Float	288
855	1856	Johnson's Alarm Safety Valve Apparatus	290
859 to 862	1856	Knowelden's Alarm Safety Valve Apparatus	290, 291
901	1867	Kenyon's Alarm Safety Valve and Apparatus	295
906	1870	Kimball's Alarm Whistle Float Apparatus	295
912	1871	Kirk's Alarm Valve and Float in Mercury	296
907	1870	Langlet's Alarm Valve, Weight Lever, and Float	295
911	1871	Lee's Alarm Safety Valve and Apparatus	296
888 and 889	1862	McCarthy's Alarm Safety Valve Float Lever	293
897	1867	Macpherson's Alarm Safety Valves and Apparatus	294
879	1858	Normandy's Alarm Valve and Apparatus	293
865 and 866	1858	Parson's Safety Piston Valve	291
867	1858	Parson's Twin Safety Piston Valves	291
868	1858	Parson's Safety Plug Valve and Rings	291
872 and 873	1858	Parson's Alarm Valve	292
874	1858	Parson's Feed Water Whistle Alarm	292
905	1870	Pratt's Alarm Valve, Whistle, Water Globe, and Valve	295
856 and 857	1856	Routledge's Alarm Flame Tube	290
896	1866	Swann's Alarm Valve	295
845, 846	1853	Tayler's Alarm Safety Valve	288
846 to 850	1853	Tayler's Alarm Safety Valve Apparatus	289
890	1863	Turner's Alarm Safety Valve Apparatus	294
863	1856	Walley's Alarm Safety Valve Apparatus	291
875 and 876	1858	Wright's Water Float Rod	292
884	1859	Walker's Alarm Safety Valve	293
858	1856	York's Alarm Safety Valve Apparatus	290

BOILER FEED PUMPS AND ENGINES.

Fig.	Date.	NAME.	Page
944	1870	Baumann's Boiler Feed Pump and Engine	302
936	1866	Brown's Boiler Feed Pump and Engine	300
965 and 966	1870	Bishop's Boiler Feed Oscillating Piston Pump	305
968	1871	Burgh's Pump Valves	305
913	1852	Cameron's Boiler Feed Pump and Engine	296
919	1866	Cameron's Boiler Feed Pump and Engine	297
917	1862	Cowan's Boiler Feed Pump and Engine	297
932	1855	Cowburn's Boiler Feed Pump and Engine	300
947 and 948	1872	Cope's Feed Pump and Engine	302
928	1872	Cope's Boiler Feed Pump and Engine	299
957	1869	Clark's Boiler Feed Apparatus	304
927	1871	Clarkson's Boiler Feed Pump and Engine	299
925	1871	Davy's Boiler Feed Pump and Engine	298
951	1867	Davy's Boiler Feed Rotary Pump	303
926	1871	De Bergue's Boiler Feed Pump and Engine	298
918	1864	Duprey's Boiler Feed Pump and Engine	297
952	1870	Faure's Boiler Feed Rotary Pump	303
955	1869	Friedman's Boiler Feed Injector	304
934	1860	Gargan's Boiler Feed Pump	300
953 and 954	1858	Giffard's Feed Injector	303
959	1871	Grindrod's Boiler Feed Water Apparatus	304
960 and 961	1867	Holman's Boiler Feed Pump	304
962	1867	Holman's Boiler Feed Pump	305
914	1852	Johnson's Boiler Feed Pump and Engine	296
930	1853	Johnson's Boiler Feed Pump and Engine	299
970	1858	Johnson's Feed Water Supply and Non-return Valves	305
935	1861	Knowelden's Boiler Feed Pump	300
915	1861	Knowelden's Boiler Feed Pump and Engine	297
937	1866	Kittoe's Boiler Feed Pump	300
938 to 940	1866	Kittoe's Boiler Feed Pump	301
967	1869	Kittoe's Feed Pump Valves and Seats	305

3 D

INDEX OF ILLUSTRATIONS.

Fig.	Date.	NAME.	Page
933	1856	Mellor's Boiler Feed Pump and Engine	300
943	1868	Macabie's Boiler Feed Pump and Engine	301
958	1870	Macabie's Boiler Feed Water Apparatus	304
963	1868	Macabie's Boiler Feed Apparatus	305
964	1868	Maxwell's Boiler Feed Engine	305
921	1868	Maxwell's Boiler Feed Pump and Engine	298
931	1855	Mellor's Boiler Feed Pump and Engine	299
915	1853	Newton's Boiler Feed Pump and Engine	297
945	1870	Pearn's Boiler Feed Pump	302
923 and 924	1869	Ramsbottom's Boiler Feed Pumps and Engine	298
946	1871	Ramsbottom's Boiler Feed Pump and Engine	302
941 and 942	1868	Samuël's Boiler Feed Pump	301
920	1868	Tijon's Boiler Feed Pump and Engine	297
969	1863	Turner's Boiler Feed Valves and Casing	305
949	1856	Whitaker's Boiler Feed Rotary Pump	303
950	1867	Wilson's Boiler Feed Rotary Pump	303
956	1863	Wagstaff's Feed Tank and Steam Lever Gear	304
922	1872	Walker's Boiler Feed Pump and Engine	298
929	1872	Wolstenholme's Boiler Feed Pump and Engine	299

CHAPTER IX.

SECURING AND CONNECTING TUBES.

Fig.	Date.	NAME.	Page
976	1868	Allibon's Boiler Stay Tube	306
972	1853	Dunn's Method of Securing the Ends of Tubes	306
982	1833	Field's Water Circulating Branch Piece	307
975	1868	Howard's Central Water Circulating Branch Connection	306
983	1858	Hopkinson's Stay Tube Connection	307
984	1861	Harlow's Set Screw Connection	307
973	1855	Johnson's Method of Closing Tubes in Boiler Plates	306
974	1867	Langlois' Method of Securing Stay Tubes	306
977 and 978	1870	Langley's Method of Fitting Boiler Tubes in Plates	306
979	1870	Langley's Method of Securing Boiler Tubes at an Angle	306
971	1839	Wahl's Method of Securing the Ends of Tubes	306
980 and 981	1871	Watt's Method of Fitting Boiler Tubes to Plates	306

SECURING WATER CIRCULATING TUBES.

Fig.	Date.	NAME.	Page
1001	1869	Desvigne's Vertical Syphon Tubes	309
996	1866	Feyh's Water Circulating Branch Piece	308
998	1868	Howard's Water Circulating Tubes	309
988	1857	Joly's Water Circulating Tubes	307
1002	1870	Lee's Cylindrical Boiler	309
1003	1870	Lee's Globular Evaporating Cup	309
990	1862	Merryweather's Hanging Tubes	308
991 and 992	1864	Marshall's Hanging Syphon Water Tube	308
995	1866	Miller's Vertical Water Circulating Tubes	308
1004	1868	Nason's Circulating Tubes	309
1006	1868	Nason's Circulating Tubes	309
985	1831	Perkins' Semi-globular Twin Land Boiler, fitted with hanging tubes	307
986	1831	Perkins' Wagon Boiler, fitted with water circulating plates	307
997	1868	Perkins' Boiler Tube Connections	309
993	1865	Smith's Hanging Tubes	308
1005	1868	Smith's Water Circulating Tubes	308
1007 to 1010	1871	Todd's Water Circulating Tubes	310
1000	1869	Thirion's Syphon Tube	309
989	1859	Varley's Cylindrical Boiler fitted with Water Pockets	307
994	1865	Wise's Hanging Tubes	308
999	1868	Wiegand's Vertical Water Circulating Tubes	309

SECURING BRANCH PIECES TO CIRCULATING TUBES.

Fig.	Date.	NAME.	Page
1011 and 1012	1865	Belleville's Water Circulating Branch Piece	310

INDEX OF ILLUSTRATIONS.

Fig.	Date.	NAME.	Page	Fig.	Date.	NAME.	Page
1014	1867	Carville's Tubes	310			CONNECTION OF TUBES AT RIGHT ANGLES, ETC.	
1027 to 1029	1871	Howard's Stay Tubes	312	1044	1872	Allison's "Deflectors"	313
1030 to 1033	1871	Howard's Trough Water Circulating Tubes	312	1034	1865	Belleville's Water Circulating Tubes	312
				1040	1871	Bartlott's Method of Connecting Tubes	313
1026	1871	Mirchin's Water Circulating Branch Piece	312	1036	1866	Howard's Method of Connecting Tubes	313
				1037	1866	Howard's Water Circulating Branch Piece	313
1013	1867	Root's Water Circulating Branch Connections	310	1038	1868	Howard's Water Circulating Tubes	313
				1039	1868	Howard's End Water Circulating Branch Connections	313
1015 to 1018	1870	Root's Water Circulating Branches	310 311	1041	1842	Lewthwaite's Water Circulating Sheet Tubes	313
1019	1871	Westerman's Water Circulating Branches	311				
1020 1021	1871	Watt's Water Circulating Branches	311	1042	1864	Marshall's Water Circulating Sheet Tubes	313
to 1024	1871	Watt's Flange Stay Bolts	311	1043	1870	Parman's "Deflectors"	313
1025	1871	Watt's Method of Fitting Boiler Tubes to Plates	312	1035	1865	Turbill's Water Circulating Branches	312

CHAPTER X.

PERFORATED FIRE BARS.

Fig.	Date.	NAME.	Page	Fig.	Date.	NAME.	Page
1057	1869	Brown's Fire Bars with Twin Sides	314	1049 and 1050	1860	Stratford's Fire Bars, with single and double row of perforations	314
1059	1870	Broughton's Fire Bars, with angular holes through the centre	314				
1058	1870	Cone's Fire Bar, with vertical openings	314			SOLID AND HOLLOW FIRE BARS.	
1061	1872	Dilnut's Fire Bars, with vertical grooves at the sides	314	1053	1860	Blackwood's Fire Bar, with longitudinal passage through the bar	315
				1068	1870	Batchelor's Combined Fire Bars	315
1055 and 1056	1868	Fletcher's Fire Bar, with grooves at side and top	314	1062	1844	Chanter's Fire Bar, with grooves at side and top, and holes through sides	315
1048	1859	Harden's Fire Bar, with vertical grooves	314	1067	1869	Fletcher's Fire Bars, having central openings	315
1047	1859	Harden's Bars, with vertical lugs for air	314	1064	1865	Groen's Fire Bars, with angulated bars, laid across the fire box	315
1053	1866	Harrison's Fire Bars, with air spaces	314				
1051	1861	Jackson's Fire Bar, with vertical holes	314	1065 and 1066	1868	Jordan's Fire Bars, with top steps and side lugs	315
1054	1868	Lewis's Fire Bars, with transverse perforations	314	1069	1871	Whitelaw's Fire Bars, with longitudinal grooves on each side of the bar	315
1045	1852	Moreau's Fire Bar	314				
1046	1859	Martin's Fire Bars, with vertical grooves	314			WATER FIRE BARS.	
1052	1863	Mylrea's Fire Bar, with openings at sides	314	1072	1867	Barlow's Water Tube Fire Bars	316
1060	1871	Raper's Fire Bar, with vertical holes	314	1077	1872	Ellis' Fire Bars, with a feed water pipe cast in the bar	316

INDEX OF ILLUSTRATIONS.

Fig.	Date.	NAME.	Page
1076	1869	Gaze's Fire Bars, with a feed water pipe passing through the bar	316
1071	1856	Haywood's Water Tube Fire Bars	316
1070	1864	Miguet's Water Tube Fire Bars	315
1073 to 1075	1868	Vicar's Fire Bars, consisting of two bars, one within the other, the outer bar containing water	315
		MOVABLE FIRE BARS.	
1083	1860	Annan's Fire Bars and lever motion	317
1079	1844	Chanter's Fire Bar, lever motion	316
1084	1861	Colquhoun's Fire Bar, lever motion	317

Fig.	Date.	NAME.	Page
1078	1838	Drew's Fire Grate, which is raised and lowered by levers	316
1088	1871	Holt's Fire Bars, with perforated bars placed across the fire box	317
1081	1857	Johnson's Fire Bars, with eccentric motion	317
1082	1857	Johnson's Fire Bar, with catch and pinion motion	317
1086	1868	Jordan's Fire Bar, with eccentric motion	317
1087	1868	Lewis's Fire Bar, lever motion	317
1080	1856	Mash's Fire Bar, lever motion	317
1085	1863	Shillito's Fire Bars, with vertical motion	317

CHAPTER XI.

MECHANICAL FEED FUEL APPARATUS.

Fig.	Date.	NAME.	Page
1089	1819	Brunton's Feed Fuel Apparatus	318
1096	1870	Burnley's Feed Fuel Apparatus	320
1104	1870	Butterworth's Feed Fuel Apparatus	322
1095	1869	Crosland's Feed Fuel Apparatus	319
1101	1829	Church's Feed Fuel Apparatus	321
1108 to 1110	1869	Crosland's Feed Fuel Apparatus	323
1111	1871	Heginbottom's Feed Fuel Apparatus	323
1090	1848	Jucko's Feed Fuel Apparatus	318
1105 to 1107	1866	Leigh's Feed Fuel Apparatus	322
1099	1866	Ripley's Feed Fuel Apparatus	321
1097	1868	Shillito's Corkscrew Fuel Feeder	320
1092 1093 and 1094	1868	Taylor's Feed Fuel Apparatus	318
	1868	Taylor's Feed Fuel Apparatus	319
1098	1866	Turner's Corkscrew Fuel Feeder	320
1091	1867	Vicar's Feed Fuel Apparatus	318
1102 and 1103	1867	Vicar's Feed Fuel Apparatus	{321 322}
1100	1870	Waller's Feed Fuel Apparatus	321

FIRE-BOX DOORS.

Fig.	Date.	NAME.	Page
1122	1870	Auld's Fire-box Door and Frame	325
1115 to 1118	1858	Beattie's Fire-box Doors and Frames	324
1121	1855	Cliff's Fire-box Door and Frame	325
1112	1857	Common Fire-box Door	323
1129	1873	Martin's Pendulous Fire Door	326
1119	1853	Prideaux's Fire-box Door and Frame	324
1120	1855	Prideaux's Fire-box Door and Frame	324
1123	1870	Prideaux's Fire Door, Shutter, and Regulator	325
1124 and 1125 1126	1870	Prideaux's Fire Doors and Frames	325
	1872	Prideaux's Inverted Regulators	325
1127 and 1128	1872	Prideaux's Fire Doors and Frames	326
1113	1857	Wye-Williams' Fire-box Door	323
1114	1857	Wye-Williams' Fire-box Doors and Frames	323

INDEX OF ILLUSTRATIONS. 389

CHAPTER XII.
IGNITION OF COAL.

Fig.	Date.	NAME.	Page	Fig.	Date.	NAME.	Page
1130	1873	Normal Condition	327	1134	1873	Fourth Stage of Ignition	327
1131	1873	First Stage of Ignition	327	1135	1873	Fifth Stage—Flaming Coke	327
1132	1873	Second Stage of Ignition	327	1136	1873	Sixth Stage—Dead Coke	327
1133	1873	Third Stage of Ignition	327				

CHAPTER XIV.
ACTION OF FLAME AND RAISING OF STEAM.

Fig.	Date.	NAME.	Page	Fig.	Date.	NAME.	Page
1137	1873	Action of flame sliding along the plate	352	1145	1873	Action of flame in Burgh's cylindrical tubular boiler	354
1138	1873	Action of flame against an angular surface	352	1146	1873	Action of flame in a modern cylindrical marine boiler	355
1139	1873	Action of flame against a plate at right angles	352	1147	1873	Action of flame in a pair of marine boilers	355
1140	1873	Action of flame in a tube	352				
1141 and 1142	1873	Action of flame amongst tubes	353	1150	1868	Bailey's Pyrometer	355
1143	1873	Action of flame around syphon tubes (Moy and Shill's patent)	353	1148	1872	Casartelli's Pyrometer	355
1144	1873	Action of flame in Burgh's treble combustion chamber tubular boiler	354	1149	1866	Wood's Pyrometer	355

CHAPTER XV.
CAUSES OF BOILER EXPLOSIONS.

Fig.	Date.	NAME.	Page	Fig.	Date.	NAME.	Page
1151	1873	Illustration of a burnt plate	356	1155	1873	Illustration of the result of proper drilling and rivoting plates	359
1152	1873	Illustration of the sediment in a marine boiler	356	1156 and 1157	1869	Illustrations of the collapse of a fire-box	360
1153	1873	Illustration of the corrosion of a boiler plate	358	1158	1867	Illustration of the collapse of a fire-box of a Cornish boiler	360
1154	1870	Illustration of the result of defective punching and rivoting plates	359				

CHAPTER XVI.
BOILER-MAKING.

Fig.	Date.	NAME.	Page	Fig.	Date.	NAME.	Page
1159	1851	Expansion Joint for flue tube of boiler	361	1161	1860	Hill's Expansion Ring for flue tubes of boilers	361
1160	1850	Adamson's Expansion Joint for flue tube of boilers	361	1162 and 1163	1858	Beattie's Strengthening Plates for boilers	363

INDEX OF SUBJECT-MATTER.

	PAGE
ACTION OF FLAME	351
in Cornish and Lancashire Boilers	351
in Locomotive Boilers	351
in Marine Boilers	351
along a Plate	352
at an Angle of a Plate	352
at Right Angles to a Plate	352
in a Tube	352
amongst Tubes	353
to cause the utmost Evaporation of Water	354
in a Vertical Boiler	354
in a large Combustion Chamber	354
in a small Combustion Chamber	355
in a Pair of Marine Boilers	355
BOILER EXPLOSIONS,	
Causes of	359
Collapsing of Tubes	360
Corrosion of Boiler Plates	358
Deposit in Boilers	356, 357
Defective Riveting	359
Good Workmanship, Prevention of	359
Grease causing Galvanic Action	357
Properties of Sea Water	357
Properties of Deposit	357
Pitting, Cause of	358
BOILER-MAKING	361 to 362
Illustrative Plates	362
Cornish and Lancashire, Defects of	361
Correct Method of	362
Expansion Joints for Flue Tubes	361
Laps of Plates	363
Lagging of Boilers	363
Repairing of Boilers	363

	PAGE
COMBUSTION.	
Average Proportion of the Chemical Constituents of Coal	328
Amount of Air required per lb. of Fuel	333
Chemical Constituent of	329
Combustion 328, 329, 331, 332, 334, 335, 336, 340	
Combustible Substances	330
Carbon	330
Heat in Carbon	333, 343
Heat in a Furnace	343
Latent Heat	332
Sensible Heat	332
Smoke, Cause of 329, 335, 336, 337, 338, 339, 340, 341, 342, 343, 347	
Smoke, Composition of	338
Smoke Numbers	347
COMPOSITION OF	
Air	329
Ammonia	329
Bisulphurate of Carbon	329
Carbonic Oxide	329
Carbonic Acid	329
Sulphurous Acid	329
Sulphuretted Hydrogen	329
Water	329
TABLES OF	
Chemical Constituent of Coal	328
Chemical Constituent of Combustion	329
Combustible Substances	330
Total Evaporative Powers of Fuel	332
Units of Heat	333
Compound Ingredients of Fuel	333
Amount of Air per lb. of Fuel	334

INDEX OF SUBJECT-MATTER.

TABLES OF
	PAGE
Specific Heat	343
The Duty of Coal (Experiments at Newcastle-on-Tyne, 1857)	344
Average Result of North Country and Welsh Coals	344
Quality of Land Engine and Smithery Coal (Keyham Yard, 1870)	345
Quality of Land Engine and Smithery Coal (Keyham Yard, 1871)	345
Respecting the Coals Consumed on Board Ship	346
Smoke Numbers	347
Trial of H.M. Ship "Northumberland," 1871	347
Trial of H.M. Ship "Northumberland," 1871	348
Trial of H.M. Ship "Topaze," 1871	348
Trial of H.M. Ship "Monarch," 1871	349
Trial of Mixed Coal in H.M. Ship "Active," 1870	349
Evaporative Power in Locomotive Engines, 1860	350
Use of Coal in H.S. Ships	345

BOILER-MAKING.
RULES FOR
	PAGE
Diameter of Rivets	364
Pitch of Rivets	365
Laps of Plates	365
Collapsing Pressure of Boiler Flue Tubes	366
Bursting Pressure of Cylindrical Boilers	366
Working Pressure of Cylindrical Boilers	366
Area of Stays and Gussets for the Ends of Cylindrical Boilers	366

CORNISH AND LANCASHIRE BOILERS,
	PAGE
Length of	367
Diameter of	367
Flue Tubes of	367
Flues in Brickwork for	367
Grate or Fire Bar Surface for	367
Safety Valves (Land Boilers)	367

MARINE BOILERS,
	PAGE
Tube Surface for	367
Diameter of Tubes	367
Length of Tubes	367
Rake of Tubes	367
Number to one Fire Box	367

MARINE BOILERS—continued.
	PAGE
Position of Stays in	367
Fire Bar of Grate Surface	367
Proportions of Fire Box	367
Fire Door Opening	367
Steam Room Capacity	368
Water Line above Tubes	368
Proportions of Smoke Box	368
Proportions of Uptake	368
Area of Chimney or Funnel	368
Proportions for Telescopic Chimney	368
Air Casings around Telescopic	368
Safety Valves	368
Fire Bars	368
Coal Bunkers	368
Coal Capacity	369
Stoking Space between Boilers	369

[Proportions of Locomotive Boilers can be obtained from Plate 22, which is perfectly complete for that purpose.]

RULE TABLES OF THE
	PAGE
Algebraic Signs	369
Bursting and Safe Pressures of Boilers	365
Gravity of Water	369
Heat Conducting Power of Metals	369
Melting Temperatures of Metals	369
Properties of Steam	370
Properties of Steam	371
Proportions of Cornish and Lancashire Boilers	366
Proportions of Weights of Marine Boilers	371
Specific Gravities	369
Strength of Materials	364
Strength of Materials	365
Strength of Riveted Joints	364
Strength of Riveted Joints	365
Strength of Flue Tubes	366
To find the Working Pressure of Cylindrical Boilers	366
Weights of Angle Iron of Equal Sides	369
Weight in lbs. per Foot of different Materials in Plates	369
Weight of Marine Boilers in Details	372
Weight of Round and Square Bars	369
Weight of Wrought Iron Pipes	372

www.ingramcontent.com/pod-product-compliance
Lightning Source LLC
Chambersburg PA
CBHW022113290426
44112CB00008B/653